Lecture Notes in Physics

Edited by H. Araki, Kyoto, J. Ehlers, München, K. Hepp, Zürich
R. Kippenhahn, München, D. Ruelle, Bures-sur-Yvette
H. A. Weidenmüller, Heidelberg, J. Wess, Karlsruhe and J. Zittartz, Köln
Managing Editor: W. Beiglböck

327

K. Meisenheimer   H.-J. Röser   (Eds.)

Hot Spots in Extragalactic Radio Sources

Proceedings of a Workshop
Held at Ringberg Castle, Tegernsee, FRG, February 8–12, 1988

Springer-Verlag
Berlin Heidelberg GmbH

Editors

K. Meisenheimer
H.-J. Röser
Max-Planck-Institut für Astronomie
Königstuhl 17, D-6900 Heidelberg, FRG

Scientific Organizers

Klaus Meisenheimer
Richard A. Perley
Hermann-Josef Röser

ISBN 978-3-662-13715-4   ISBN 978-3-540-46126-5 (eBook)
DOI 10.1007/978-3-540-46126-5

2158/3140-543210 – Printed on acid-free paper

P r e f a c e

Only a few percent of all galaxies are strong radio emitters. Of these again only a fraction show what we now call classical double structure, with giant lobes of radio emission straddling the central galaxy, and prominent bright spots of emission near the outer ends of the lobes. In the widely accepted standard picture, these hot spots terminate the supersonic or even relativistic jets, transporting energy, mass and momentum out from an active galactic nucleus to the giant radio lobes.

"Hot Spots in Extragalactic Radio Sources" - why such a specialized topic for a workshop ? Hot spots are an excellent "laboratory" to study the interaction of the jet with the surrounding medium and to test our understanding of the overall source dynamics. Furthermore, soon after the first detailed radio maps of Cygnus A had been obtained in 1973, the origin of the relativistic particles and magnetic fields giving rise to the intense synchrotron emission in the hot spots was a matter of intense debate. From an observational point of view, the hot spot radiation is our only clue for understanding the acceleration and propagation of these particles. On the other hand, the observed properties of hot spots constrain the parameter space of jet models, e.g. their densities and velocities.

New observational data and unprecedented computational tech-niques have recently become available. Summarizing our current knowledge and coordinating efforts for future work was our prime motivation for this " specialists' " meeting. It was made possible by the Max-Planck-Gesellschaft, which not only provided facilities at

Ringberg Castle but also supplied a generous grant, for which we would like to express our thanks on behalf of all participants. Our special thanks go to Dr. Dietmar Nickel (MPG) for his efficient help, to Mr. A. Hörmann, the castle manager, and his staff, who made our stay at Ringberg so comfortable, and to Mr. Axel Quetsch, who took over organizational matters during the workshop.

We would like to thank Miss Rita Wagner for her patience in preparing the manuscripts of the invited papers, Miss Rachel Blythe for a careful checking of the whole text and Mrs. Doris Anders, Karin Dorn, Martina Weckauf and Mr. Werner Neumann for their skilled preparation of the camera-ready manuscripts.

Heidelberg and Socorro   Klaus Meisenheimer
December 1988   Rick Perley
                                   Hermann-Josef Röser
                                   (Scientific Organizing Committee)

Contents

Radio Observations

Optical Observations

Models and Simulations

Particle Acceleration

List of Participants

Stefan Appl	31	Landessternwarte, Heidelberg, F.R.G.
Peter D. Barthel	21	Caltech, Pasadena, U.S.A.
Stefi Baum	34	Radio Sterrenwacht, Dwingeloo, Netherlands
Peter L. Biermann	12	Max-Planck-Institut für Radioastronomie, Bonn, F.R.G.
Dieter Biskamp		Max-Planck-Institut für Plasmaphysik, Garching b. München, F.R.G.
Wil van Breugel	20	Radio Astronomy Lab., UC Berkeley, Univ. of California, Berkeley, U.S.A.
Max Camenzind		Landessternwarte, Heidelberg, F.R.G.
Chris Carilli	29	NRAO/MIT, Socorro, New Mexico
Wayne A. Christiansen	26	University of North Carolina, Chapel Hill, U.S.A.
Philippe Crane		European Southern Observatory, Garching b. München, F.R.G.
John W. Dreher	28	MIT, Cambridge, U.S.A.
Luke Drury	19	Dublin Institute for Advanced Studies, Dublin, Ireland
Jean Eilek	11	Physics Department, New Mexico Tech., Socorro, U.S.A.
Robert A.E. Fosbury	13	ST-ECF, European Southern Observatory, Garching b. München, F.R.G.
Klaus-Dieter Fritz	1	Max-Planck-Institut für Radioastronomie, Bonn, F.R.G.
Philip E. Hardee	23	Department of Physics & Astronomy, University of Alabama, Tuscaloosa, U.S.A.
Alan Heavens	35	Dept. of Astronomy, Univ. of Edinburgh Royal Observatory, Edinburgh, U.K.
Peter Hiltner	27	Max-Planck-Institut für Astronomie, Heidelberg, F.R.G.
John Kirk		Max-Planck-Institut für Physik und Astrophysik, Garching b. München, F.R.G.
Dieter Kössl	17	Max-Planck-Institut für Astrophysik, Garching b. München, F.R.G.
Wolfgang Kundt	37	Institut für Astrophysik, Universität Bonn, F.R.G.
Robert Laing	16	Royal Greenwich Observatory, Hailsham, U.K.

J. Patrick Leahy   15 National Radio Astronomy Observatory,
                            Socorro, New Mexico, U.S.A.

Colin L. Lonsdale   25 Haystack Observatory, Westford, U.S.A.

Alan Matthews   32 Mullard Radio Astronomy Observatory,
                            Cavendish Laboratory, Cambridge, U.K.

Klaus Meisenheimer   4 Max-Planck-Institut für Astronomie,
                            Heidelberg, F.R.G.

Friedrich Meyer   6 Max-Planck-Institut für Astrophysik,
                            Garching b. München, F.R.G.

Tom Muxlow   10 University of Manchester, Macclesfield,
                            Chesire, U.K.

Michael L. Norman   7 National Center. for Supercomputing
                            Applications, Univ. of Illinois,
                            Urbana-Champaign, Champaign, U.S.A.

Chris O'Dea   30 Radio Sterrenwacht, Dwingeloo, NL

Frazer Owen   24 National Radio Astronomy Observatory,
                            Socorro, New Mexico, U.S.A.

Rick A. Perley   8 National Radio Astronomy Observatory,
                            Socorro, New Mexico, U.S.A.

Axel M. Quetsch   38 Max-Planck-Institut für Astronomie,
                            Heidelberg, F.R.G.

Hermann-Josef Röser   3 Max-Planck-Institut für Astronomie,
                            Heidelberg, F.R.G.

Lawrence Rudnick   14 School of Physics & Astronomy, University
                            of Minnesota, Minneapolis, U.S.A.

Peter A.G. Scheuer   33 Mullard Radio Astronomy Observatory,
                            Cavendish Laboratory, Cambridge, U.K.

Reinhard Schlickeiser   2 Max-Planck-Institut für Radioastronomie,
                            Bonn, F.R.G.

Martin Schlötelburg   22 Max-Planck-Institut für Astronomie,
                            Heidelberg, F.R.G.

Sperello di Serego Alighierie,   ST-ECF, European Southern
                            Observatory, Garching b. München, F.R.G.

Clive Tadhunter   18 ST-ECF, European Southern Observatory,
                            Garching b. München, F.R.G.

Edoardo Trussoni   36 Istituto di Fisica Generale
                            dell'University', Torino, Italy

Paul J. Wiita   9 Dept. of Phys. & Astronomy, Georgia State
                            University, Atlanta, U.S.A.

Michael J. Wilson   5 University of Leeds, Dept. of Applied
                            Mathematics, Leeds, United Kingdom

Rick Perley   Robert Laing

Hot Spot Radio Galaxies - An Introduction

Richard A. Perley

National Radio Astronomy Observatory
Socorro, N.M. 87801, USA

1. A Brief History of Extragalactic Radio Astronomy

Review talks frequently begin with a brief historical survey of the subject. I imagine that this is done because the reviewers are frequently those who have been involved with the evolution of the subject, and because they feel that such a survey is useful to introduce the subject. In my case, I can make no claim concerning being involved with the history of extragalactic radio astronomy. However, I do feel that a brief historical review is useful as an introduction to the subject in general, and to 'hot spots' in particular. I also feel that although this meeting is very much a specialists' meeting, the backgrounds of the participants are sufficiently diverse that an introductory talk should start with the observational basics. And a good way to introduce the subject is from an historical perspective.

In reviewing the literature for this review, I noticed a clear watershed divide in the understanding of the basic physics giving rise to the phenomenon of 'hot spot' radio galaxies. This occured in 1974 with the publication of the five papers cited below. In the simplest terms, this was the year that the phenomenon of 'hot spots' was clearly recognized. Prior to this year, surveys with the low-resolution synthesis interferometers at Cambridge had established the well-known general characteristics of extragalactic radio morphology:

(1) They had a 'double' sided structure, extending in some cases hundreds of kiloparsecs on each side of the optical galaxy. It was recognized that the radio brightness commonly increased towards the extrema of the structure.

(2) Assuming the radio emission was via the synchrotron process, (and that the redshifts were cosmological), the observed structures contained minimum total energies up to 10^{61} erg with radio luminosities of up to 10^{45} erg/sec. Arguments based on synchro-

1

tron ageing combined with observations of the spectra led to source ages of 10^7 to 10^8 years.

These basic characteristics were established by extensive imaging of the 3C survey using, principally, the Cambridge aperture synthesis instruments. The results of this extensive program are in many papers found in the Monthly Notices from 1968 onwards. However, the resolution of these instruments was generally insufficient to resolve any small-scale structure, and it is thus not surprising that the theories advanced to explain the observed large-scale structure were centered around 'one-shot' models - the observed structure is a result of a single massive explosion which was somehow directed in a fairly narrow cone. The brightening at the ends of many sources was then interpreted as due to deceleration of the launched blob by an extragalactic medium. (De Young and Axford, 1967, Mitton and Ryle, 1969, Longair and Macdonald, 1969). Note that by 1970, indications of ongoing activity were emerging (Mitton, 1970, Graham, 1970).

All this changed quite dramatically after the construction of the 5-km telescope (Ryle, 1972). This instrument gave, for the first time, resolutions of 2" combined with complete aperture plane coverage. (The latter point is especially important in imaging large and complicated objects. Long-baseline interferometers had detected fine-scale structure in some extended objects by this time (Donaldson et al., 1971, Miley and Wade, 1971), but the lack of phase and sufficient aperture plane coverage hindered interpretation). The first results on an extragalactic source from this telescope (Cygnus A) completely altered the theories of the day. The outstanding result was the clear detection of 'hot spots' located at the ends of the previously known 'radio lobes'. Many other prominent sources were quickly imaged by this telescope. The result of these new revelations quickly showed up in the following series of papers:

(1.) Longair, Ryle and Scheuer (1973). This seminal paper summarizes the theoretical problems with existing radio source theories, and clearly advances the need for continuing particle acceleration.

(2.) Hargrave and Ryle (1974). This paper presents the 5-km results of Cygnus A. Hot spots are clearly detected, and the consequences of their existence stated. The need for continuing acceleration in these hot spots is shown.

(3.) Scheuer (1974). Peter Scheuer presents models for high-luminosity extragalactic radio sources including the effects of 'jets' or 'beams', the conduits through which the energy required to power the hot spots flows.

(4.) Fanaroff and Riley (1974). This short (3 pages) paper is perhaps the best known of all those mentioned here. I will expand upon the fundmental relation they discovered below.

(5.) Blandford and Rees (1974). This seminal work discusses the origin and evolution of hot spots, and their relation to high luminosity radio sources. They predict the existence of jets in all high luminosity radio sources.

These papers completely changed the way we looked at, and understood, extragalactic radio sources. I would say that the picture that evolved during that time is alive and well today. Since 1974, instrumental advances have allowed the detection of hundreds of radio jets (predicted by Blandford and Rees) so that there is no longer any doubt that radio galaxies are 'fed' by the nuclear regions through outflow in narrow channels (Bridle and Perley, 1984). In recent years, supercomputers have been used to model the outflow of supersonic jets, in order to study the formation of hot spots, and the stability of the flow. And, advances in theoretical understanding allow closer examination of the acceleration processes thought to be at work in the hot spots which the jets feed.

Indeed, it is this remarkable progress in all three aspects of astronomy which should make this meeting a fruitful one. This is, to my knowledge, the first meeting devoted entirely to the subject of 'hot spots'. And, I think it is timely to have this meeting now, since we have many imaged hot spots, many computational models, and (no doubt) many theories of particle acceleration at these hot spots.

2. The Fanaroff-Riley Relation

As the Cambridge synthesis instruments better resolved extragalactic radio sources, it became clear that there were fundamental morphological differences between radio sources, and that these differences were related to radio luminosity. A hint of what was to come is found in Hooley (1974), followed shortly by the well-known paper published by Fanaroff and Riley. They defined a parameter, R, as the ratio of the distance between the brightest features in a radio source to the maximum total extent of that radio source. They then noted that nearly all sources with R < 0.5 had spectral lumino-sities less than $P_{178} = 2 \times 10^{25}$ watt Hz^{-1} $ster^{-1}$, ($H_o = 50$), while nearly all sources with R > 0.5 had luminosities above this value.

This relation has stood very well the test of time and much improved imaging instruments. There really is a fundamental morpholo-gical difference between most high luminosity radio sources (R > 0.5), and low luminosity radio sources (R < 0.5). The low luminosity objects ('edge-darkened doubles') generally contain prominent, often 2-sided jets, with the lobe emission trailing off into intergalactic space. They give the strong impression of a rising plume, and indeed, current theories so represent them. The high luminosity objects ('edge brightened doubles', 'classical doubles', 'hot spot radio galaxies') contain hot spots in one or both lobes. The radio emission is always sharply bounded on the outer extremi-ties. They give the appearance of dynamically expanding volumes. In the radio astronomer's jargon, the low luminosity sources are called 'FR-I', the high luminosity objects 'FR-II'.

So, the fundamental question must be why there is such a clear morphological difference which is (almost) uniquely specified by luminosity. I believe that another way of putting the question in perhaps simpler terms is to ask why most of the most luminous sources have hot spots with sharply bounded lobe emission while their less energetic brethren really do look like gentle plumes statically con-fined by an external medium.

At this stage I wish to point out some recent work of Parma and DeRuiter. Using the 3C and B2 surveys, they have assembled the fol-lowing statistical summary of the fraction of sources which contain hot spots, as a function of radio luminosity.

Table 1. Frequency of Occurence of Hot spots

log P_{1400} [watt Hz^{-1}]	Fraction with Hot spots
21-22	No Doubles
22-23	0/4
23-24	0/17
24-25	13/46 (28%)
25-26	16/23 (70%)
26-27	4/6 (67%)

By way of comparison, the spectral luminosity of Cygnus A is about 10^{28} watt Hz^{-1}.

There are two main points to note here. The first is that hot spots *only* occur in high luminosity radio sources (although this statement is not likely to be universally accepted due to the wide spread in opinion of what a hot spot is). The second is that not all highly luminous radio sources have hot spots. A significant fraction (~30%) of such sources do not contain hot spots, and are classified as FR-I objects. Perhaps the best known luminous radio galaxy which has FR-I structure is Hercules A, 3C348. (Dreher and Feigelson, 1984). This observation adds an interesting new dimension to the problem stated above, namely:

Why are there a significant number of luminous radio sources which have FR-I structure - i.e., *don't* have hot spots?

3. Just What IS a Hot Spot?

I have perhaps erred in my order of presentation of this review by talking about hot spots without attempting to define them. From what I have already said, it might be concluded that hot spots are 'bright' and 'small'. How bright? How small? This is the problem. We observers have clearly failed to provide quantitative values on the brightness and physical size of hot spots. I have never seen a table listing the physical properties of hot spots of a wide variety of objects. And, I believe it is time to do so. Further, it is also time to attempt a clear definition of what constitutes a hot spot. After all, if we are assembled here to discuss these things, it seems reasonable that we should all have a clear understanding of what they are. So, in the following sections, I am going to attempt such a definition, and then let the chips fall where they may.

First, we must make clear what is *not* a hot spot. The radio emission nearly always seen which is coincident with the galaxy nucleus is *not* a hot spot. This emission is surely connected with the 'central engine', generally has an inverted (optically thick) spectrum, and, when resolved by VLB techniques, is found to contain a small (pc-scale) one-sided jet pointing to one of lobes. The nuclear emission falls outside the scope of the this review (and meeting), and I will not discuss it further.

Before attempting a definition of hot spots, it is worthwhile surveying the range of structures which might conceivably fall within such a definition. My intention in doing so is to suggest, and give evidence, that it is not easy to cleanly define hot spots. In fact, I believe that it may not even be possible to define hot spots in terms of observed parameters alone. Arranged on the following pages are contour maps of seven high luminosity objects, each containing high brightness regions. Shown in Fig. 1 is the classical high luminosity radio galaxy Cygnus A (about which we will have much to say later),

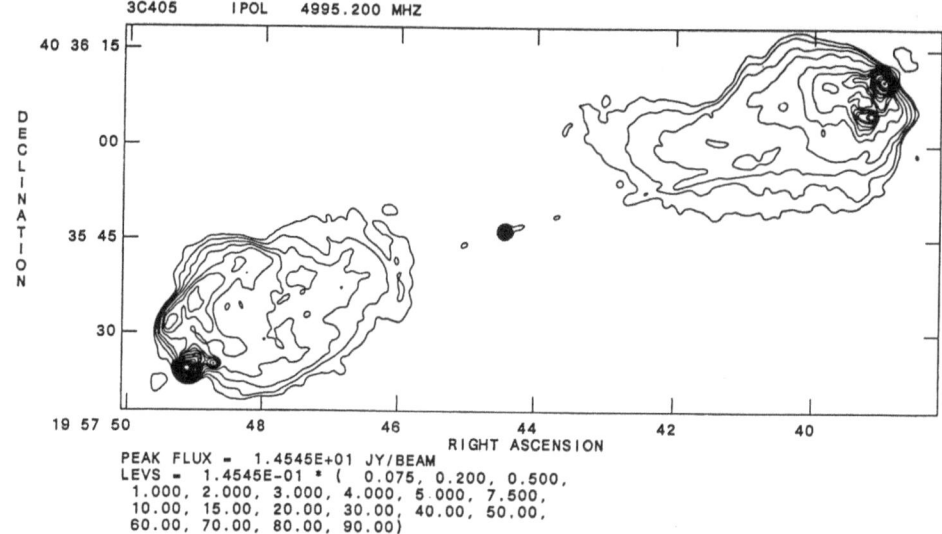

Figure 1

and in Fig. 2 a much less luminous radio galaxy, Pictor A. In both of these, note the presence of very bright, sharply bounded regions of emission located at the extremities. For both of these sources, the peak brightness of the hot spots is a factor of 20 to 200 or more times the typical brightness of the extended regions lying between the hot spots. (The '200 or more' is for Pictor A, as we have not yet resolved the hot spots). In physical units, the peak brightness of the hot spots for these sources is something like 20 Jy/arcsec^2 at 6 cm wavelength; the typical size less than 5 kpc with gradients on the scale of 100 pc or less (and probably less than 10 pc for Pictor A).

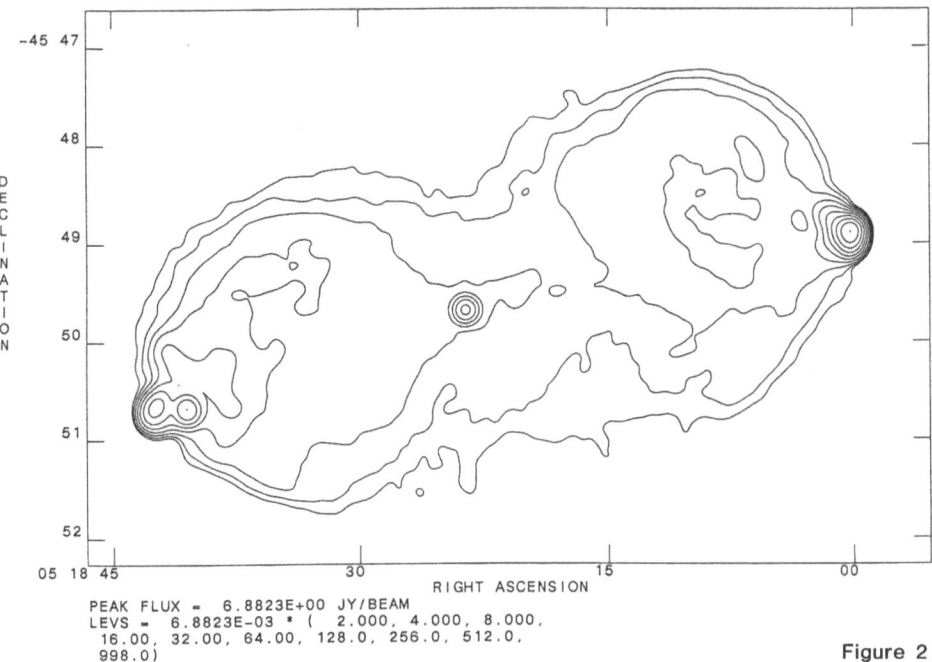

PEAK FLUX = 6.8823E+00 JY/BEAM
LEVS = 6.8823E-03 * (2.000, 4.000, 8.000,
16.00, 32.00, 64.00, 128.0, 256.0, 512.0,
998.0)

Figure 2

On the other hand, a very different kind of hot spot is shown in Fig. 3, an image of 3C33. Here the southern 'hot spot' is much more resolved, and much less bright, somewhat less than 1 Jy/arcsec^2 at 6 cm. Strong optical synchrotron emission is seen in both Pictor A and 3C33, but, curiously, not from Cygnus A. See the review by Röser in these proceedings.

Now peruse the following images, showing variants on the hot spot phenomenon. Unfortunately, in most of these the hot spots are not resolved, so I cannot give brightnesses or sizes. Fig. 4 shows 3C337. Here, the hot spots clearly protrude from the 'lobe'. This is rather uncommon - I don't know of a more extreme example than this one. In general, the maximum protrusion is about the diameter of the hot spot itself, a few kiloparsecs. The process which creates hot spots seems to know when the hot spot is 'exposed', and manages to prevent excessive 'exposure'. In Fig. 5 is 3C219, where the hot spots are deeply embedded within the lobe. In this source, the northern hot spot is actually a ring, with a circumferential magnetic field (Perley et al., 1979). A source with hot spots that both protrude and are embedded is 3C303, shown in Fig. 6. Note that there are additional 'hot spots', located between the nucleus and the prominent double

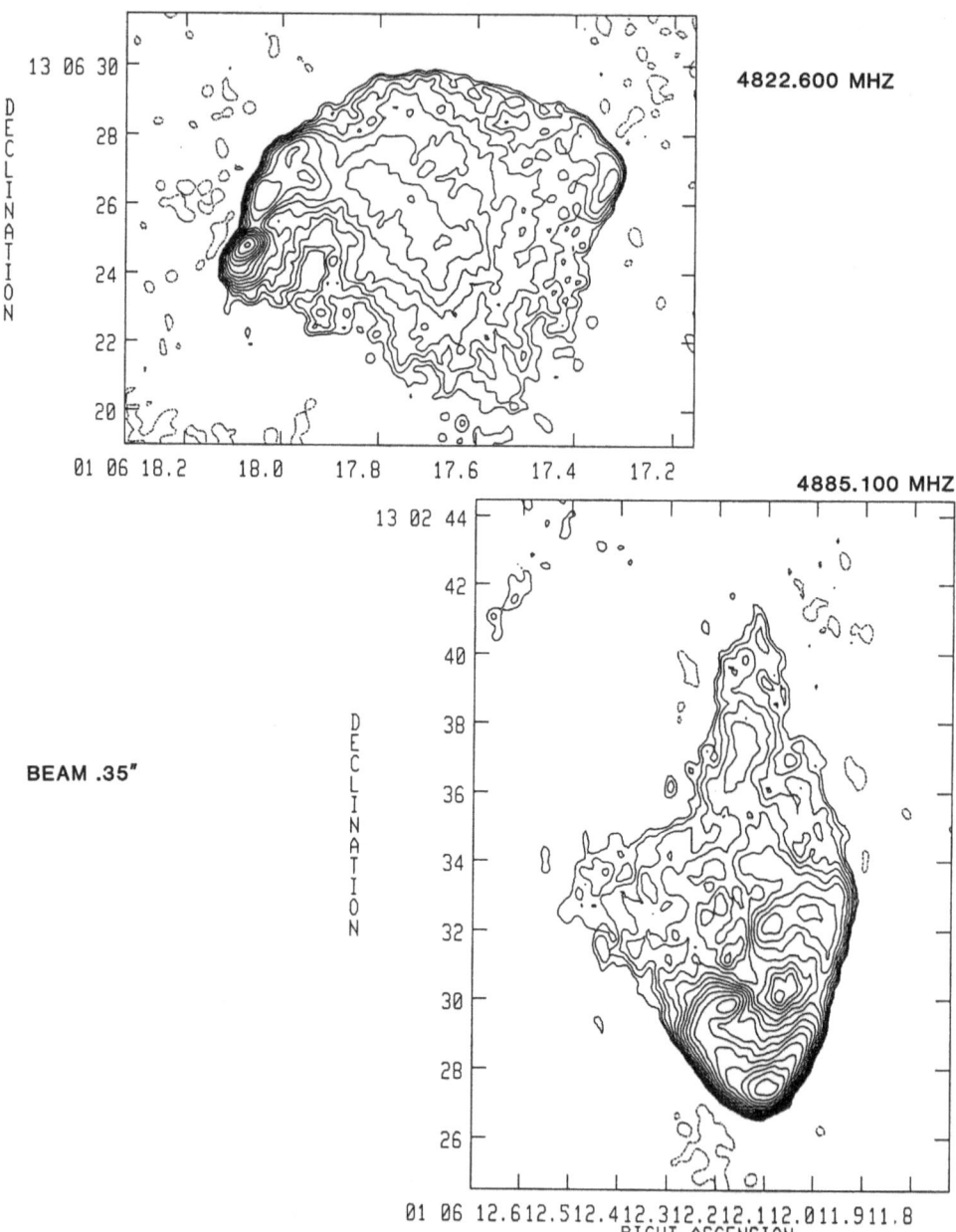

4822.600 MHZ

4885.100 MHZ

BEAM .35"

RIGHT ASCENSION

PEAK FLUX = 2.1278E-02 JY/BEAM
LEVS = 2.5000E-04 * (-1.00, 1.000, 1.400,
1.960, 2.744, 3.842, 5.378, 7.530, 10.54,
14.76, 20.66, 28.93, 40.50, 56.69, 79.37,
111.1, 155.6, 217.8, 304.9, 426.9)

PEAK FLUX = 9.2205E-02 JY/BEAM
LEVS = 5.0000E-04 * (-1.00, 1.000, 1.400,
1.960, 2.744, 3.842, 5.378, 7.530, 10.54,
14.76, 20.66, 28.93, 40.50, 56.69, 79.37,
111.1, 155.6, 217.8, 304.9, 426.9)

Figure 3: 3C 33

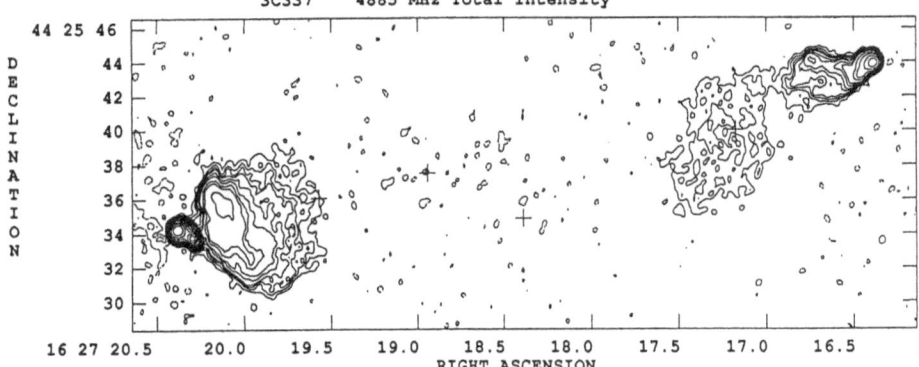

3C337 4885 MHz Total Intensity
DECLINATION
44 25 46
44
42
40
38
36
34
32
30
16 27 20.5
20.0
19.5
19.0
18.5
18.0
17.5
17.0
16.5
RIGHT ASCENSION

Figure 4

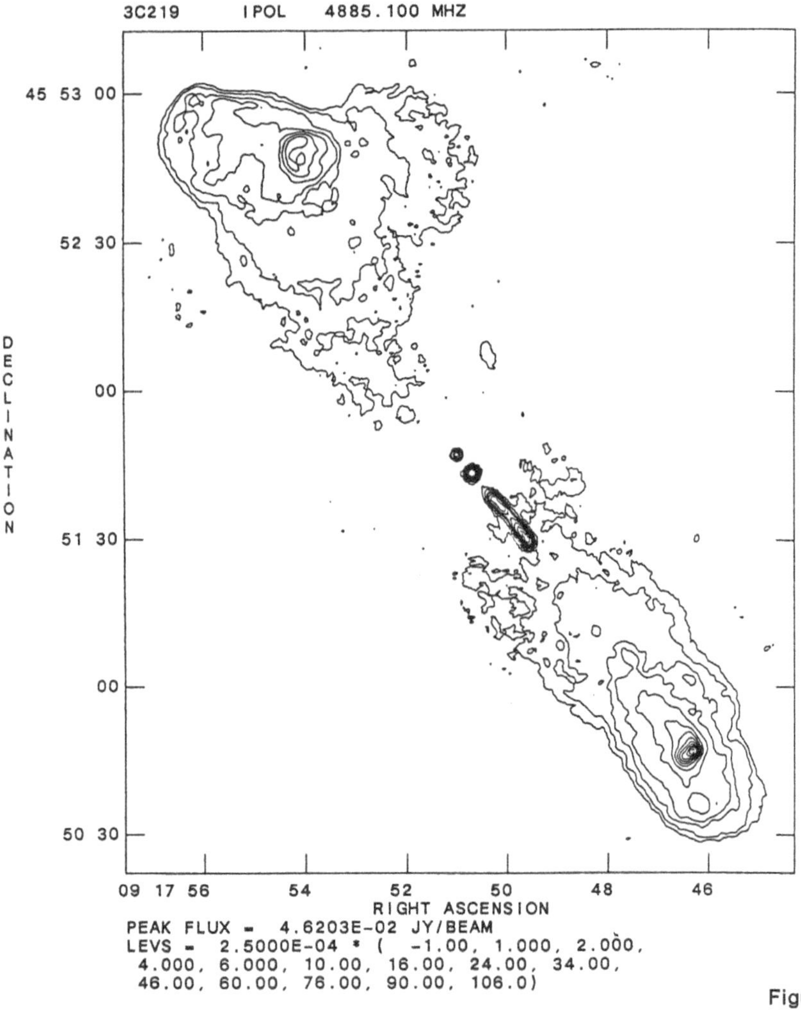

3C219 IPOL 4885.100 MHZ
DECLINATION
45 53 00
52 30
00
51 30
00
50 30
09 17 56
54
52
50
48
46
RIGHT ASCENSION
PEAK FLUX = 4.6203E-02 JY/BEAM
LEVS = 2.5000E-04 * (-1.00, 1.000, 2.000,
4.000, 6.000, 10.00, 16.00, 24.00, 34.00,
46.00, 60.00, 76.00, 90.00, 106.0)

Figure 5

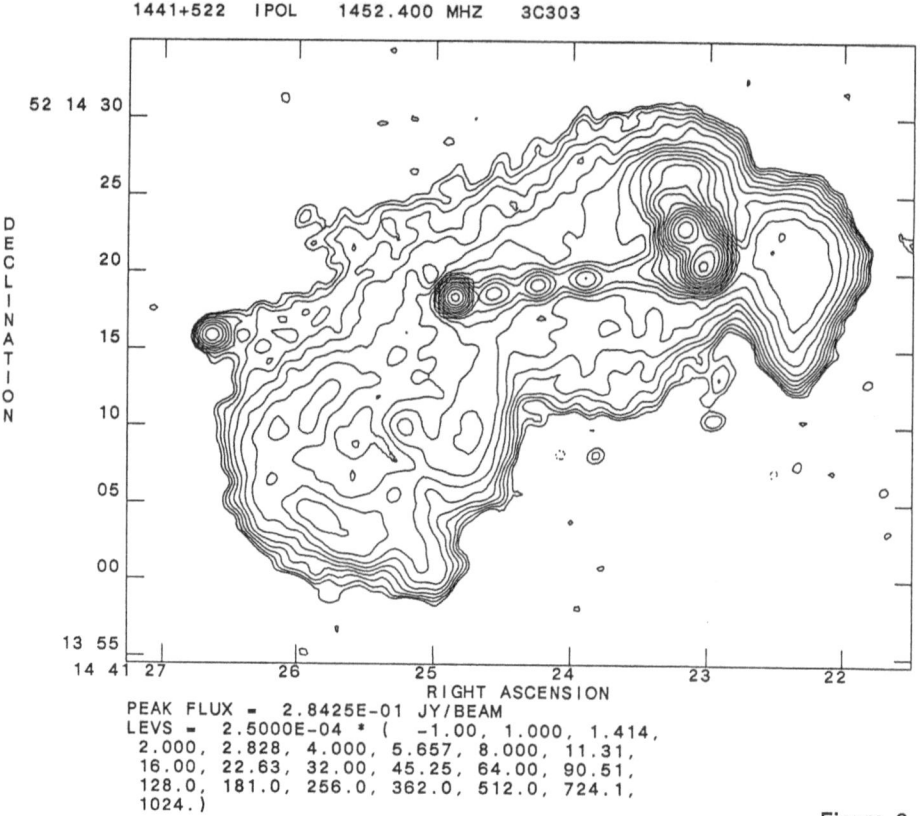

1441+522   IPOL   1452.400 MHZ   3C303

PEAK FLUX = 2.8425E-01 JY/BEAM
LEVS = 2.5000E-04 * (-1.00, 1.000, 1.414,
2.000, 2.828, 4.000, 5.657, 8.000, 11.31,
16.00, 22.63, 32.00, 45.25, 64.00, 90.51,
128.0, 181.0, 256.0, 362.0, 512.0, 724.1,
1024.)

Figure 6

hot spot in the western lobe. These are believed to be 'knots' in the jet feeding the double hot spots. Not all hot spots can be called 'spots' - 3C390.3 (Fig. 7) has in the northern lobe a 'hot line' (a wall jet?). Note that the opposite lobe of this source contains a very bright hot spot of normal appearance. Hot spots seen on one side are not uncommon. Finally, Fig. 8 shows an object which is almost all hot spot - the quasar 0800+608. The northern half is a long, knotty jet. Which of these brightness enhancements is the 'hot spot', if any?

Those foolish enough to attempt to define hot spots should keep these variants of the phenomenon in mind. It is, in my opinion, not enough to simply say that hot spots are bright spots. Nor can we say that they are bright spots at the far edges of radio sources. It is definitely preferable to make the definition in terms of observed parameters alone (so as not to bias the physics), but it may be

10

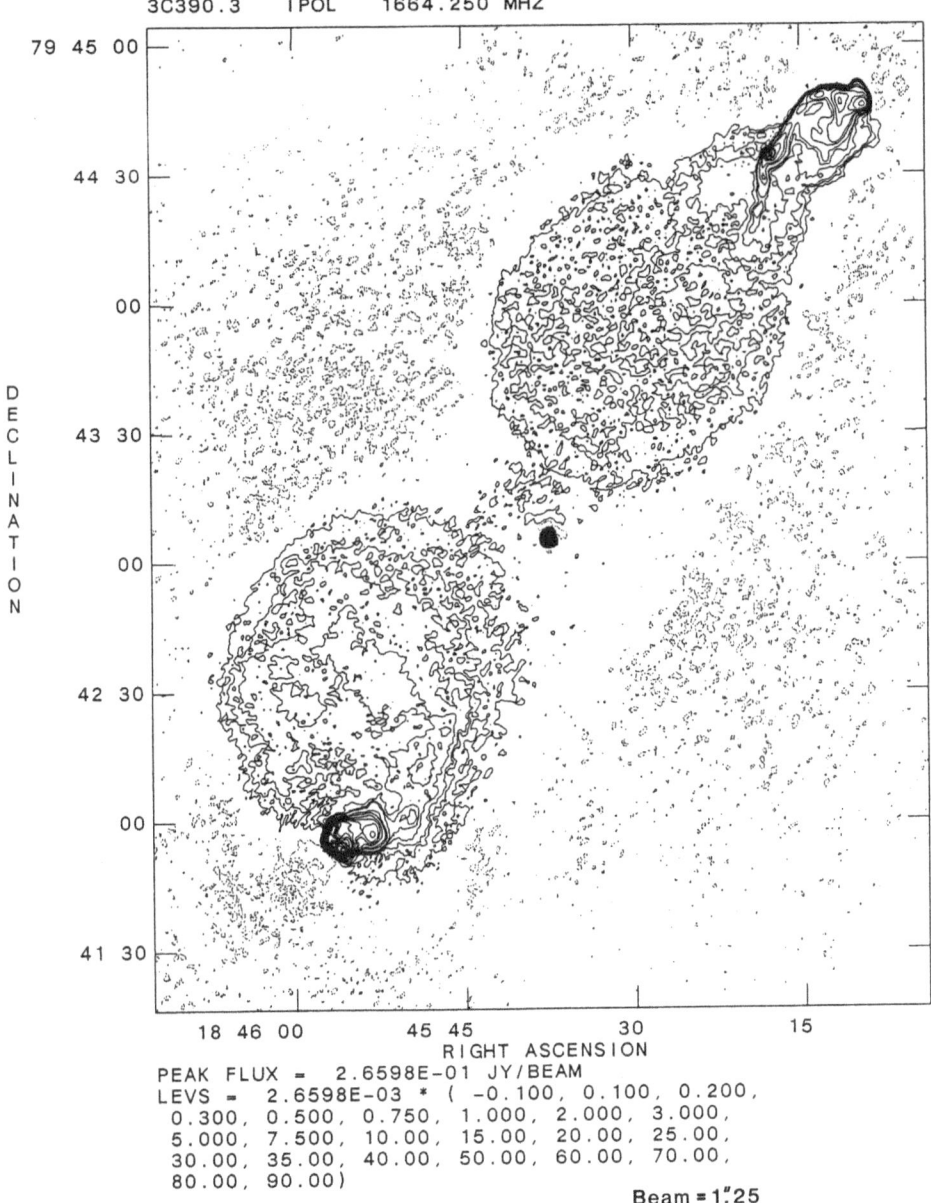

PEAK FLUX = 2.6598E-01 JY/BEAM
LEVS = 2.6598E-03 * (-0.100, 0.100, 0.200,
0.300, 0.500, 0.750, 1.000, 2.000, 3.000,
5.000, 7.500, 10.00, 15.00, 20.00, 25.00,
30.00, 35.00, 40.00, 50.00, 60.00, 70.00,
80.00, 90.00)

Beam = 1.ʺ25

Figure 7

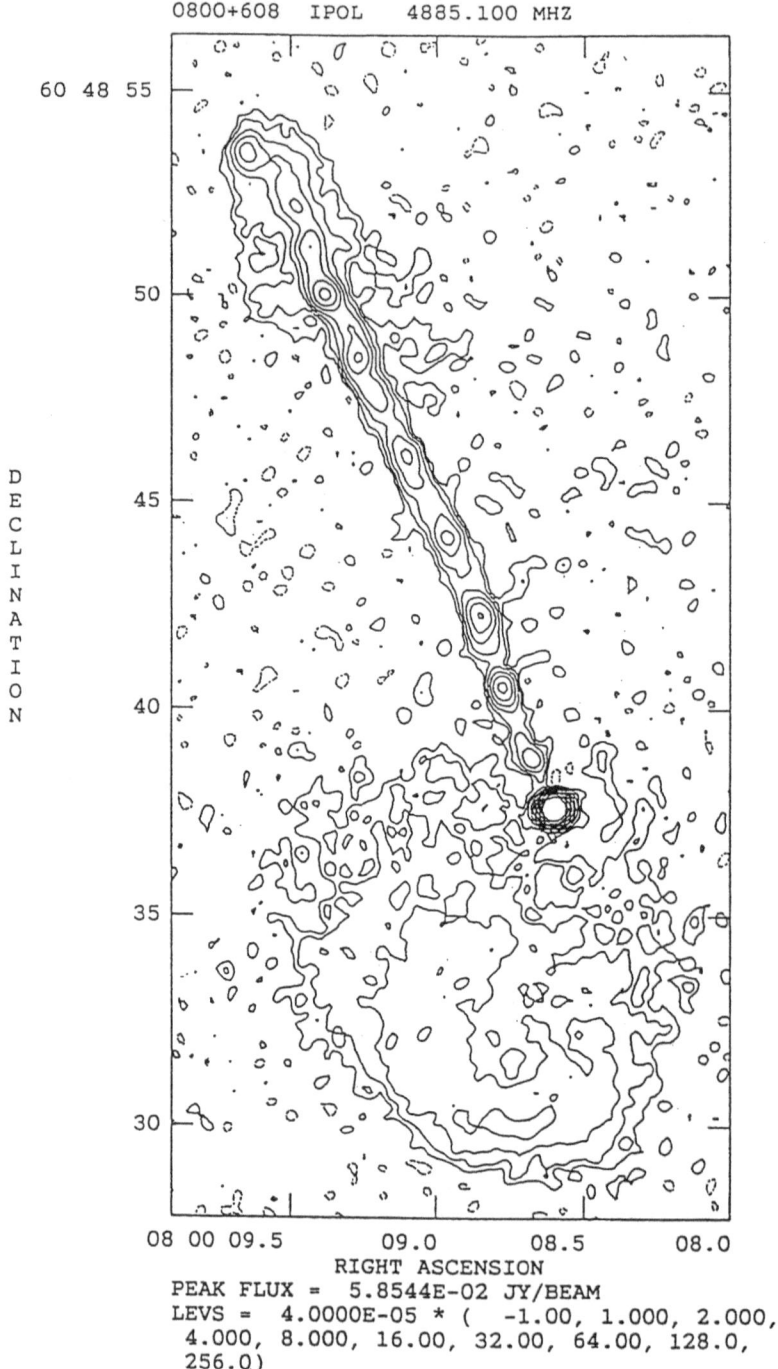

0800+608   IPOL   4885.100 MHZ

PEAK FLUX = 5.8544E-02 JY/BEAM
LEVS = 4.0000E-05 * (-1.00, 1.000, 2.000,
4.000, 8.000, 16.00, 32.00, 64.00, 128.0,
256.0)

Figure 8

12

necessary to inject some 'current wisdom' into the definition to allow discrimination between hot spots and objects of different physical origin, but similar appearance. I now give my zeroth order attempt at the definition of a hot spot:

A hot spot is a sharply bounded region of high surface brightness exceeding xxx mJy/arcsec^2 whose physical extent is less than yyy kpc and which is located within or in close proximity to the radio emitting lobes of extragalactic radio sources.

I have not put in values for 'xxx' or 'yyy' because I don't know what they should be. I would hazard a guess that 'xxx' will be some tens to perhaps 100 millijanskys per square arcsecond at 20 cm observing wavelength, and that 'yyy' will be a few hundred parsecs. As an afterthought, I will add that perhaps a relative brightness definition ('a hot spot is a compact region some zzz times as bright as the lobe surrounding it') might be more appropriate. Again, I will stress that more imaging of a large number of objects with sufficient resolution to resolve the hot spots will be needed.

This definition (when the numbers are put into it) should cover almost all examples of what we call hot spots. But, what should we do with an object such as 0800+608 (Fig. 8)? It is likely that the entire jet is a sort of a hot spot, with each 'knot' contributing to the deceleration of the flow, such that the jet, upon reaching the end, has been so weakened that a strong hot spot cannot be formed. There will be many examples of exceptions to the rule I have attempted, but I think my definition will cover most of the ranges of the known phenomena.

In the previous section, I stated the original definition of an FR-II radio source. This was based on observations made with low-resolution instruments. When high-resolution instruments are brought to bear, it seems that it is the presence of hot spots which made the 'R' parameter greater than 1/2. Thus, I propose that a suitable definition of an FR-II radio source should be as follows:

An FR-II radio source is one which possesses at least one (1) hot spot, as defined above.

4. The Basic FR-II Model

It hardly seems necessary to introduce, or discuss the now widely accepted model of extragalactic radio sources to such a specialist meeting. However, it may be useful to do so, in order that the story be somewhat complete.

Almost immediately after the Cambridge 5-km telescope maps of Cygnus A were published (Hargrave and Ryle, 1974) came the seminal paper of Blandford and Rees, proposing what is now widely known as the 'jet model'. Recall that the Hargrave and Ryle result showed, for the first time, the amazing brightness (and hence, short lifetimes) of the hot spots of a prominent radio source. They produced an analysis showing that energy deposition must be an ongoing phenomenon in these structures, but did not suggest any specific mechanism by which this could be happening. Blandford and Rees (1974) proposed a highly efficient 'pipeline' - a supersonic jet, produced by the nucleus, which quickly advected cold gas from the nucleus to the edges of the radio lobes. Here the jet is abruptly stopped by its impact with the undisturbed, denser intergalactic medium. This produces a strong shock, which at the least compresses the fluid and greatly increases the emissivity, and is also likely the site of relativistic particle acceleration. The shock is identified with the 'hot spot', and the 'waste' material from this shock inflates an underdense cavity which is identified with the radio lobe. This constitutes the standard model, which is widely accepted in the community.

Observations often show the presence of two hot spots in a lobe, although more than two are very rare, if any at all exist. This, plus the fact that the minimum pressures of the hot spots exceed the lobe pressures by up to two orders of magnitude, indicate that hot spots are probably transient features, forming, dying, reforming, etc. Other lines of evidence pointing towards transient hot spots are that hot spots are commonly found recessed, that they are rarely, if ever, found 'outside' of a lobe by more than a few kiloparsecs (i.e., the scale of the hot spot itself), and that the jets which feed them are commonly bent. These observations gave rise to Peter Scheuer's 'dentist drill' model, a modification of the 'standard model'. Here, the hot spots are short lived, perhaps less than 10^5 years, (this timescale being set by the free expansion timescale of the hot spot), in which time the hot spot moves forward (i.e., extending the lobe boundary) by a few kiloparsecs, then for reasons presumably due to stability, they pinch off, causing a new hot spot to be formed elsewhere.

I won't say any more on the double hot spot picture, as Robert Laing will have much to say about them in the next talk.

These standard pictures are now commonly accepted, and are well supported by computer simulations. See the contribution by Mike Norman in these proceedings.

5. The Environment of Radio Galaxies

5-1. The Optical Identifications of Radio Galaxies

It has been known since the earliest days of optical identification of radio sources that luminous radio galaxies are (nearly) exclusively identified with elliptical galaxies. Further, these galaxies are generally 'normal' in appearance. Although FR-II radio galaxies usually have strong nuclear emission lines, there is no simple one-to-one correspondence between the presence of strong radio activity and emission lines (Hine and Longair, 1979).

Recent work has allowed quantitative measure of these ideas, and has allowed a clear distinction to be drawn between the type of galaxies identified with FR-I radio sources, and those identified with FR-II sources. Longair and Seldner (1979) showed that FR-I sources are generally found in clusters, while FR-II sources are usually isolated field galaxies. However, note that this latter trend is not absolute, as the most famous FR-II radio source, Cygnus A, is located in a loose cluster which is a luminous X-ray source (Fabbiano et al., 1979). Longair and Seldner's work has been extended by Prestage and Peacock (1988), using a larger, more complete sample. They support Longair and Seldner's conclusions, but point out that the FR-Is encompass a wide range of locations and types.

These conclusions are supported by careful optical photometry done by Owen and Laing (1988). They find that FR-II galaxies are not abnormally large, and have a remarkably small spread in absolute

15

magnitude, with $M_v \sim -22$. Further, these galaxies have normal light profiles with a small spread in magnitude and radius. It seems that FR-II galaxies form a remarkably homogeneous set. On the other hand, the FR-I galaxies are very much more varied. They are generally bigger, brighter and, in agreement with Longair and Seldner, generally found in clusters. Their spread in radius and magnitude is very large. These associations are shown in Fig. 9, kindly reproduced by permission of Frazer Owen. This shows the radio-optical luminosity distribution, with the radio morphology noted by the plotted points.

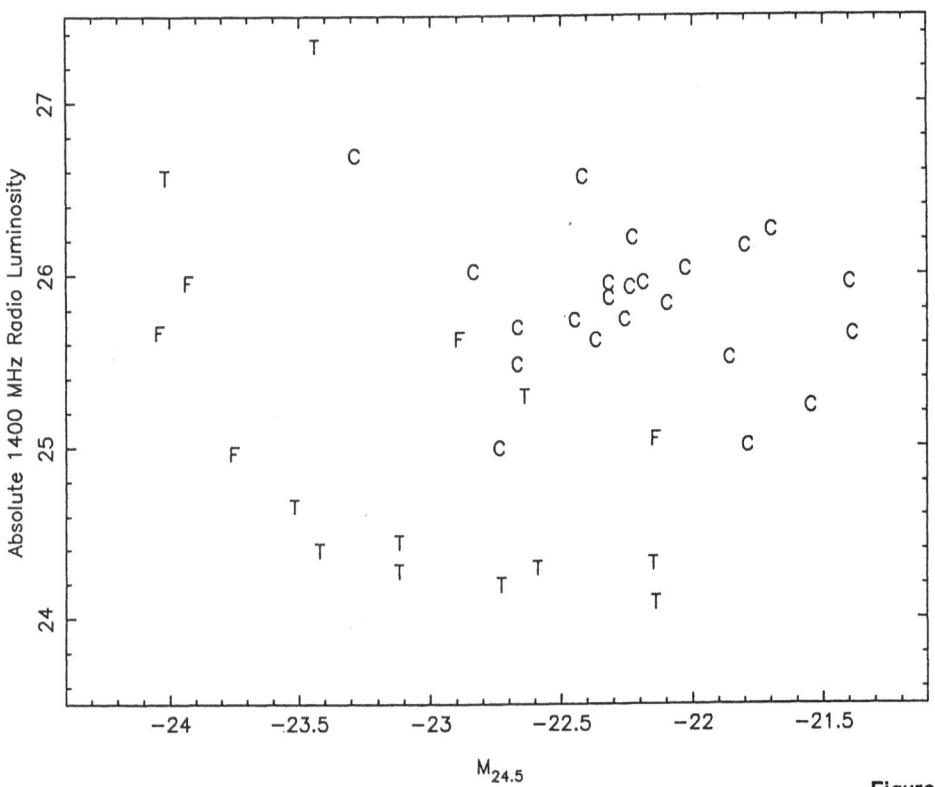

Figure 9

Note the restricted range in which classical doubles are found (marked by 'C'). All other points denote FR-Is, following a scheme given in their paper. An important point which comes out of all this is that FR-II galaxies cannot evolve into FR-Is, since the dynamical timescales for this evolution are very much greater than the lifetimes of the radio sources.

Although Owen and Laing state that the galaxies they studied have normal light profiles, Heckman, van Breugel and their co-workers

have described a rather different picture, stating that 1/3 to 1/4 of
all powerful radio galaxies have peculiar optical morphology. The
difference between these is most likely one of degree - that is, in
Owen and Laing's definition of 'normal' vs. Heckman and van Breugel's
definition of 'peculiar'.

5-2. Evidence for Gaseous Environments

Perusal of any of the high quality VLA images of luminous radio
sources should be enough to strongly encourage the thought that the
radio lobes are confined by an external medium. Indeed, in the context of the standard model, an external medium must be present in
order that hot spots be seen. I note here that adiabatic expansion of
a synchrotron-emitting cloud results in a catastrophic reduction in
flux density and brightness. The simple analysis presented by
Longair, Ryle and Scheuer (1973) demonstrates that the flux density
scales with r^{-5}, the surface brightness with r^{-7}. A simple, adiabatic
expansion of a cloud by a factor of two will certainly result in
extinction of the radio emission.

Although the general morphology of luminous radio sources lends
encouragement to ideas of confinement, or at least retardation by an
external gas, I note that some objects persist in complicating the
simple picture. For example, although 3C341 (Fig. 10) looks nicely
confined, the structure of 4C14.27 (Fig. 11) shows a distinct trough
which bifurcates the radio structure. For those who think this may be
explained by some dense galactic disk, I offer up 4C14.11 (Fig. 12)
and 0938+39, (Fig. 13) both of which show 'X-shaped' structure.
Instead of a trough, these objects offer wings. Leahy and Williams
(1984) have a model for this kind of structure involving backflow
through the lobes along with buoyancy-driven outflow. Precession of
the radio axis also comes to mind. The main point is that interpretation of lobe structure is not necessarily simple.

The most compelling evidence for confining hot gas surrounding
ellipticals comes from X-ray observations. Feigelson and Berg (1983),
analysing data from 'a representative sample' of extragalactic radio
sources report that the inferred pressures are sufficient to confine
lobes and bridges of radio galaxies, but not the hot spots. The clear
inference is that the hot spots are 'confined' by ram pressure, so
that they are likely to be short-lived phenomena, as I have argued in

17

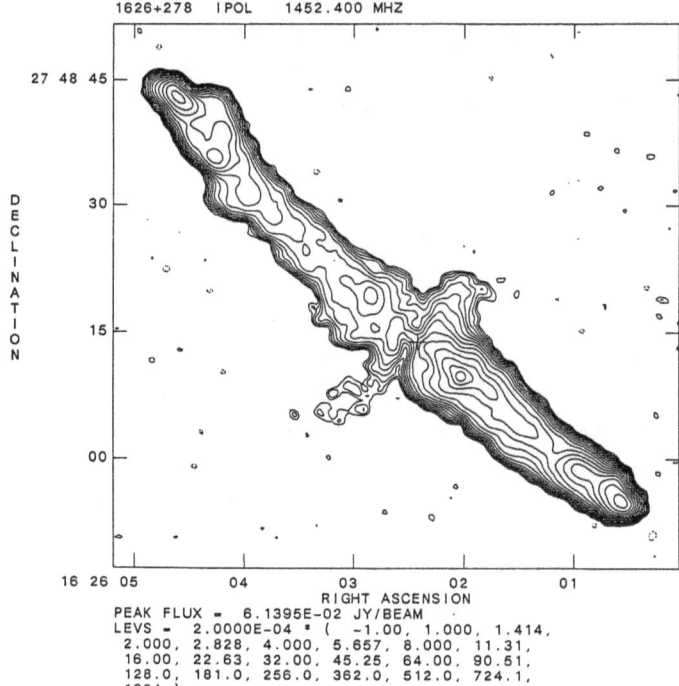

1626+278 IPOL 1452.400 MHZ

PEAK FLUX = 6.1395E-02 JY/BEAM
LEVS = 2.0000E-04 * (-1.00, 1.000, 1.414,
2.000, 2.828, 4.000, 5.657, 8.000, 11.31,
16.00, 22.63, 32.00, 45.25, 64.00, 90.51,
128.0, 181.0, 256.0, 362.0, 512.0, 724.1,
1024.)

Figure 10

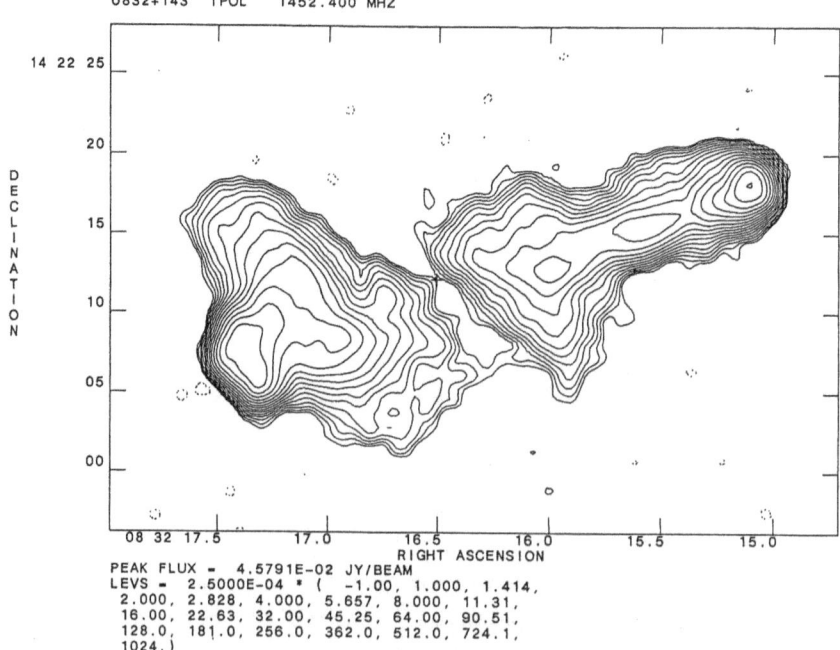

0832+143 IPOL 1452.400 MHZ

PEAK FLUX = 4.5791E-02 JY/BEAM
LEVS = 2.5000E-04 * (-1.00, 1.000, 1.414,
2.000, 2.828, 4.000, 5.657, 8.000, 11.31,
16.00, 22.63, 32.00, 45.25, 64.00, 90.51,
128.0, 181.0, 256.0, 362.0, 512.0, 724.1,
1024.)

Figure 11

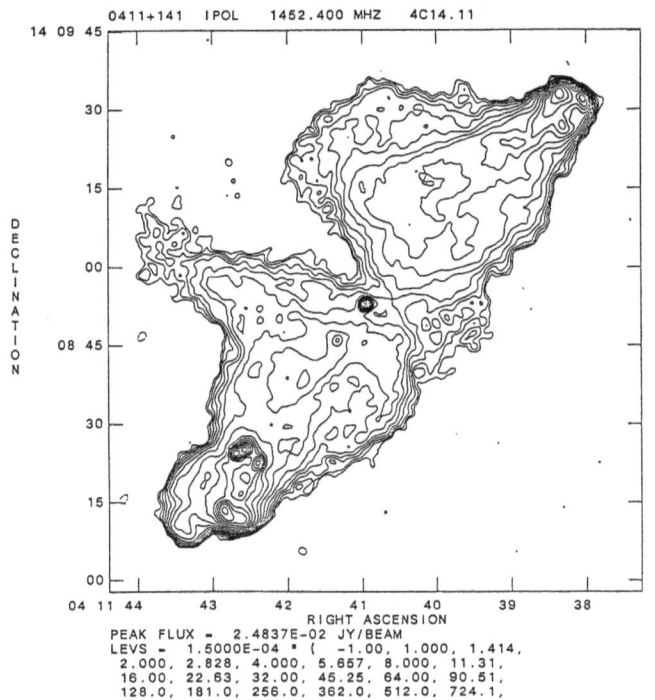

0411+141   IPOL   1452.400 MHZ   4C14.11

PEAK FLUX = 2.4837E-02 JY/BEAM
LEVS = 1.5000E-04 * (-1.00, 1.000, 1.414,
2.000, 2.828, 4.000, 5.657, 8.000, 11.31,
16.00, 22.63, 32.00, 45.25, 64.00, 90.51,
128.0, 181.0, 256.0, 362.0, 512.0, 724.1,
1024.)

Figure 12

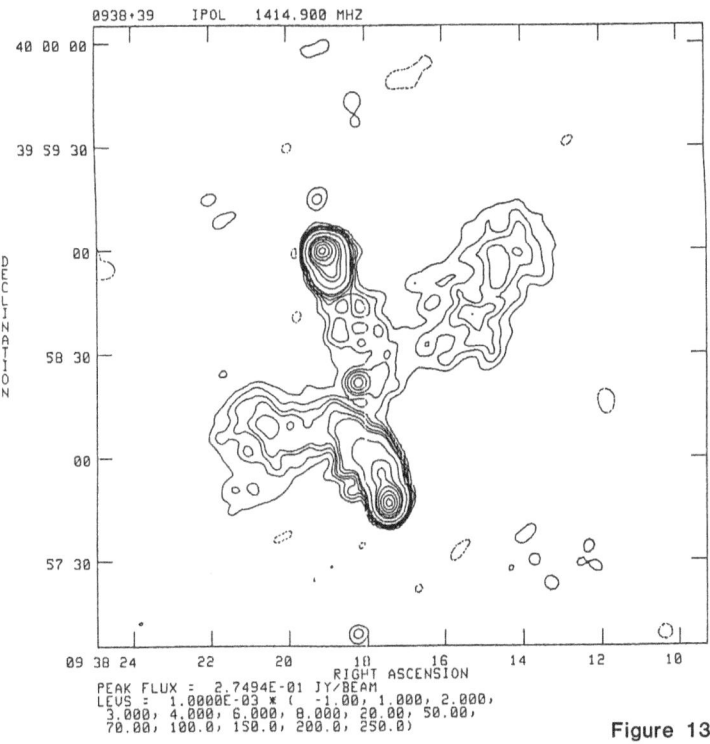

0938+39   IPOL   1414.900 MHZ

PEAK FLUX = 2.7494E-01 JY/BEAM
LEVS = 1.0000E-03 * (-1.00, 1.000, 2.000,
3.000, 4.000, 6.000, 8.000, 20.00, 50.00,
70.00, 100.0, 150.0, 200.0, 250.0)

Figure 13

the previous section. Feigelson and Berg's sample included no FR-II sources, so their conclusions are more relevant to FR-I objects.

Fabbiano et al. (1984), analysing Einstein data, studied a better-defined sample of radio galaxies concluded that the X-ray luminosity of FR-II sources is greater than that of FR-Is, and that the X-ray luminosity of these FR-II objects was correlated with the radio power of the nucleus. This indicates that the X-ray emission from luminous radio galaxies does not originate from extended gas. Miller et al. (1985) presented a careful study of the X-ray emission from FR-II sources. Their analysis was of those FR-II sources close enough to have extended emission on galactic scales to be resolvable by the Einstein instrument. Unfortunately, this necessary requirement results in a very small sample size. Miller et al.'s analysis shows no extended X-ray emission in any object, and further indicates that the upper limit in confining pressure is insufficient to statically confine the lobes of luminous radio galaxies (except possibly the faintest regions). They favor a steady, powered expansion of the radio source, noting that if the central engine turned off, a quick extinction of the radio emission is likely, due to adiabatic losses, as argued above.

Since so few FR-II radio galaxies were within reach of Einstein, another approach is to consider elliptical galaxies in general, arguing that the mean X-ray properties of the extended emission of these are unlikely to be different than for those galaxies hosting strong radio emission. Thus, one can consider the results of Forman et al. (1985) who analyzed the Einstein observations of 55 E-galaxies. They show that hot coronae are common around ellipticals, with typical central temperatures of 10 million K, central densities of 10^{-2} cm^{-3} and core radii of about 5 kpc. Using a standard iso-thermal model, these parameters predict that the hot gas, at radii of over 100 kpc is insufficient to statically confine typical radio lobes. However, I note that very little indeed is known about gas densities at these radii - in particular, a very hot, dilute gas cannot be ruled out on the basis of current observations.

It is quite possible that the radio sources themselves are, in a sense, responsible for the apparent confinement. If the radio lobes expand supersonically (which is quite possible, - certainly the hot spots move forwards supersonically), they will drive a shock wave in advance of them, which in turn heats and compresses the external medium. The resulting pressure rise is of order M^2. Put another way, the expanding source builds up a pressure wall in front of it due to the swept-up external gas, whose effective (stagnation) pressure is approximately ρv^2. The intertia of this 'wall' could be sufficient to give the sharply bounded appearance of the radio lobes - especially in the outer areas. In the older central regions, lobes are often difficult to detect, and more spread out, with boundaries much less sharp. Here we can imagine adiabatic losses taking their toll, while the expansion slows so that the inertial of the external gas becomes much less effective in 'confining' the interior gas.

Another line of evidence comes from spectroscopy of the environs of luminous radio galaxies. Van Breugel and Heckman, and their co-workers have reported abundant line emission from the environs of radio galaxies. See Baum et al. (1988) for the latest on this subject. Also, Wil van Breugel's review of the subject is in this workshop proceedings. This emission is much more commonly seen around radio sources than around radio-quiet galaxies of similar optical appearance, indicating the radio activity is somehow intimately connected with the line emission. Yet there is no clear correlation between the sites of the line emission and radio structure, although some correlations exist in a few sources. Given that the environs of galaxies are known to be hot, and thus fully ionized, it is difficult to understand how these cool spots form, how long they last, and what their relation is to the radio structure. However, their very existence some tens to hundreds of kiloparsecs from the core certainly constitutes evidence for gas at these distances.

Another important line of evidence for external gas around radio sources comes from observations of the polarized flux. If a source of polarized brightness (such as a radio galaxy) is viewed through an magnetized ionized gas, the plane of polarization will be rotated following the well known law:

$$\chi(\lambda) = \chi(0) + RM\lambda^2$$

where RM is the rotation measure, commonly expressed in radians per meter squared. This is in turn related to the physical parameters of the gas by:

$$RM = 8.1 \times 10^5 \int n_e \ [\text{cm}^{-3}] \ B \ [\text{Gauss}] \cdot dl \ [\text{parsec}]$$

Thus, the Rotation Measure is a density weighted integral of the line-of-sight magnetic field. As such, we can hope to learn something of the density of gas around radio sources, although it is more commonly used to estimate the magnetic field, since density information is independently attained through X-ray observations. Nevertheless, the very detection of an RM is evidence of an environment (here I am ignoring the Galactic contribution to the RM).

In most FR-II radio sources (meaning, all those not in dense clusters), the RM is found to be less than a few dozen radians/m^2 (Leahy et al., 1986). These values are easily explained as coming from galactic gas in conjunction with fields of 1 microgauss or less. These parameters are much the same as expected for the very extended FR-I sources. However, in two, very special FR-II sources, the RM is vastly greater. These are 3C405 (Dreher et al., 1987) and 3C295. Both these objects are extraordinarily luminous $P_{1400} \sim 10^{28}$ watt Hz^{-1}, and both are at the center of dense X-ray emitting gas. The RM images of these objects show values of plus/minus 4000 radians/m^2 for Cygnus, and 5 times greater values of 3C295. In both objects, the RM structure is not random, but shows large-scale order, suggesting ordered fields in the clusters on scales for 50 to 100 kpc, or even greater. The implied magnetic fields, using cluster-sized path lengths, and densities given by X-ray observations, run from 1 to 10 μG. It is quite possible that a significant fraction of the RM is due to a local origin - the shocked sheath surrounding the radio lobe. Within this sheath, densities are enhanced by factors of 4, as are transverse magnetic fields. This origin of the RM, which is strongly supported by the discovery of the 'Carilli Arc' (see the contribution from Chris Carilli in this volume), require magnetic fields in the sheath of some 50 to 100 μG.

In general, the subject of magnetic fields in clusters, as well as surrounding field galaxies, is quite wide open. There are few observations, and plenty of opportunity for useful work.

5-3. Depolarization of Radio Emission

The term 'depolarization' is commonly used to denote a lowering of the degree of linear polarization of radio emission with increasing wavelength. This is commonly observed in extragalactic radio sources. However, the interpretation of this phenomenon is not necessarily easy. There are two separable reasons for depolarization:

(1.) Internal Depolarization. There is thermal gas intermixed with the radiating relativistic particles, causing a rotation of the plane of polarization of the back emission w.r.t. the front. This amount of rotation depends upon frequency, and increases at lower frequencies. This causes a true depolarization. This effect has been extensively searched for, with (in the author's opinion) no credible detections. The corresponding upper limits on internal thermal gas density run from 10^{-3} to 10^{-5} per cm^{-3}. (Hargrave and Ryle, 1974, Spangler and Sakurai, 1985)

(2.) External Depolarization. If the radiation passes through a Faraday screen, the plane of polarization rotates, as described above. If it happens that the screen contains structure which is unresolved by the beam, then instrumental ('external') depolarization can occur. This is also a function of wavelength, and is difficult to separate from true internal depolarization. Another form of instrumental depolarization occurs if the rotation measure is such that the plane of polarization rotates by a significant amount across the bandpass used. Note that depolarization caused by structure of the radio source itself (the plane of the emitted radiation changes on a scale smaller than the beam) will not cause depolarization, since this will occur at all wavelengths.

Interestingly enough, depolarization by a Faraday screen may give us some interesting information about the environments of radio galaxies. Laing (1988), and Garrington et al. (1988) have noted an asymmetry in the depolarization across radio galaxies, in the sense that the 'unjetted' side of a radio source is more depolarized than the 'jetted' side. The simplest interpretation is that the 'unjetted' side points away from us, so that radiation from that end of the radio source passes through more external ionized medium than the 'jetted' side. This interpretation would provide strong support for the contention that the jets in these sources are relativistic.

5-4. Spectral Evolution

Under the standard model of FR-II radio source evolution, the source grows longer as the jet pushes back the external medium. This indicates that the central regions are 'older' than the outer regions, in the sense that particles located in the central regions were processed through the hot spot shock at a time before those currently radiating at the ends of the radio source. Provided there is no significant re-acceleration within the lobes, and no significant transport of particles since the last acceleration, there should be a significant spectral gradient in the radiation across the radio source. That is, the spectral index of the radiation from the central regions should have a steeper spectral index than that at the ends.

Indeed, this is the case in most (but probably not all) FR-II radio galaxies. This 'steepening' of the spectrum is especially noted in Cygnus A and 3C295, where the (equipartition) fields are very strong, causing short radiative lifetimes. The appearance of a bend in the spectrum can be interpreted as an age, using the familiar expression

$$\nu_{GHz} = 10^{18} \; H_{\mu G}^{-3} \; t_{yr}^{-2}$$

where ν_{GHz} is the frequency at which the 'synchrotron bend' is believed (or observed) to occur.

The approach taken by Burch (1977, 1979), Mayer (1979), Winter et al. (1980), and Alexander (Alexander 1987, Alexander and Leahy, 1987, Alexander et al, 1984) is to fit the spectra at various points across a radio lobe, using multi-frequency data, assume a magnetic field (usually the equipartition field), and so derive the age as a function of position. For Cygnus A, this yields 6 million years. Alexander and Leahy (1987) have done this for 21 objects. In doing this, they find that the lobe expansion is less than 0.2c for all sources (with a range of 1000 to 60,000 km/sec), that the lobe expansion velocity increases with source power, that acceleration in the lobes is not required to explain the spectra, and that the anomalous sources which don't fit the above trends are not 'classical doubles'. These expansion velocities are within the limits set by statistical arguments based on the brightness and angular size ratios of radio lobes by, for example, Mackay (1973), Longair and Riley (1979) and Swarup and Banhatti (1981).

6. Summary

Recent instrumental improvements have given observers the tools that accurately quantify the hot spot phenomenon. In many cases, the basic data are taken. Unfortunately, not all the images are made, and no comprehensive effort has yet been undertaken to consolidate the information in a way to make it useful to all. It seems to this reviewer that accurate images are needed before accurate descriptions can be made.

Nevertheless, the phenomenon of hot spots is well established. A few very good images are available. Recent advances in computing and theory should soon enable better understanding of the phenomenon.

REFERENCES

Alexander, P. *M.N.R.A.S.* **225**, 27, (1987)

Alexander, P., Brown, M.T., and Scott, P.F.
    M.N.R.A.S. **209**, 851, (1984)

Alexander, P. and Leahy, J.P. *M.N.R.A.S.* **225**, 1, (1987)

Burch, S.F. *M.N.R.A.S.* **180**, 623, (1977)

Burch, S.B. *M.N.R.A.S.* **186**, 519, (1979)

Baum, S.A., Heckman, T., Bridle, A., van Breugel, W., and Miley, G.
    Ap.J.Suppl.Ser. **68**, (1988)

Blandford, R.D. and Rees, M.J. *M.N.R.A.S.* **169**, 395, (1974)

Bridle, A.H., and Perley, R.A. *Ann.Rev.Astr.Astrophys.* **22**, 319 (1984)

De Young, D.S. and Axford, W.I. *Nature* **216**, 129, (1967)

Donaldson, W., Miley, G.K., and Palmer, H.P.
    M.N.R.A.S. **152**, 145, (1971)

Dreher, J.W., Carilli, C.L., and Perley, R.A. *Ap.J.* **316**, 611, (1987)

Dreher, J.W., and Feigelson, E. *Nature* **308**, 43, (1984)

Fabbiano, G., Doxsey, R.E., Johnston, M., Schwartz, D.A., and
    Schwarz, J. *Ap.J.* **230**, L67, (1979)

Fabbiano, G., Miller, L., Trinchiere, G., Longair, M. and Elvis, M.
    Ap.J. **277**, 115, (1984)

Fanaroff, B.L. and Riley, J.M. *M.N.R.A.S.* **167**, 31P, (1974)

Feigelson, E.D., and Berg, C.J. *Ap.J.* **269**, 400, (1983)

Forman, W., Jones, C., and Tucker, W. *Ap.J.* **293**, 102 (1985)

Garrington, S.T., Leahy, J.P., Conway, R.G. and Laing, R.A.
    Nature **331**, 147, (1988)

Graham, I. *M.N.R.A.S.* **149**, 319, (1970)

Hargrave, P.J., and Ryle, M. *M.N.R.A.S.* **166**, 305, (1974)

Hine, R.G. and Longair, M.S. *M.N.R.A.S.* **188**, 111, (1979)

Hooley, T. *M.N.R.A.S.* **166**, 259, (1974)

Laing, R.A. *Nature* **331**, 149, (1988)

Leahy, J.P., Pooley, G.G., and Riley, J.M.
      M.N.R.A.S. **222**, 753 (1986)

Leahy, J.P. and Williams, *M.N.R.A.S.* **210**, 929, (1984)

Longair, M.S., and Macdonald, G.H. *M.N.R.A.S.* **145**, 309, (1969)

Longair, M.S., and Riley, J.M. *M.N.R.A.S.* **188**, 625, (1979)

Longair, M.S., Ryle, M., and Scheuer, P.A.G.
      M.N.R.A.S. **164**, 243, (1973)

Longair, M.S., and Seldner, M. *M.N.R.A.S.* **189**, 433, (1979)

Mackay, C.D. *M.N.R.A.S.* **162**, 1, (1973)

Mayer, C.J. *M.N.R.A.S.* **186**, 99, (1979)

Miley, G.K., and Wade, C.M. *Astrophys. Lett.* **8**, 1, (1971)

Miller, L, Longair, M.S., Fabbiano, G., Trinchiere, G., and Elvis, M.
      M.N.R.A.S. **215**, 799, (1985)

Mitton, S. *M.N.R.A.S.* **149**, 101, (1970)

Mitton, S., and Ryle, M. *M.N.R.A.S.* **146**, 221, (1969)

Owen, F.N., and Laing, R.A. preprint (1988)

Perley, R.A., Bridle, A.H., Willis, A.G. and Fomalont, E.B.
      A.J. **85**, 499 (1980)

Prestage, R.M., and Peacock, J.A. *M.N.R.A.S.* **230**, 131, (1988)

Ryle, M. *Nature* **239**, 435, (1972)

Scheuer, P.A.G. *M.N.R.A.S.* **166**, 513, (1974)

Spangler, S.R. and Sakurai, T. *Ap.J.* **297**, 84 (1985)

Swarup, G. and Banhatti, D.G. *M.N.R.A.S.* **194**, 1025, (1981)

Winter, A.J.B., Wilson, D.M., Warner, P.J., Waldram, E.M.,
      Routledge, D., Nicol, A.T., Boysen, R.C., Bly, D.W.J., and
      Baldwin, J.E. *M.N.R.A.S.* **192**, 931, (1980)

Radio Observation of Hot Spots

R. Laing

Royal Greenwich Observatory
Herstmonceux Castle
Hailsham
East Sussex, BN27 1RP
United Kingdom

Abstract

A simple morphological description of radio hot spots is given, based on recent observations of high linear resolution. It is stressed that hot spots are intrinsically non-axisymmetric and the consequences for theoretical modelling are indicated. It is suggested that flow velocities in powerful sources are relativistic on all scales up to and including that of the hot spots.

1. Introduction

As Perley (these proceedings) has indicated, it became clear in the early 1970's (e.g. Miley & Wade 1971; Hargrave & Ryle 1974) that powerful radio sources of Fanaroff & Riley (1974)'s Class II (hereafter FR-II) contained components of high surface-brightness at their ends. The picture which emerged, primarily from observations with the Cambridge 5-km Telescope (Jenkins, Pooley & Riley 1977 and references therein), was that such "hot spots" were roughly axisymmetric structures, located at the outer extremities of the radio lobes. They had typical sizes of a few kpc and merged into more diffuse "tails" which pointed back towards the central galaxy or quasar. Their theoretical interpretation as the "working surface" at the end of a jet was developed by Blandford & Rees (1974) and Scheuer (1974). In the 1970's, it was already clear that hot spots often had complicated sub-structure (e.g., Hargrave & Ryle 1974, 1976; Hargrave & McEllin 1975), but the underlying regularities have only become

This paper is concerned mainly with the developments in the study of hot spots which have taken place since 1980, and the review by Miley (1980) provides a convenient starting-point. The main technical advances have been made by MERLIN and the VLA and I shall stress the importance of the increased linear resolution provided by these instruments. Section 2 discusses the definition of a hot spot. In Section 3, I summarise the results of a survey of about 60 powerful sources at a resolution of 0.1 arcsec and use them to develop a simplified description of hot spot morphologies. In Section 4, the spectra and magnetic field configurations of hot spots are outlined. Section 5 is concerned with the sidedness of compact hot spots and their relation to jets and Section 6 describes a simple physical picture of the flow. Finally, Section 7 summarises the conclusions.

2. Definition

There is no accepted definition of a hot spot. Qualitatively, a hot spot is a region of high surface-brightness close to the outer end of a radio lobe. Whilst it is often possible to identify discrete subcomponents, the distinction between "hot spot" and "lobe" is subjective and it is not clear to me that it has any physical validity. An additional difficulty is caused by the confusion between hot spots and knots in jets. In their study of the correlation of fractional flux in hot spots with luminosity, Jenkins & McEllin (1977) defined a hot spot to be a region of diameter 15 kpc, centred on the brightest point at the end of a radio lobe. This suggestion is clearly unsatisfactory as it stands, if only because some sources have an overall size less than this limit. A more useful definition might be made by replacing the fixed linear diameter by a fraction of the total size of the source but for the remainder of this review, I shall use the term "hot spot" in its qualitative sense.

3. Morphology

3.1 Observations

The generalisations in this section are based mainly on observations of a sub-sample of the 3CR catalogue (as revised by Laing,
Riley & Longair 1983) at a resolution of 0.1 arcsec, using the VLA at
a frequency of 14.9 GHz (Laing, in preparation). The selection criteria, in addition to those of the parent sample, were: $z > 0.3$;
largest angular size > 4 arcsec and surface-brightness > 40 mJy/beam
area with a beam of FWHM = 0.4 arcsec at a frequency of 4.9 GHz.
69 sources satisfy these conditions, of which 62 were observed.

Examples are shown in Figures 1-7 below. In each case, the
overall structure of the source at 4.9 GHz is shown in panel (a),
and the remaining illustrations show details of the hot-spot
structure at higher resolution. The frequency of observation is
marked on each map. The shaded circles or ellipses indicate the FWHM
beamshape and the cross denotes the position of the optical
identification. Polarization vectors, where plotted, indicate the E-
vector direction. Their lengths are proportional to the percentage
polarization (100% linear polarization is indicated by a horizontal
line).

3.2 Primary hot spots

It has been recognised for some time (e.g. Hargrave & Ryle 1976;
Kapahi & Schilizzi 1979) that some of the sub-structure within hot
spots is very small. The properties of the compact components (which
will be referred to as primary hot spots) have been discussed by
Laing (1981, 1982), Valtaoja (1984) and Lonsdale & Barthel (1986).
The following summary of their properties is based on these references and on the observations described in Section 3.1.
(a)   The sizes of primary hot spots are small (< 0.5-2 h^{-1} kpc, where
      $h = H_o$ / 100 kms^{-1} Mpc^{-1}, for the present sample). Their sizes
      may scale with the overall size of the source.

(b)   The majority of the present sample have components of this size
      in at least one of their hot spots.

(c)   There is rarely more than one bright, compact component in each
        hot spot, although fainter knots (usually associated with jets)
        are sometimes present (e.g. 3C 204; Figure 4a).

(d)   A primary hot spot may contribute very little of the flux den-
        sity of the source (1 and 2%, respectively, for the two compo-
        nents in 3C 427.1; Figure 2) and may therefore by very difficult
        to detect. Observations of high sensitivity with a FWHM <
        (largest angular size)/100 are the most effective in identifying
        faint primary components.

(e)   The primary components are not in general located at the extreme
        ends of their radio lobes (Figures 1-7). It is likely that they
        are mostly at the sides of the lobes, set back slightly from the
        leading edges, and that projection causes them to appear along
        the centre-lines of the lobes in some cases (e.g. Figure 2). A
        substantial minority are at the ends of their lobes (e.g.,
        3C 204 East, Figure 4).

(f)   Where a primary hot spot is well resolved, it is extended along
        the source axis and/or towards the brightest parts of the
        surrounding, more extended emission. Good examples are 3C 20
        East (Laing 1982) and 3C 133 East (Laing 1988).

(g)   The internal energy densities calculated from synchrotron theory
        for primary hot spots can be extremely high. Values in the range
        5×10^{-9} to 2×10^{-7} Jm^{-3} have been estimated for the sources in
        the sample of Section 3.1, the largest being that determined by
        Lonsdale & Barthel (1984) for 3C 205 South. As with the sizes of
        hot spots, the minimum energy densities are clearly correlated
        with source size. There are serious uncertainties in the estima-
        tion of minimum energy densities and the assumptions that elec-
        trons in hot spots radiate at frequencies as low as 10 MHz in
        the observed frame and that their spectral indices at low fre-
        quencies remain steep are especially open to question.

3.3 Secondary structure

        The more diffuse emission surrounding the primary components can
have a variety of morphologies, often complex, but two characteristic
structures are:

(a) Double hot spots (e.g. 3C 268.4 South; Figure 5). The identifi-
     cation of primary and secondary components is straightforward
     in this case and has been described (using a variety of equiva-
     lent names) by Laing (1981, 1982), Kronberg & Jones (1982),
     Valtaoja (1984) and Lonsdale & Barthel (1986).

(b) Flaring hot spots (e.g. 3C 441 North; Figure 1). Instead of
     being separated into two discrete components, this structure
     consists of a primary with a curved emitting region expanding
     away from it.

4. Spectra and Magnetic Field Structures

4.1 Spectra

     Spectral indices of hot spot (defined in the sense $S_\nu \propto \nu^{-\alpha}$) are
typically in the range $0.5 < \alpha < 1.0$, steepening somewhat at frequencies above 5 GHz (e.g., Laing 1981; Alexander & Leahy 1987; Stephens
1987). There is evidence that some spectra flatten at low frequencies
($\nu < 300$ MHz; Muxlow & Leahy, private communication), possibly as a
result of synchrotron self absorption. The spectrum of a primary hot
spot is almost always flatter than that of its associated secondary
component (Laing 1981; Lonsdale & Barthel 1986).

4.2 Magnetic fields

     The projected magnetic field direction in primary hot spots
tends either to be orthogonal to the source axis (e.g. Figure 1b) or
along the line joining the primary and secondary components. In the
secondary components themselves, it follows the edges of the sources
and the contours of total intensity, as in the extended lobes.

5. Sidedness: The Connection Between Hot spots and Jets

5.1 Jets and primary hot spots

If a jet and a primary hot spot are both present in a radio
lobe, then the jet invariably connects the nucleus and the primary.
This is not to say that all powerful jets terminate in hot spots
(some flare without brightening), but there is clearly a close physical connection. Examples are 3C 204, 275.1 and 245 (Figures 4, 6 and
7, respectively) and 3C 133 (Laing 1988). If the primary is resolved,
it tends to be extended along the jet (cf. Section 3.2f).

5.2 Relative sidedness of jets and hot spots

There is a strong correlation between the sidedness of jets and
primary hot spots. In the sample described in Section 2.1, there are
30 sources having either a continuous jet or compact knots on one
side of the nucleus (jets in powerful sources are almost invariably
one-sided; Bridle & Perley 1984). In 26 cases, the brighter (or only)
primary hot spot is on the same side as the jet/knot. This result is
considerably stronger than that obtained by Bridle & Perley (they
found that the sidedness of jet and hot spots were strongly correlated only for sources with very bright cores), almost certainly
because of the higher resolution of the present data, which effectively isolates the primary components. In fact, 2 of the 4 sources
which go against the trend are very small, and it may be that the
resolution is as yet insufficient to locate their primary components.
There is one clear counter-example, however (3C 200; Burns et al.
1984).

The conclusion is that the mechanisms which cause side-to-side
asymmetries in kpc-scale jets and in primary hot spots are related.
In addition, jets on pc and kpc scales have the same sidedness (e.g.
Bridle & Perley 1984; Browne 1987). Although direct connections have
only been traced in a few cases, the supposition is that pc and kpc
jets and primary hot spots form a continuous flow, whose sidedness is
determined in the same way on all scales.

6. Non-Axisymmetric Flows

6.1 Formation of the primary hot spot

Given the close connection between jets and primary hot spots, models such as those of Kronberg & Jones (1982) in which compact components result from instabilities in the flow are ruled out. It seems clear that a primary hot spot is the end of a jet. The position of the primary hot spot within a lobe suggests that it is formed by the impact of a jet on the side wall of the cavity. The details of the flow pattern have not yet been determined: possibilities include the formation of a "splatter spot" (Williams & Gull 1985) or a "wall jet" (Wilson, these proceedings) and the disruption of the jet by instabilities (Norman, these proceedings). In any case, a shock will form at the point of impact of the jet on the wall, and it will be oblique to the flow. The post-shock flow may still be supersonic in this case. The primary hot spot is then identified with the immediate post-shock flow, whilst the secondary structure is the region of high pressure formed as the flow slows down. A problem with this picture, stressed by Lonsdale & Barthel (1986), is that the estimated pressures within some of the primary hot spots are so high that they would be expected to move much faster than the surrounding secondary emission. This may not present serious difficulties: firstly, the pressures may be overestimated because of incorrect extrapolation to low frequencies (see above), secondly, the primary may move around the lobe due to motion of the beam or instabilities (Scheuer 1982; Norman, these proceedings) and finally, relativistic beaming may be important, as is discussed in the next section.

6.2 Relativistic flow in hot spots

Bulk relativistic motion on pc scales has been inferred to explain superluminal motion, rapid variability and the absence of Inverse-Compton X-rays in compact radio sources. It is therefore natural to ask whether the one-sided nature of kpc jets and primary hot spots is due to Doppler beaming. The argument that the flow in powerful sources is intrinsically two-sided has received some support recently as a result of the discovery of a second jet in Cygnus A (Carilli, these proceedings) and the tentative detection of counter-jets in several quasars (Bridle et al., in preparation). Several

sources in the sample of Section 3.1 have hot spots on both sides, each with synchrotron lifetimes which are less than the light-travel time from the nucleus (one of the original argument for continuous, two-sided energy supply quoted by Hargrave & Ryle 1974).
The hypothesis that one-sided jets and hot spots are a result of Doppler boosting of an intrinsically symmetrical flow also gives a natural explanation of the depolarization asymmetry found by Laing (1988) and Garrington et al. (1988; see also Perley, these procee-dings).

The fact that jets and compact hot spots have the same sidedness was previously taken as evidence against the relativistic flow model, because it was assumed that the flow velocities in the hot spots were similar to the advance speed of the radio lobe as a whole, which is much less than c (see below). The change from an axisymmetric flow model to a non-axisymmetric one is crucial in allowing the post-shock flow to remain relativistic, so that beaming can occur. An additi-onal consequence is that the energy densities of some hot spots will be overestimated, so the ram-pressure balance problem may be less severe than previously thought. The deceleration and decollimation of the flow at the primary hot spot will change the beaming angle of the flow, so we should not expect the brightnesses of the jet and the hot spot to be perfectly correlated, but the sidedness should usually be the same and in cases which are close to the line of sight we should expect both the jet and the primary hot spot structure to be very one-sided (e.g., 3C 245, Figure 7). The velocity of the lobe as a whole is likely to be much slower than c if the jet is lighter than its surroundings, so the constraints on the distribution of separa-tion ratios (v < 0.3c; Longair & Riley 1979) and from observations of spectral steepening along the lobes (0.02-0.2 c; Alexander & Leahy 1987 and references therein) will not be violated.

7. Summary

The main points of this review are as follows:

Many hot spots have the characteristic structure of a compact, primary component at the side of the radio lobe, set back from the leading edge and associated with a more diffuse secondary structure.

Primary hot spots have the same sidedness as small- and large-scale jets. In many cases, jets from the nucleus terminate in primary hot spots. The cause of the asymmetry must be the same on all scales.

Flow in FR-II radio sources is intrinsically non-axisymmetric. Primary hot spots may be identified with the flow emerging from an oblique shock, formed where the jet hits the wall of the cavity.

A plausible working hypothesis is then that FR-II sources are powered by light, symmetrical jets which remain relativistic up to and including the post-shock flow in the hot spots, the asymmetries on all scales being caused by Doppler boosting.

Acknowledgements

I am grateful to the Max-Planck-Society for providing travel expenses.

References

Alexander P. & Leahy, J.P., 1987, Mon.Not.R.astr.Soc. **225**, 1.
Blandford, R.D. & Rees, M.J., 1974, Mon.Not.R.astr.Soc. **169**, 395.
Bridle, A.H. & Perley, R.A., 1984, Ann.Rev.Astr.Astrophys. **22**, 319.
Browne, I.W.A., 1987. Superluminal Radio Sources, p. 129, eds.
        Zensus, J.A. & Pearson, T.J., Cambridge University Press.
Burns, J.O., Basart, J.P., De Young, D.S. & Ghiglia, D.C., 1984,
        Astrophys. J. **283**, 515.
Fanaroff, B.L., Riley, J.M., 1974. Mon.Not.R.astr.Soc. **167**, 31P.
Garrington, S.T., Leahy, J.P., Conway, R.G. & Laing, R.A. 1988,
        Nature **331**, 147.
Hargrave, P.J. & McEllin, M., 1975, Mon.Not.R.astr.Soc. **173**, 37.
Hargrave, P.J. & Ryle, M., 1974, Mon.Not.R.astr.Soc. **166**, 305.
Hargrave, P.J. & Ryle, M., 1976, Mon.Not.R.astr.Soc. **175**, 481.
Jenkins, C.J. & McEllin, M., 1977, Mon.Not.R.astr.Soc. **180**, 219.
Jenkins, C.J., Pooley, G.G.&Riley, J.M., 1977, Mem.R.astr.Soc. **84**,61.
Kapahi, V.K. & Schilizzi, R.T., 1979, Nature **277**, 610.
Kronberg, P.P. & Jones, T.W., 1982. IAU Symp. 97, Extragalactic
        Radio Sources, eds Heeschen, D.S. & Wade, C.M., p. 157, D.
        Reidel, Dordrecht.
Laing, R.A., 1981, Mon.Not.R.astr.Soc. **195**, 261.

Laing, R.A., 1982, IAU Symp. 97, Extragalactic Radio Sources, eds. Heeschen, D.S. & Wade, C.M., p. 161, D. Reidel, Dordrecht.

Laing, R.A., 1988, Nature **331**, 149.

Laing, R.A., Riley, J.M. & Longair, M.S., 1983, Mon. Not. R. astr. Soc. **204**, 151.

Longair, M.S. & Riley, 1979, Mon. Not. R. astr. Soc. **188**, 625.

Lonsdale, C.J. & Barthel, P.D., 1984, Astr. Astrophys. **135**, 45.

Lonsdale, C.J. & Barthel, P.D., 1986, Astron. J. **92**, 12.

Miley, G.K., 1980, Ann. Rev. Astr. Astrophys. **18**, 165.

Miley, G.K. & Wade, C.M., 1971, Astrophys. Lett. **8**, 11.

Scheuer, P.A.G., 1974, Mon. Not. R. astr. Soc. **166**, 513.

Scheuer, P.A.G., 1982, IAU Symp. 97, Extragalactic Radio Sources, eds Heeschen, D.S. & Wade, C.M., p. 163, D. Reidel.

Stephens, P.W., 1987, Ph.D.Thesis, Victoria University of Manchester.

Valtaoja, E., 1984, Astron. Astrophys. **140**, 148.

Williams, A.G. & Gull, S.F., 1985. Nature **313**, 34.

Figures

The following VLA maps are displayed with contour levels at C x (-1, 1, 2, 3, 4, 5, 10, 15, 20, 25, 50, 75, 100, 125, 250, 375, 500, 1000) mJy/beam. If present, a horizontal bar represents 100% polarization. The maps show:

1) 3C 441   Radio galaxy, z = 0.708
   a) C = 0.4, peak = 78.3 mJy/beam, b) C = 0.5, peak = 11.0 mJy/beam

2) 3C 427.1   Radio galaxy, z = 0.572
   a) C = 0.3, peak = 42.1 mJy/beam, b) C = 0.75, peak = 4.8 mJy/beam
   c) C = 0.75, peak = 1.96 mJy/beam.

3) 3C 41   Radio galaxy, z = 0.794
   a) C = 0.5, peak = 124.5 mJy/beam, b,c) C = 0.5, peak = 12.6 mJy/beam,
   d) as c) but some contours omitted.

4) 3C 204   Quasar, z = 1.112
   a) C = 0.25, peak = 48.0 mJy/beam, b,c) C = 1.0, peak = 8.4 mJy/beam (core).

5) 3C 268.4   Quasar, z = 1.400
   a) C = 0.25, peak = 254 mJy/beam, b,c) C = 2.5, peak = 61.0 mJy/beam,
   d) C = 4.0, peak = 26.3 mJy/beam, uppermost contour: 24 mJy/beam.

6) 3C 275.1   Quasar, z = 0.557
   a) Peak = 107.7 mJy/beam, contours at 0.3, 0.5, 0.8, 1.0, 2, 5, 10, 15, 20,
      30, 40, 50, 60, 80, 99% of peak.
   b,c) C = 0.7, peak = 222.9 mJy/beam (core).

7) 3C 245   Quasar, z = 1.029
   a) C = 1.5, peak = 985 mJy/beam, b) C = 0.75, peak = 605 mJy/beam.

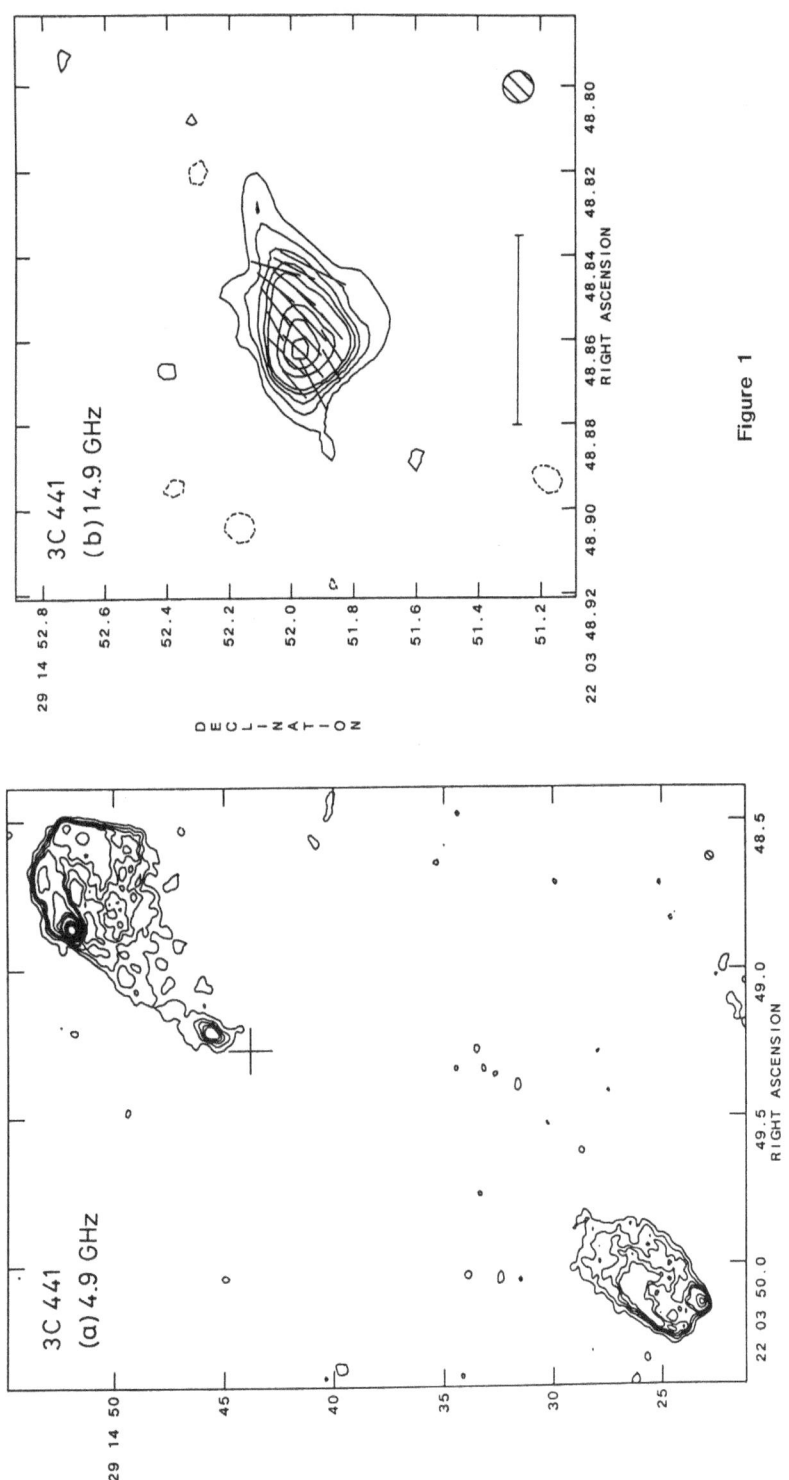

3C 441
(a) 4.9 GHz

3C 441
(b)14.9 GHz

Figure 1

37

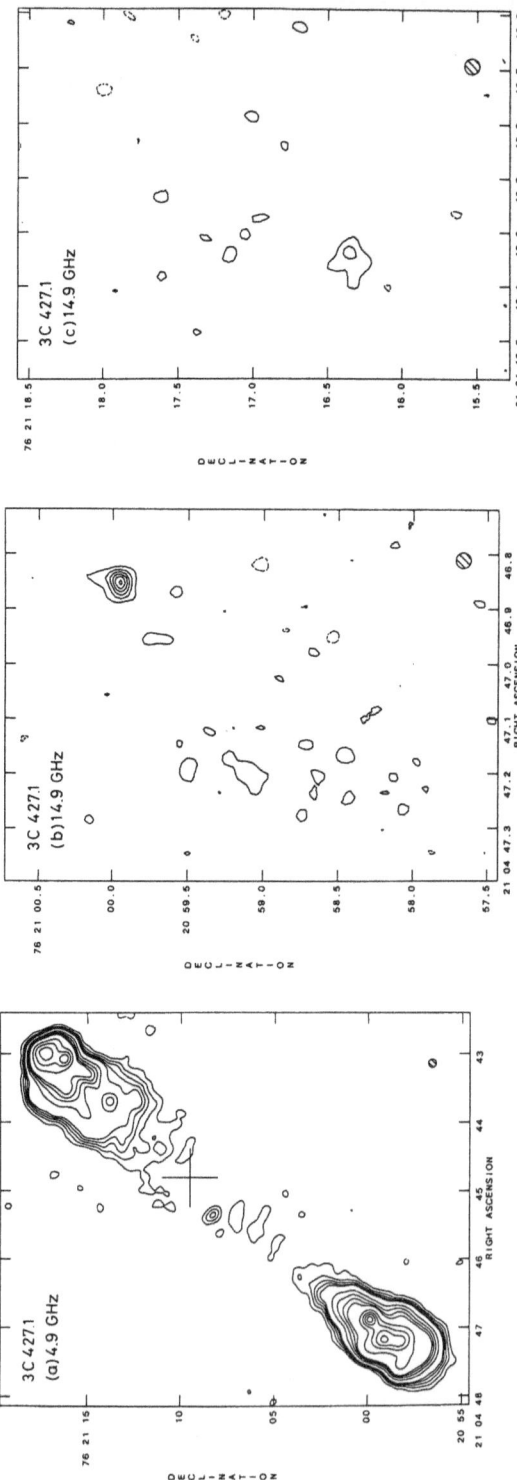

Figure 2

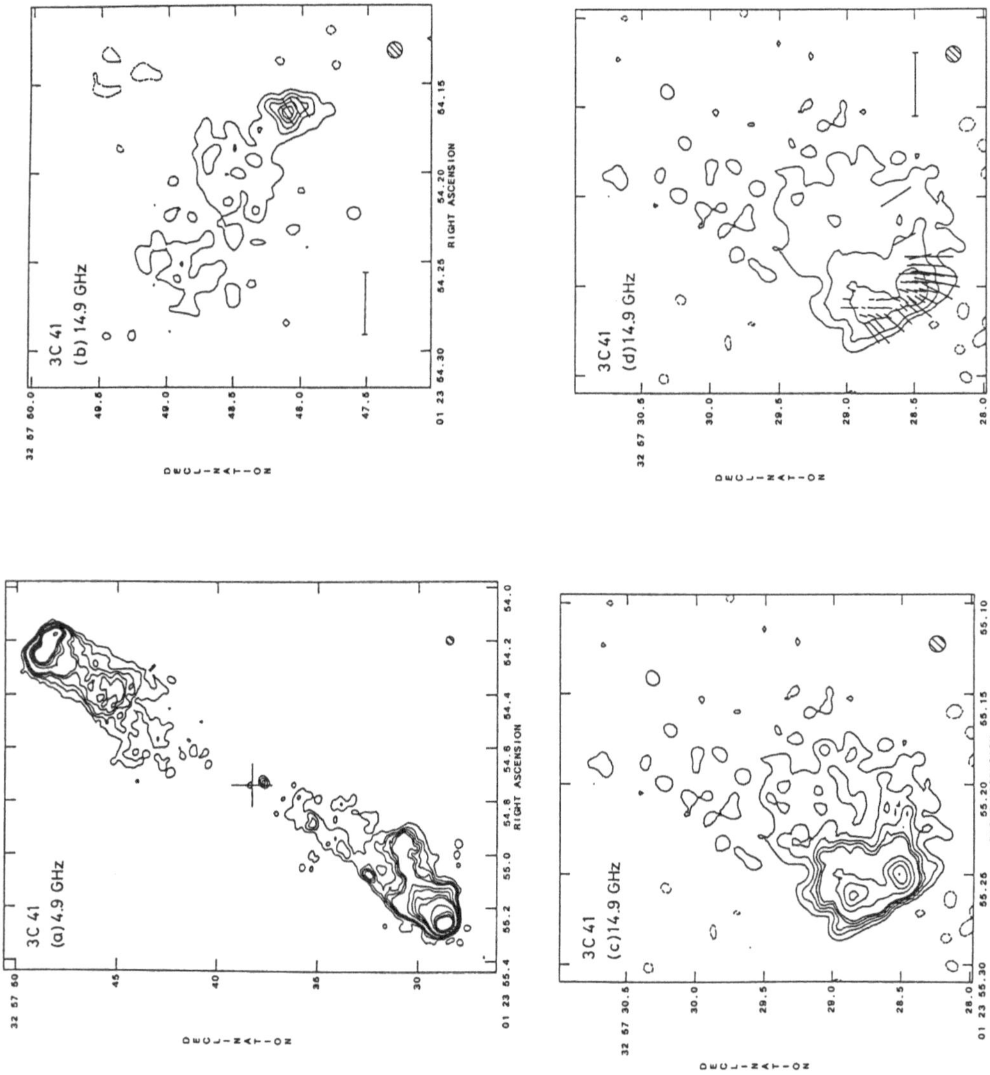
3C 41
(a) 4.9 GHz
(b) 14.9 GHz
(c) 14.9 GHz
(d) 14.9 GHz
DECLINATION
RIGHT ASCENSION

Figure 3

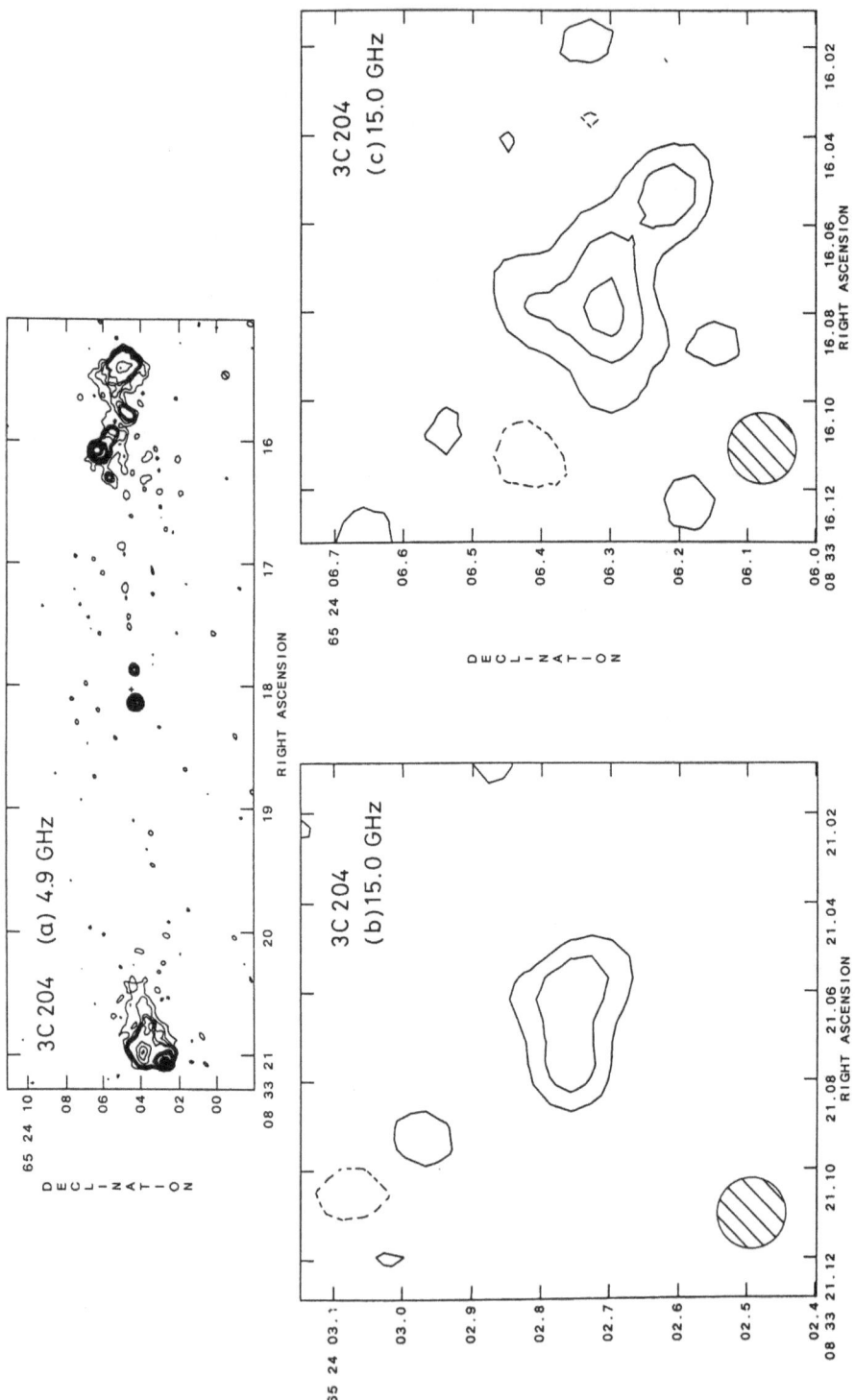

Figure 4

Figure 5

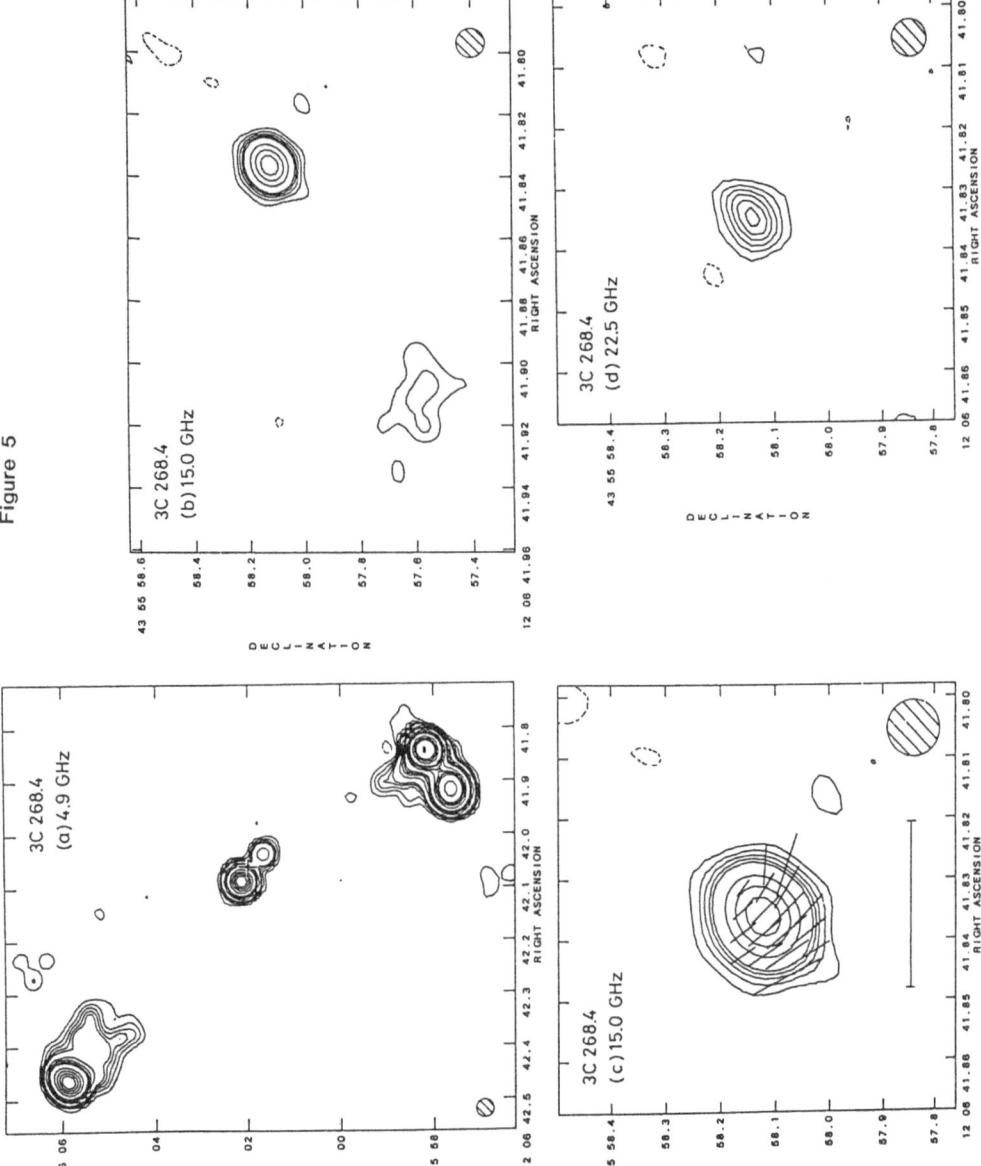

3C 268.4
(a) 4.9 GHz

3C 268.4
(b) 15.0 GHz

3C 268.4
(c) 15.0 GHz

3C 268.4
(d) 22.5 GHz

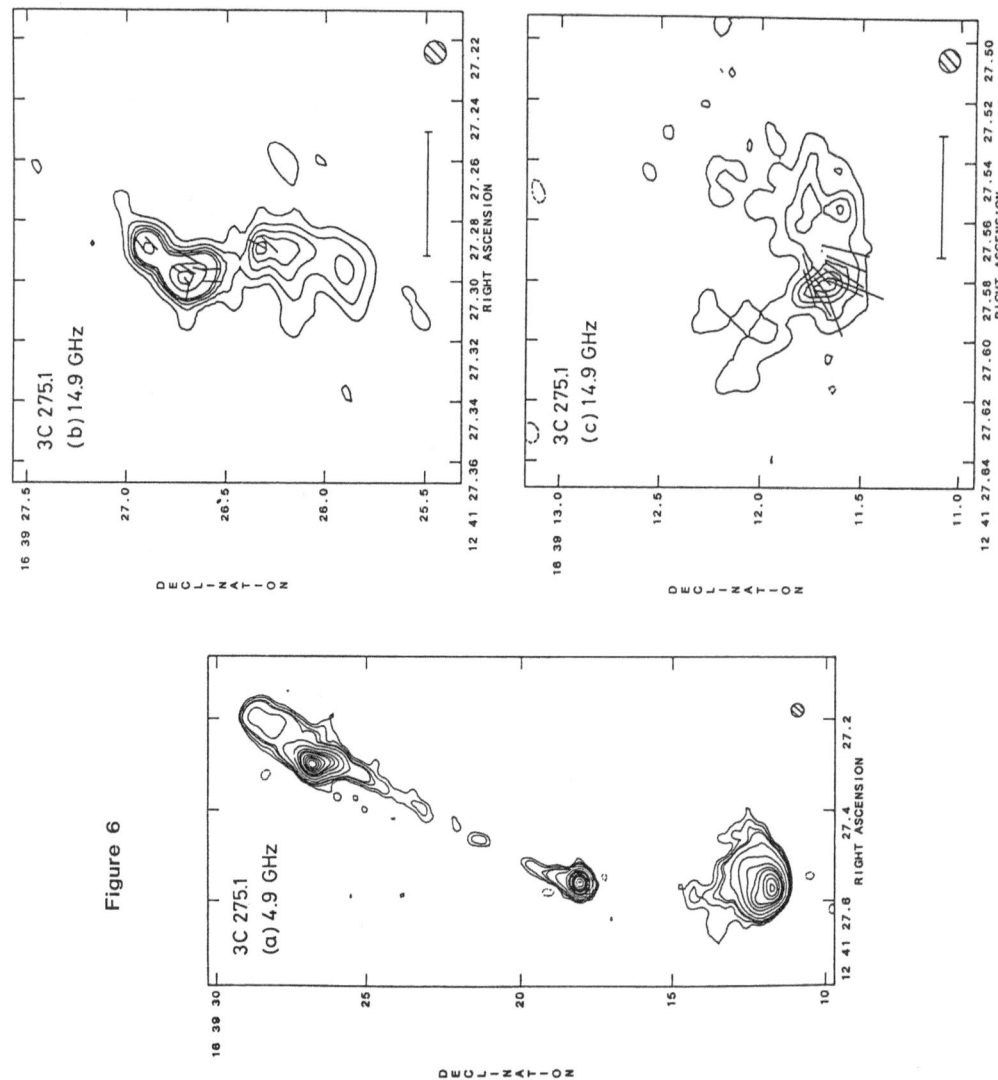

Figure 6

3C 275.1
(a) 4.9 GHz

3C 275.1
(b) 14.9 GHz

3C 275.1
(c) 14.9 GHz

42

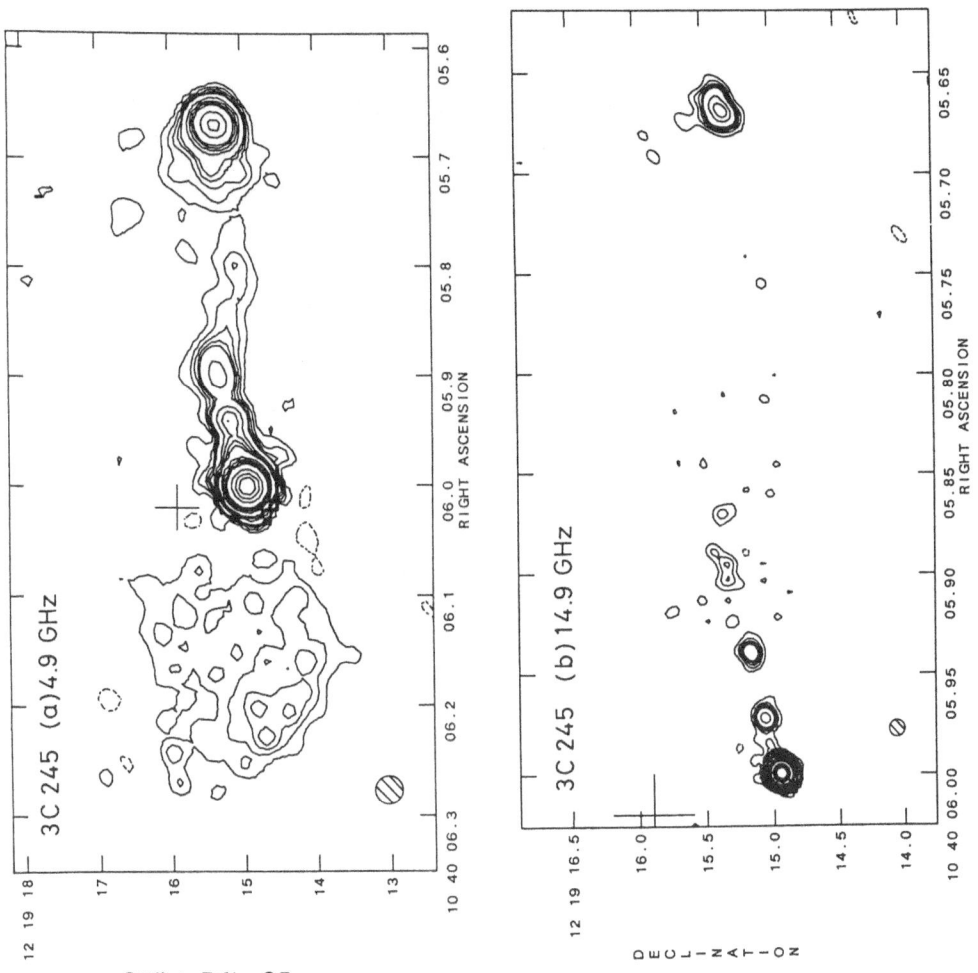

3C 245 (a) 4.9 GHz

3C 245 (b) 14.9 GHz

Figure 7

43

Colin Lonsdale

Robert Laing   Chris Carilli   John Dreher

           Rick Perley   Peter Barthel

44

COMPACT HOTSPOTS, DOUBLE HOTSPOTS, AND SOURCE ASYMMETRY

Colin J. Lonsdale
Haystack Observatory
Westford, Massachusetts 01886, USA

This contribution is divided into three parts. First, some recent VLBI results on an extremely compact hotspot in 3C205 will be briefly discussed. Next, preliminary results of a new study of double hotspots will be presented and placed in the context of previous work by myself and collaborators. Finally, the subject of side-to-side asymmetries in radio sources will be mentioned, with particular regard for the apparently strong tendency for double hotspots to lie on the same side as a jet (if both are present). All this work stems from collaborative efforts with others, principally Peter Barthel of Caltech and Tom Muxlow of Jodrell Bank.

In November of 1986, we observed the southern lobe of 3C205 with a global VLBI network at 18cm wavelength. The fringe spacing on the longest baselines was around 4 milliarcseconds. We used the MkIII recording system, and this experiment represented the first sensitive search for very compact emission more than a few tens of milliarcseconds away from the core of an extragalactic radio source. The southern hotspot of 3C205 is 8 arcseconds distant from the core, which corresponds to 40 kiloparsecs (projected) for $H_o = 75 km\,s^{-1}\,Mpc^{-1}$ and $q_o = 0.05$.

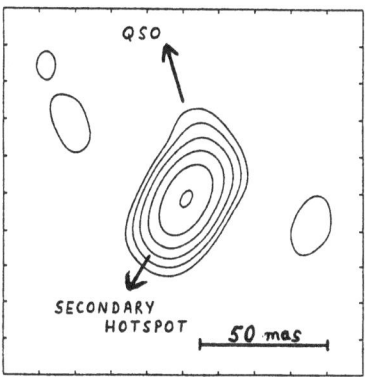

Fig. 1. VLBI map of the southern hotspot of 3C205. The resolution is 15 mas.

The hotspot was clearly detected on the longest baselines (Bonn–VLA, Bonn–Greenbank, Greenbank–VLA), although for only a short period. The visibility on these baselines is consistent with a ridgelike structure in p.a. $\sim 40°$, in close agreement with the elongation already known from lower resolution studies (Lonsdale and Barthel 1984). The width of this ridge can be no greater than 4 milliarcseconds, which corresponds to about 20 parsecs. The axial ratio of the ridge appears to be at least 5. Efforts are continuing to produce a hybrid map at the full resolution of the array, but global fringe–fitting is mandatory and significant software development has been necessary. A map using only the shorter baselines, at 15 milliarcseconds resolution, is shown in Fig. 1. This hotspot contains structure on a finer scale and at a higher brightness temperature than any other known feature of a radio source beyond the self–absorbed core. The extreme values one obtains for minimum internal pressure, amongst other things, place severe constraints on steady–state models of "working surface" type hotspots, whilst the hypothesis that the feature is transient and in free expansion suffers from the problem that adiabatic expansion would snuff out the emission very quickly. Compact hotspots appear to be too common for that to work. Such arguments are discussed at length in our two previous papers on 3C205 (Lonsdale and Barthel 1984, 1986). The new VLBI results were presented in much the same form at the IAU symposium 129 in Cambridge, Massachusetts, in 1987. Additional searches for similarly compact emission from hotspots in other sources are planned.

The second part of this contribution concerns a new VLA study of 15 radio sources with double hotspots in at least one lobe. The sources were selected by examining the literature, and requiring that the sources be suitable for high-fidelity mapping. This translates into the following selection criteria:

- At least one lobe has at least two distinct peaks in surface brightness separated by a region of distinctly lower surface brightness, as determined from a published map of resolution ~ 2 arcsec or better (our loose definition of a double hotspot).

- Declination δ greater than $\sim 20°$ (to permit adequate u-v coverage in 2×20-minute scans).

- Flux density greater than perhaps 100mJy at λ6cm (to ensure a reasonable chance of mapping the hotspots in both total intensity and linear polarization at λ2cm).

- Total angular extent less than $\sim$1 arcminute (to keep the computational burden within reasonable bounds).

As can be seen from this list, the source selection is driven more by practical considerations than by a desire to define a statistically meaningful sample. The conclusions of this study must therefore be accompanied by strong cautions regarding selection effects. The study consists of matched-array VLA observations at λ6cm and λ2cm, with angular resolution 0.35 arcseconds. The double hotspots are mapped in total intensity and linear polarization at both wavelengths, yielding (in principle) spectral index, Faraday rotation and depolarization distributions across the hotspots, in addition to the disposition of the double hotspots in relation to the rest of the source. The aim is to confirm or identify trends in these properties, which constitute clues to the mechanism of double hotspot formation. The strategy is to compare these trends with the predictions of various models, and identify areas in which selection effects may be confusing the issue.

As of this writing, the primary data reduction effort is roughly 50% complete, and some sample maps are shown in Fig. 2. The results so far are consistent with the trends noted (using much inferior data) in Lonsdale and Barthel (1986). In particular, the tendency for the primary (most compact) hotspot of the double to have a flatter spectral index is confirmed, as is the frequent presence of "tails" of emission attached to the primary pointing towards both the core and the secondary. In addition, the tendency of the primary to lie closer to the source axis (defined by the opposite lobe and the core) than the secondary is apparent in the new data.

The interpretation of these trends is controversial as yet. In the abovementioned paper, we made a case that double hotspots are formed by a deflection of the energy supply beam at the primary towards the secondary. The main cause of the deflection was argued to be a dense, massive intergalactic cloud rather than the side of the cavity excavated by the coccoon (Williams and Gull 1985) or other mechanisms which rely on changes in the orientation of the energy supply beam (hereinafter referred to as *"beam jitter"*) as the fundamental departure from axial symmetry. This is because the tendency of the primary to lie on the source axis indicates that the flow remains axisymmetric to that point. Beam jitter models postulate departure from axisymmetry *before* the primary is formed.

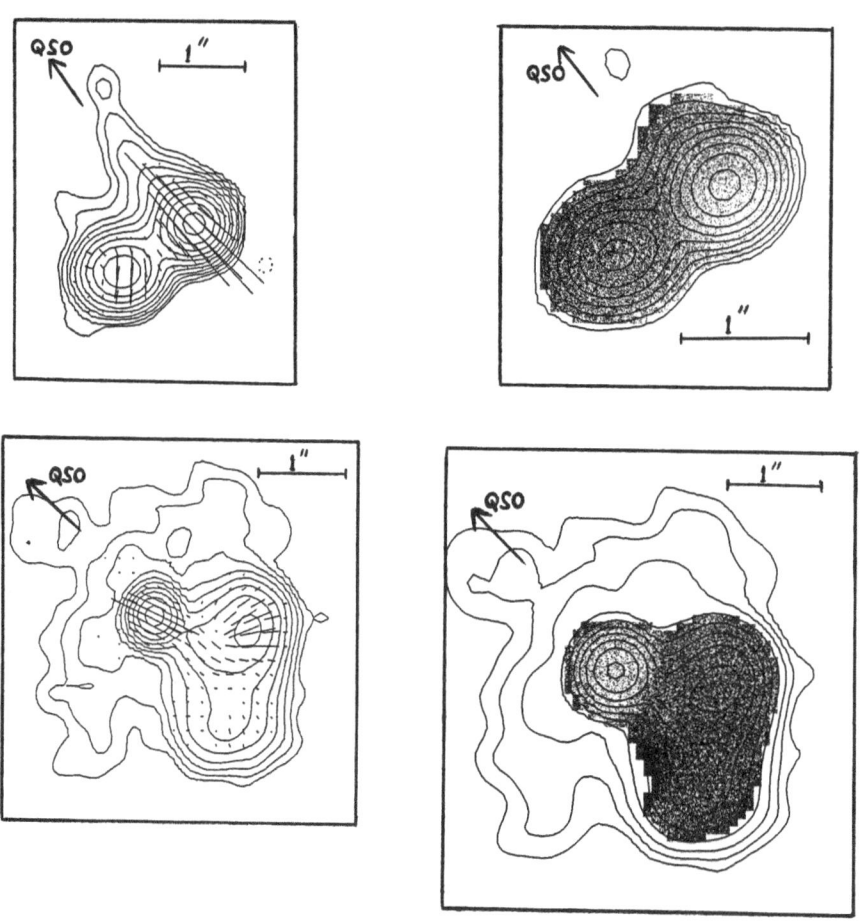

Fig. 2. Maps of two double hotspots. The upper source is 1206+439 (3C268.4), of which the southern hotspot complex is shown. The lower source is 1140+223 (3C263.1), again with the southern hotspot complex in view. On each map the lowest contour is between 0.2 and 0.3 mJy/beam, and contours are separated by a factor of 2. On the left is the λ6cm map with E-vectors of polarized emission superimposed. On the right, the grey scale represents spectral index, with darker shading indicating a steeper spectrum. The total intensity contours on the right are at λ2cm for 1206+439, and λ6cm for 1140+223.

47

A second key point from our studies is that the secondary hotspots contain large amounts of internal energy, and therefore are not recently formed. Ignoring radiation and expansion energy losses, it must take typically $10^5 - 10^6$yrs for an energy supply mechanism delivering $10^{44} - 10^{45}$erg/sec to accumulate the observed internal energy, constraining the longevity of the entire double hotspot configuration. The evidence for energy supply from the primary to secondary, whether by deflection or other mechanism, is powerful, implying that the conditions necessary for such supply are similarly long-lived. In a deflection model, this places severe lower limits on the mass and density of whatever is doing the deflecting, and virtually rules out such tenuous obstacles as the inner surface of the coccoon (the beam would rapidly deform and then penetrate such a surface).

These far-reaching conclusions depend on the validity and generality of the observational trends noted above, and the VLA project is a first step towards putting these trends on firmer ground. There remain, however, obstacles to the acceptance of our data as being representative of the double hotspot phenomenon, due to possible selection effects.

One problem is that our sources were previously known to possess double hotspots, constituting a selection effect favouring similar luminosities in the primary and secondary hotspots (very weak secondaries will not be recognized as hotspots). For sources with very weak secondaries, the longevity constraints for the double hotspot complex mentioned above will be less stringent. However, double hotspots of the type we are studying are found in a significant fraction of all edge-brightened double radio sources, so these large secondary internal energies must still be explained as a frequent element of double hotspot characteristics.

Another potential difficulty, noted by R. Laing, concerns the crucial observation that primaries lie closer to the source axis than secondaries, thus greatly weakening models which rely on beam jitter to break axial symmetry. If secondaries are randomly oriented with respect to primaries *and* there is no mirror symmetry in the source (contrary to naive expectations in a beam jitter model), then the alignment we observe will result from a selection effect caused by our requirement that the source possess a previously discovered double hotspot. Briefly, the line joining primary and secondary will be biased in orientation, whilst the corresponding line in the opposite lobe will not be biased. Especially if Doppler boosting hides the opposite primary, the opposite secondary will tend to line up better with the primary than with the secondary, as observed. The necessary circumstances are contrived, but plausible.

Unfortunately, definitive tests of physical models for phenomena in radio sources are rarely possible (witness the still unresolved debate over flow speeds in large-scale jets), and the question of the mechanism of double hotspot formation is likely to remain open for some time. At present, however, the weight of evidence favours a cloud-collision model, and the current VLA project continues to generate evidence supporting that model.

My final topic also concerns double hotspots, with particular regard to the side-to-side asymmetry in radio source morphology they represent. Despite the lack of a generally accepted definition of a hotspot, there is a definite class of objects which most present radio source scholars would call "double hotspots". The situation with radio jets is more straightforward, with a reasonably unambiguous definition in the literature (Bridle and Perley 1984). The interesting observation I wish to draw people's attention to here is the fact that if a double hotspot appears on one side of a source, and a one-sided jet appears in the same source, *they always appear on the same side.*

There is still much debate about whether jet one-sidedness is primarily due to Doppler boosting or intrinsic asymmetry, with advocates of the former hypothesis generally preferring to postulate that there exist twin, symmetrical beams in all sources. In this case, all significant manifestations of source asymmetry which are correlated with jet sidedness would be consequences of orientation effects in one guise or another. An excellent recent example of such an asymmetry is the so-called "depolarization asymmetry" (Garrington *et.al.* 1988, Laing 1988), in which unjetted lobes systematically depolarize more rapidly than jetted lobes. Those who favour intrinsically symmetric beams can find a natural explanation for this asymmetry in the form of differing path lengths through a circumgalactic magnetoionic medium for the radiation from the approaching and receding lobes. This effect will obviously be strongly correlated with the Doppler boosting that is postulated as the main cause of the apparent jet one-sidedness. Other plausible explanations can be formulated within the framework of an intrinsic asymmetry model, however, and whilst an observational effort directed at finding the origin of the depolarization could discriminate between the models decisively, such observations are extremely challenging. The data required to convincingly eliminate the intrinsic asymmetry hypothesis consists of evidence in several sources that the depolarization is due to a foreground screen enveloping, but *not* spatially correlated with, the radio source.

The general point here is that a sidedness correlation between jet asymmetry and some other type of asymmetry can discriminate between models, provided we can decide whether the correlated asymmetry is an orientation effect or not. Since the presence of double hotspots seems to correlate with jet sidedness, we can apply the above test to try and settle the asymmetry question. There is really only one plausible orientation effect capable of changing the appearance of a double hotspot into that of a typical opposite lobe, and that is Doppler boosting of the primary due to relativistic motion of the hotspot material. The differential boosting between the approaching and receding primary hotspots must be large enough to effectively hide the latter, and since primary hotspots are small, bright and luminous, the required boosting ratios are frequently several hundred. The resulting constraints on the *hotspot* flow speed (i.e. post-shock flow speed) are much more extreme than the constraints on symmetric *jet* flow speeds, which are derived from jet intensity ratios of typically no more than 10 or 20 to 1. If these constraints become too severe for beam and hotspot models, the double hotspot asymmetry would constitute conclusive evidence that jet one-sidedness is an observational manifestation of an intrinsic source asymmetry.

I am grateful to Bob Phillips for a critical reading of this manuscript.

REFERENCES

Bridle A.H. and Perley R.A. (1984), *Ann. Rev. Astron. Astrophys.*, **22**, 319.
Garrington S.T., Leahy J.P., Conway R.G. and Laing R.A. (1988), *Nature*, **331**, 147.
Laing R.A. (1988), *Nature*, **331**, 149.
Lonsdale C.J. and Barthel P.D. (1984), *Astron. Astrophys.*, **135**, 45.
Lonsdale C.J. and Barthel P.D. (1986), *Astron. J.*, **92**, 12.
Williams A.G. and Gull S.F. (1985) *Nature*, **313**, 34.

Chris Carilli

Larry Rudnick

Cygnus A and the Williams Model

C.L. Carilli

MIT and NRAO *

P.O. Box 0 Soccoro, NM USA 87801

J.W. Dreher

Massachusetts Institute of Technology

R.A. Perley

National Radio Astronomy Observatory

April 1, 1988

Abstract

We present radio observations of the hot spots in Cygnus A. The observations show both a compact, 'primary' hot spot, and a larger, more diffuse 'secondary' hot spot in each lobe, as well as the jets which feed the primaries. We interpret our observations along the line of the Williams and Gull (1986) 'splatter spot' models for double hot spots is extragalactic radio sources.

1   Introduction

The most familiar and well studied 'classical double' radio source is Cygnus A (*cf.* Hargrave and Ryle 1974, Alexander *et al.* 1984). It is the closest of the powerful double sources (distance = 230 Mpc using $H_o = 75$ kms^{-1}Mpc $^{-1}$), and it is associated with a large cD galaxy in the center of a dense 'cooling flow' X-ray cluster (see Dreher, Carilli , and Perley 1987a, DCP, for a review). Cygnus A has the distinction of being the first source in which radio 'hot spots', or brightness enhancements at the lobe extremities, were identified (Swarup *et al.* 1963). In order to better understand the physics governing objects such as this, we have undertaken a long series of high dynamic range, high resolution, polarimetric radio observations of the source using multiple configurations of the NRAO's Very Large Array.

Following the precepts of the conference title, we concentrate on the regions at the ends of the radio lobes which include the hot spots. We present images derived from both total and polarized flux from Cygnus A. Though a number of details remain to be resolved, we believe that our results are most consistent with the emerging theoretical and numerical results depicting jets which vary their direction on time scales short compared with the lifetime of the radio source (*cf.* Scheuer 1982). Such jet deviations may be responsible for the double hot spots observed in the lobes of many powerful radio galaxies (including Cygnus A). Specifically, we interpret our data along the lines of the 'splatter spot' model of Williams and Gull (1986), in which the compact 'primary' hot spot is the oblique shock formed at the point of first jet impact with the external medium, the outflow from which feeds the larger, more diffuse 'secondary' hot spot.

---

*The National Radio Astronomy Observatory (NRAO) is operated by Associated Universities, Inc., under contract with the National Science Foundation.

2   Observations and Interpretations

We start with a 2cm, 0.4" resolution, total intensity image of the entire radio source (Figure 1). Due to difficulties with the display of high dynamic range images, the brighter structures at the ends of the lobes are saturated to better display the jets feeding the radio source. Still, the fainter parts of the jets, close to the radio core, are lost on this plot (see Dreher, Carilli, and Perley 1987b). For reference, we have labeled the positions of the two hot spots in each lobe (P for primary and S for secondary, see also figure 2), as well as the jets (J), and a compact feature in the northwest (NW) lobe, close to the primary, designated 'F'. In the southeast (SE) lobe, one can clearly trace the jet through a projected bend of $\approx 21°$, all the way into the primary hot spot. In the NW lobe the situation is not as clear. There are, however, two observations supporting a similar picture: 1) before the jet is lost in the lobe, it points much more closely towards the primary than the secondary, and 2) feature F is probably a knot in the jet as it approaches the primary hot spot. We base this on a number of facts: A) feature F is extended along a line pointing from the inner jet towards the primary hot spot (see figure 2), B) it is 35% polarized, with the projected magnetic field also pointing towards the primary, and, C) it is flatter in spectral index than the surrounding lobe emission (see figure 3). At this point the jet diverts by $\approx 17°$ on its path to the primary. Hence, these images suggest that the jets do alter direction along their course, and that they first impact the IGM at the recessed primary hot spots. We now turn to high resolution images of the regions immediately surrounding the hot spots.

Figure 2 displays the hot spots at 2cm, 0.1" resolution. At this resolution all features are essentially resolved, though we have lost sensitivity to the diffuse lobe emission. Overall, we see that the morphologies of the two hot spot regions are rather similar, especially the bright, parallel 'ridges' of emission in the secondaries. This suggests that a systematic fluid mechanical process dominates the structural developement, as opposed to local 'weather'. The main difference between the fields is the relative brightness of the primaries and secondaries, with the NW lobe having a more well developed primary and weaker secondary, comparatively. This difference could be a boosting effect (see R. Laing, this volume), or just a time delay caused by source orientation in the sky plane. Although it is difficult to display using contours, there are features in both hot spot regions with fluxes well above the background, which connect the compact primaries with the diffuse secondaries. These we label as 'tails' in the figure, and, as we shall see, they may in fact be the flows between the two hot spots. The complex morphology of these hot spots clearly demonstrates the difficulty of defining a 'hot spot' within the plethora of structures often seen at the ends of the lobes. In fact, in these 2cm images, the tails and bright filaments contribute about as much to the total flux as do the bright 'heads' and ridges of the hot spots.

The spectral index for the emission between 2 and 6 cm at 0.4" resolution is shown in figures 3 a and b. We see that the hot spot heads and ridges have much flatter spectral indicies than the rest of the radio source, indicating local particle acceleration. This is seen most clearly at the heads of the primaries, where the flattest spectral indicies in both lobes are found (values as large as -0.6[1]). The tails of the hot spots also have rather flat spectral index values of $\approx$ -0.8, as would be expected if the electron population has been recently accelerated. The surrounding diffuse lobe emission generally has much steeper spectral index values of $\leq$ -1.2.

Figures 3 c and d display the fractional polarization at 2cm, 0.1" resolution. The fractional polarizations are high, reaching up to near the theoretical maximum ($\approx 70\%$) in a few places. Values are largest along the hard edges and filaments, which is consistent with the random field compression and shear models of Laing (1980). This occurs along both the outer and inner hard

---

[1]The dark patches at some of the lobe and hot spot edges are probably not real, but may be a result of small positional offsets between the two frequencies.

edges of the hot spots, which may be an indication of hot spot expansion both forward, into the IGM, and backward, into the lobes. The hot spot tails and most of the filamentary structures are also highly polarized ($\approx 40\%$), but, curiously, the regions between the ridges in the secondaries are considerably less polarized ($\leq 10\%$). This could be a resolution or a line of sight effect. Another curious feature is the 'spur' of polarized emission protruding from the secondary in the NW lobe. The feature is very long and narrow (3 kpc x 0.3 kpc), with fractional polarizations as high as 55%. Also, the magnetic field vectors project along the feature's extent, and the spectral index along the feature is considerably flatter than in the surrounding material (see figure 3 b and f). Perhaps we are seeing an internal lobe shock or collimated backflow from the secondary.

The projected magnetic fields are plotted in figures 3 e and f at 0.1" resolution. The field direction along knot F has been drawn next to the feature for clarity. As expected, the fields generally lie parallel to the filaments and hard edges, as well as parallel to the ridge lines in the secondaries. Minimum energy field strengths are ≈ 300 μG in the hot spot heads and ridges, 100-200 μG along the tails and filaments, and 80 μG for the surrounding, diffuse emission. An interesting feature to follow is the tail of the NW primary hot spot. At the head of the hot spot the fields lie parallel to the hard front edge. As the tail sweeps back, the vectors rotate $\approx$ 90° to an orientation along the tail's length, as would be expected for shear flow along the tail, or compression in a nozzle. This continues until the tail begins to widen as it approaches the secondary.

Lastly, figures 3 g and h shows the rotation measure (RM) at 0.4" resolution in the two hot spot regions. As was pointed out in great detail in DCP, the RM must be caused by gas external to the emitting regions, and that this RM screen is not within the Milky Way galaxy. The logical site for the screen is to associate it with the dense cluster gas surrounding the radio source. The overall morphology of the screen shows large RM's and RM gradients uncorrelated with total intensity. However, a close inspection of the hot spots in the NW lobe reveals a curious hemispherical feature of large RM, concentric with the primary hot spot. Rotation measures rise gradually from near zero in the feature's interior, up to over 1000 rad m^{-2} at a distance of $\approx$ 3 Kpc west of the hot spot, and then abruptly back to zero further out (note: pixels along the edge of the feature were blanked due to depolarization caused by large RM gradients). A sharp discontinuity in RM such as this is exactly what one would expect at the location of a non-radio emitting bow shock associated with the supersonically advancing hot spot. The bow shock compresses, heats, and accelerates the external IGM before the two shocked fluids (Jet and IGM) meet in pressure balance along a contact discontinuity (see Blandford and Rees 1974). The geometry of the system dictates that the primary hot spot lies at the near edge of the radio lobe. This allows the bow shock, induced as the incoming jet pushes into the lobe's side wall, to project onto the backlighting lobe. Assuming strong shocks, and using the X-ray derived external densities, we calculate a field strength of 5 μG in the unshocked IGM in order to creat the RM values observed. In this scenario, the secondary is futher away from us than the primary. If the tail of the primary is actually outflow to the secondary, then this geometry suggests that the sharp curve of the tail of the primary may be a much softer curve, seen in projection. We do not see a similar RM feature in the other (SE) lobe, which is as expected if the primary hot spot is situated on the far edge of the lobe.

3 Conclusions

We have presented detailed observations of the hot spots in Cygnus A. The points which most strongly support the splatter spot model of Williams and Gull are: 1) Both lobes have double hot spots, including a compact, recessed primary and a more diffuse secondary. 2) The jets in the two lobes seem to bend along their course, and can be traced into the primary hot spots. 3) The primaries have the flattest spectral indicies in the lobes. 4) Morphological structures connect the primaries and secondaries, along which spectral indicies are flatter than in the surrounding material and fractional polarizations are high, with field vectors parallel to the 'flow'. 5) The RM image reveals the bow shock associated with the primary hot spot in the NW lobe, which yields information on the three dimensional structure of the radio source.

Though alterations to this scenario are possible and probably likely, we feel that, overall, it represents a reasonable and consistent picture of the basic flow patterns in the source. Specifically, there is direct evidence on our images indicating that the jets alter direction on relatively short timescales, and that they first enter the compact primary hot spots in each lobe. There is also circumstantial evidence for flow from the primaries to the secondaries, though this is by no means as certain, and a number of important questions remain to be answered. Primarily, how can one get reasonably collimated and energetic flow from the primary in order to power the secondary? This point is accentuated by the short electron synchrotron lifetimes in the secondaries of $\leq$ 10^6 years (Wright and Birkenshaw 1984), and the problem becomes more accute when deflection angles become large, as in the NW lobe of Cygnus A. For hydrodynamical oblique shocks in a Newtonian fluid, supersonic flow can be maintained for deflection angles $\leq 37°$ ($49°$ for $\gamma = 4/3$), and the immediate post-shock flow will be parallel to the deflecting wall (see M. Wilson, this volume). However, the effect of a non-ridged wall (i.e. the shocked IGM), on this process must also be considered. For larger deflection angles the flow must become subsonic. The fluid can then be reaccelerated back into the lobes by pressure gradients, perhaps forming a de Laval nozzle as it re-attains supersonic velocities (Williams 1986). Another question is whether the observed alteration of jet direction is a result of processes within the central engine, or is it a result of the 'heavy weather' caused by backflowing waste jet material (see M.Norman, this volume)? Lastly, what physical processes drive the various fine scale structures observed in the hot spots (e.g. the ridges in the secondaries)? We hope further analysis of Cygnus A and similar sources, as well as more sophisticated numerical simulations, will help to resolve some of these questions.

References

Alexander, P., Brown, M., Scott, P., 1984, M.N.R.A.S., 209, 851

Blandford, R., and Rees, M., 1974, M.N.R.A.S., 165, 395

Dreher, J.W., Carilli, C., and Perley, R. 1987a, Ap.J., 316, 611 (DCP)

Dreher, J.W., Carilli, C., and Perley, R. 1987b, B.A.A.S., 19, 731

Hargrave P.J. and Ryle, M. 1974, M.N.R.A.S., 166, 305

Laing, R. 1980, M.N.R.A.S., 193, 439

Swarup, G., Thompson, A., and Bracewell, R., 1963, Ap.J., 138, 305

Scheuer, P.A., 1982, IAU Sym. 97, ed. Heeschen and Wade, p. 163

Williams, A.G., 1986, PhD Thesis, Cambridge University

Williams, A.G., and Gull, S., 1986, Nature, 313, 34

Wright, M., and Birkenshaw, M., 1984, Ap.J., 281, 135

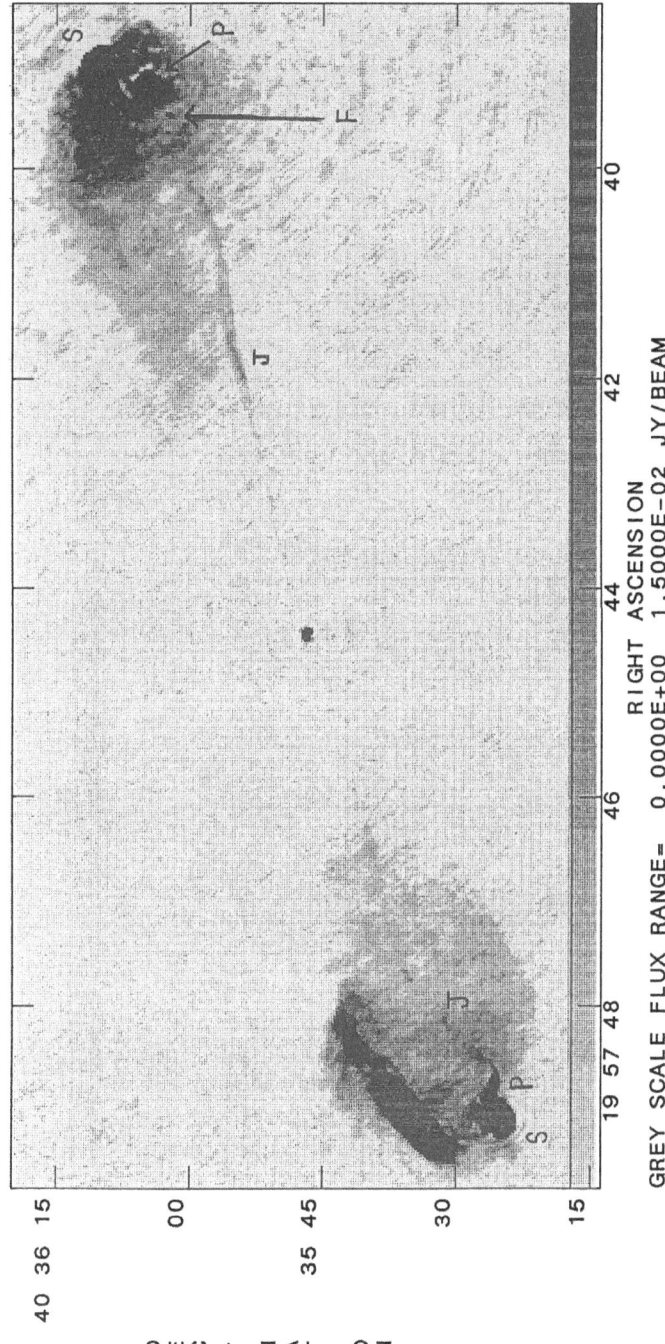

GREY SCALE FLUX RANGE= 0.0000E+00 1.5000E-02 JY/BEAM

Figure 1) A grey scale display of the full source in total intensity at 2cm, 0.4" resolution. Bright structures at the lobe ends are saturated to better show the jets feeding the lobes. Notice the bend in the southern jet as it enters the lobe.

Figure 1) A grey scale display of the full source in total intensity at 2cm, 0.4" resolution. Bright structures at the lobe ends are saturated to better show the jets feeding the lobes. Notice the bend in the southern jet as it enters the lobe.

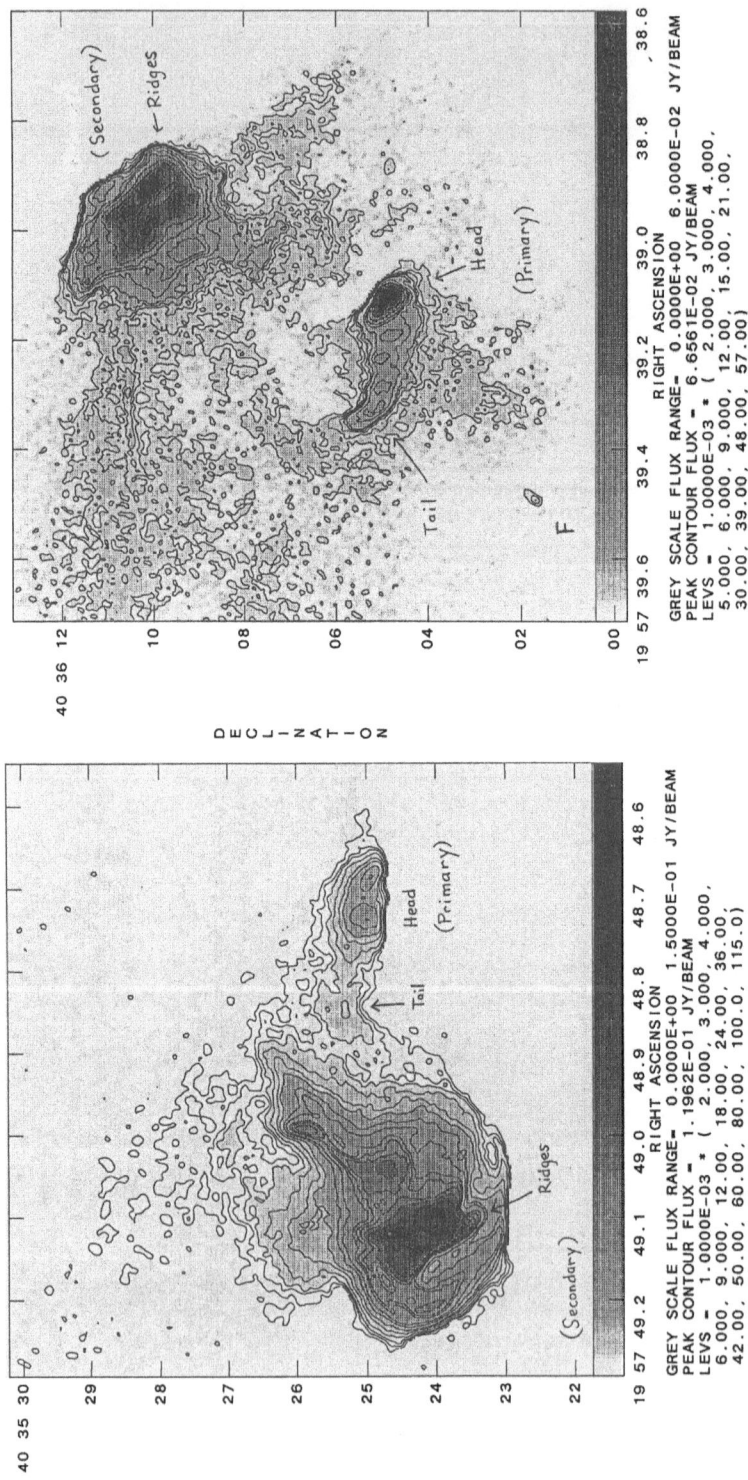

Figure 2) A contour and grey scale plot of the two hot spot regions, in total intensity at 2cm, 0.1" resolution. This shows the two hot spot regions at the ends of the lobes (i.e. the saturated regions from figure 1). The southeast hot spot region is on the left, and the northwest is on the right. Both lobes show two hot spots (primary and secondary), as well as a wealth of other structures.

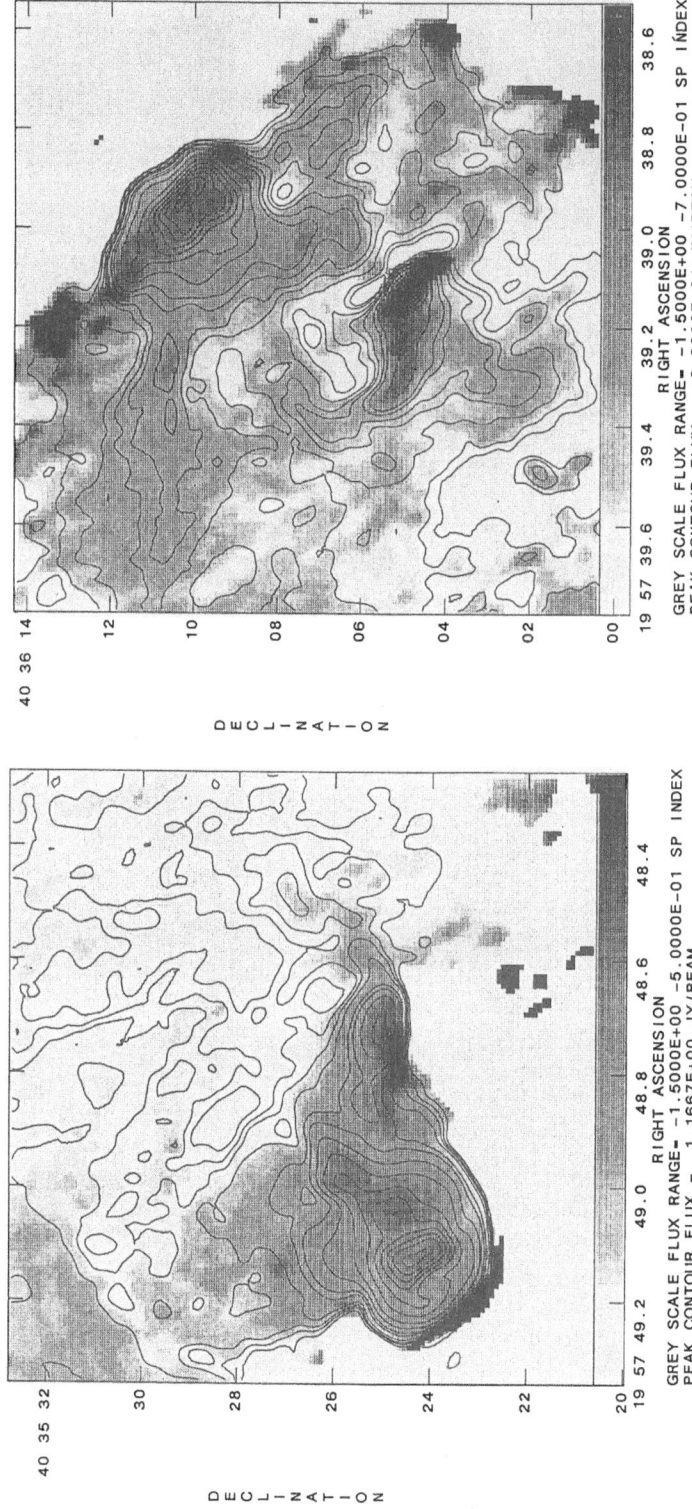

Figure 3

a, b) A grey scale plot of spectral index between 2 and 6cm, on top of 2cm total intensity contours at 0.4" resolution. Blanking was used for all points below 10σ for all the grey scale plots shown.

57

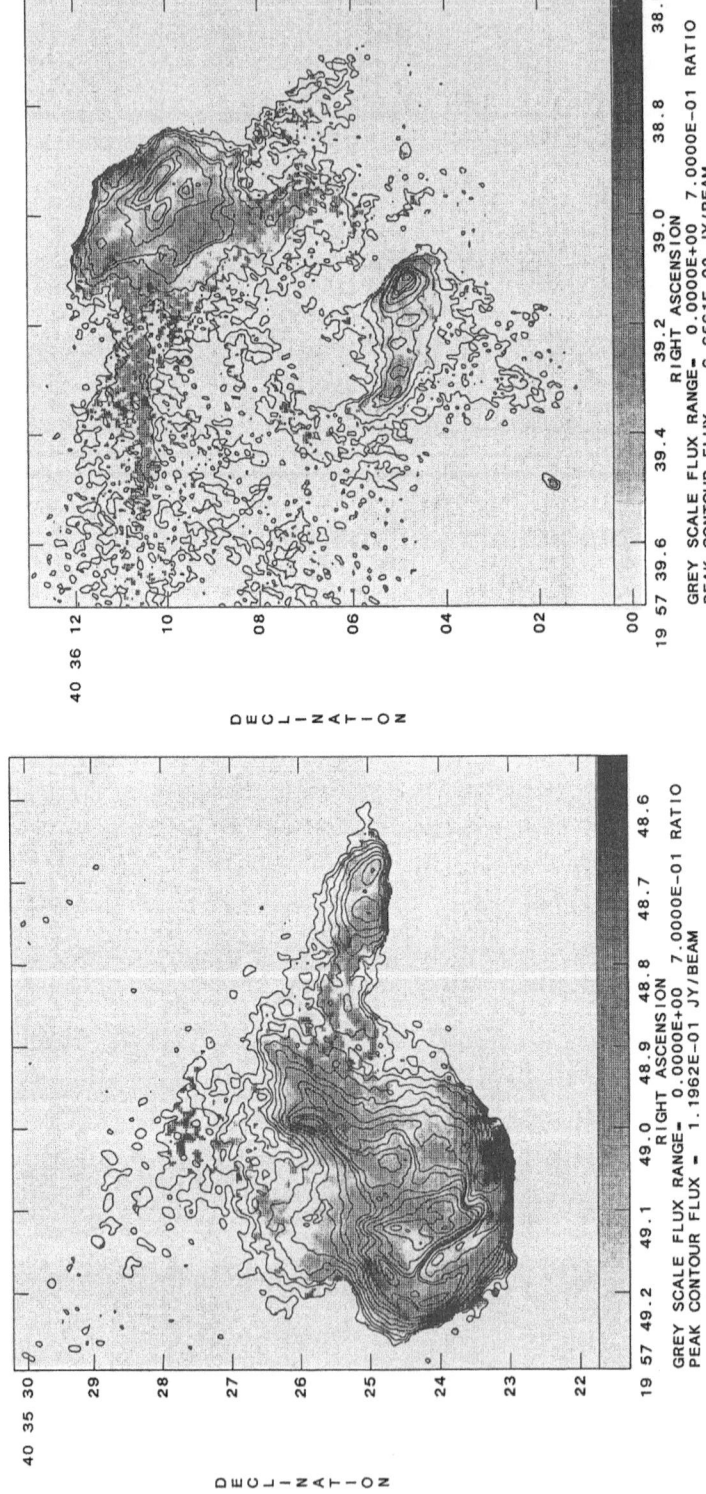

Figure 3

c,d) A grey scale plot of fractional polarization, on top of 2cm total intensity contours, at 0.1" resolution.

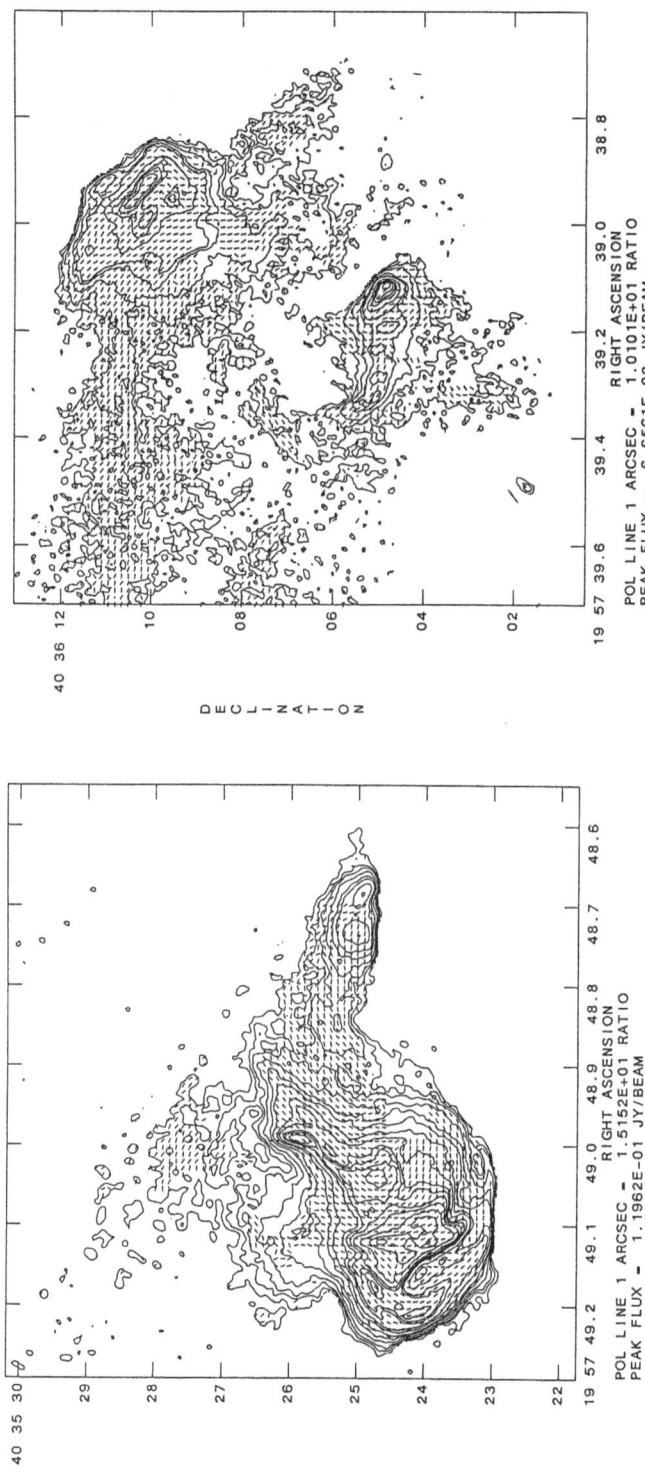

Figure 3

e,f) A vector plot of the projected magnetic fields on top of 2cm total intensity contours, at 0.1" resolution. The observed position angles have been corrected for rotation measure, and all vectors are of equal length.

59

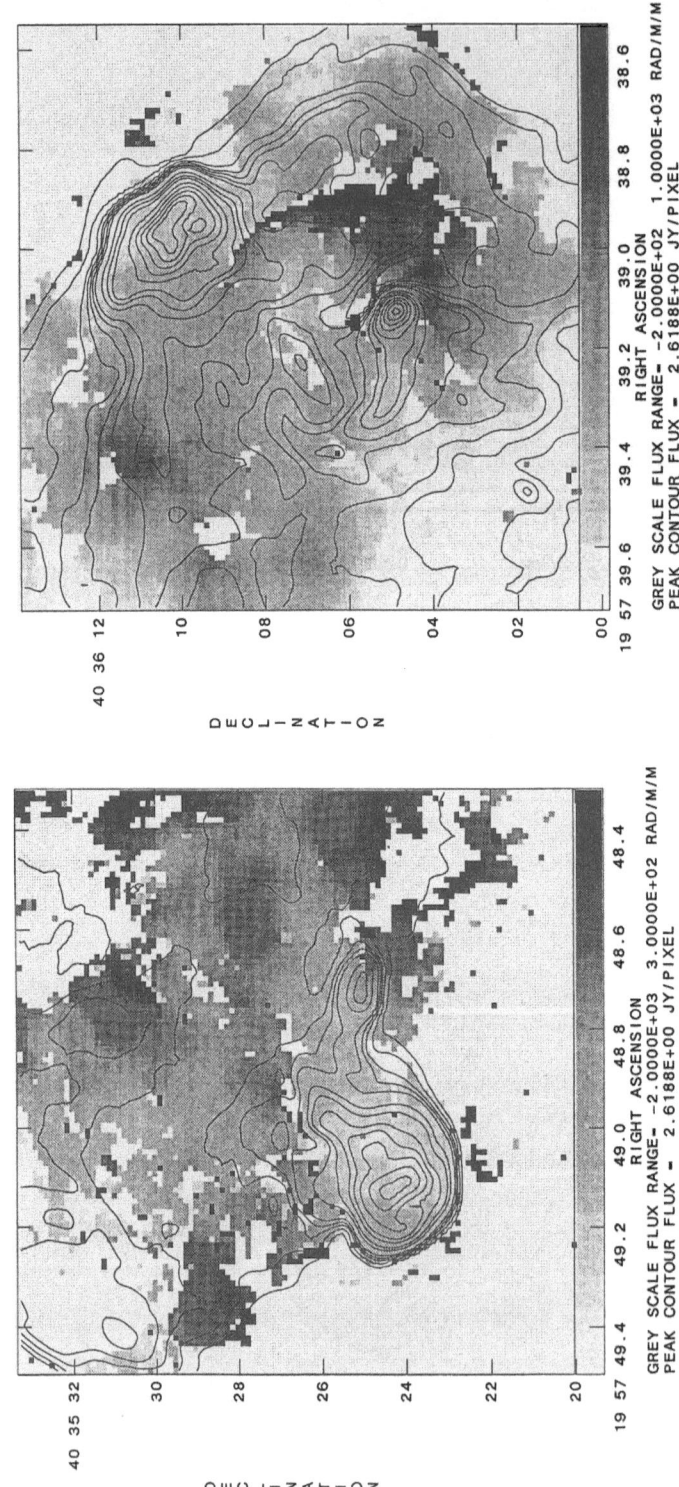

Figure 3

g,h) A grey scale plot of rotation measure, on top of 6cm total intensity contours, at 0.4" resolution. Notice the hemispherical structure, concentric with the primary hot spot in the NW lobe.

A TALE of TWO HOT SPOTS

Lawrence Rudnick, Department of Astronomy, University of Minnesota
116 Church St. SE., Minneapolis, MN 55455 USA

ABSTRACT: I use the two hot spots in 3C33 to discuss the critical hydrodynamic and particle acceleration assumptions in our standard fluid flow models of extragalactic radio sources.Details of the northern and southern hot spots in 3C33 are completely different despite strong similarities in the large scale structures. I argue that both of the hot spots appear "simple", *i.e.*, not complicated by wandering beams or external inhomogeneities, and I explore possible reasons for their differences. A key question is whether the relativistic particles and fields originate only within the contact discontinuity, or can be accelerated/compressed behind the bow shock as well. Coupled with this question is our ignorance of which physical parameters or features in the fluid flow correlate with the synchrotron emissivity.

I. A Preliminary Caution

I do not believe it is useful at this stage of our understanding to develop a "definition" of hot spots. At low resolution there are usually distinct regions of higher surface brightness which used to be called hot spots. When observed at higher resolution these may show no structure or one of a pantheon of morphologies that has so far resisted useful classification. I think it is useful to designate some sub-classes, such as "splash spots", for which distinct models can be proposed. However, almost any generic description is bound to exclude some interesting objects, such as those discussed here.

II. Individualizing Hot Spots - the Menu

Following are some basic ways in which hot spots (either of the same or different sources) could differ, although originating from the same physical situation:

Manageable differences - (i.e., those which could conceiveably find support from other lines of evidence, or from building models of classes of sources) -

   Overall Beam Properties - Mach number, density ratio to external medium, cross sectional area, seed relativistic particle density,...;

   Age- especially if initial ejection times differ by a substantial fraction of the dynamic evolution or particle lifetimes;

   Gross Environmental Characteristics- density gradients in hot (10^6 - 10^8 °K) or cold gas, large scale shear motions, gravitational potential;

Unmanageable differences - (i.e. - those for which no confirming evidence is likely to be found and which will require one explanation per observed feature) -

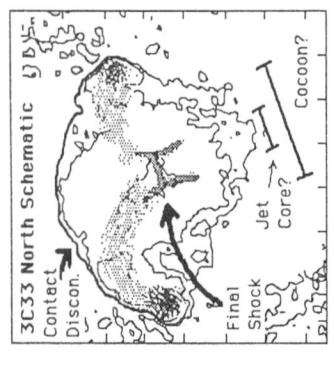

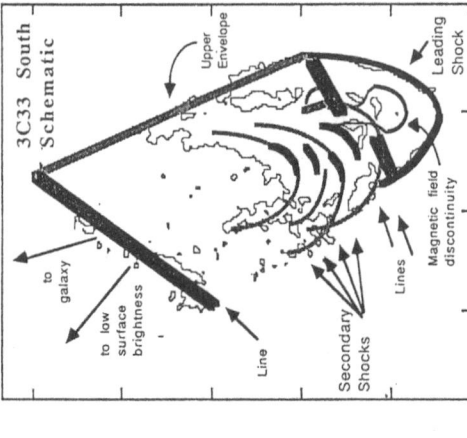

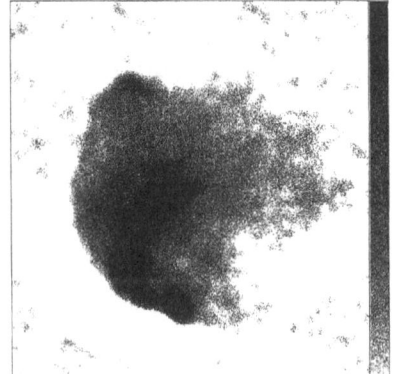

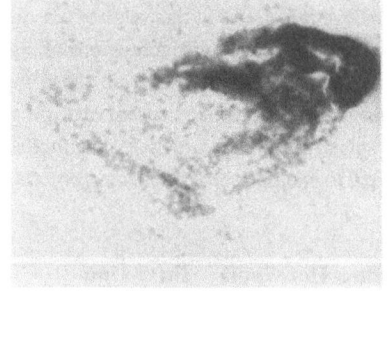

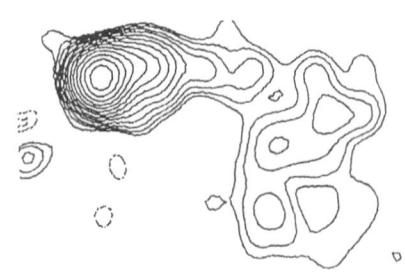

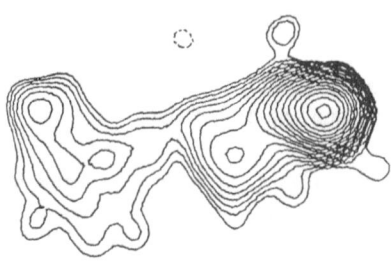

Figure 1. The two hot spots of 3C33. Left - low resolution contours. Middle - detailed gray-scales of hot spot regions, blown up by a factor of ≈5. North - total intensity; South - polarized intensity. Right - model structures. Part of this diagram is from Rudnick, 1988.

Flow Variations (short time-scale) - including a) the natural evolution of shock, turbulent and other structures on time scales comparable to the jet crossing time, and b) true, growing instabilities which may strongly distort the flow;

Environmental Inhomogeneities , clouds, relics of past activity, or small neighboring galaxies which introduce perturbations on scales ≤ 10 kpc.

III. 3C33's Hot Spots

A. Overall Characteristics

A low resolution map of 3C33 at λ6cm (Figure 1, left) shows high surface brightness features at the two leading edges. These are slightly resolved at a resolution of ≈ 10 kpc (H ≡ 100 km/sec/Mpc) and will be called the "hot spots" or "hot spot regions". Their monochromatic luminosities are North - $3x10^{23}$ W/Hz and South - $6x10^{23}$ W/Hz. Higher resolution maps of these hot spots are shown in the middle, with a suggested classification of their structures on the right.

From the low resolution observations there is no reason to expect very different structures in the two hot spots. The southern lobe is ≈ 30% closer to the nucleus, well within the normal distribution of arm length asymmetries. Beyond the first 10 kpc, the integrated luminosities of the lobes are within ≈10% of each other. The two hot spot peaks and the nucleus are collinear within ≈1°, although the low surface brightness bridge heads into left field. The alignment and the similar diffuse luminosities limit the role of distortions from the external medium.

B. Key Features / Differences

Each of 3C33's hot spot regions is characterized by a simple outline within which more complicated brightness and polarization structures appear. There is currently no detectable low surface brightness emission ahead of their leading edges. The southern hot spot is described in detail in Rudnick (1988). Work on the northern hot spot (Rudnick and Anderson, 1988) is still in progress.

The southern hot spot, in outline, is an almost perfect parabola, with a symmetry axis along a position angle of 11° (19° position angle to nucleus). It is strongly edge brightened (see Fig. 2) for the first 7" (west) and 3" (east) where transverse features mark the transition to low brightness regions.

The northern hot spot, in outline, has a very symmetric "mushroom" shape, with an axis along a position angle of 12° (19° position angle from nucleus). Its emission slopes gradually to the leading edge (not very different from a uniform emissivity hemisphere, Fig. 2) except at the position of the two symmetrically placed bright regions at the trailing end of the "mushroom cap." The center of curvature of the leading edge falls on one leg of an H - shaped feature.

The magnetic field of the southern hot spot is extremely well-ordered, aligned with the leading edge, and yields polarizations of 30% - 40%. Behind the leading edge there is a discontinuity in the field structure (see Rudnick, 1988, Fig. 7a).

The magnetic field of the northern hot spot is also well-ordered, yielding polarizations of 20% - 50%, but its complete orientation pattern is not yet clear. There is a distinct transition in field direction across a line perpendicular to the symmetry axis, at the position of the **H**.

C. Different Models?

The very simple shapes of these hot spot regions have led me to identify them with shapes that are seen in the physically simplest (but high resolution) types of numerical simulations which have been done to date. I am therefore asking whether we can describe each of the observed structures in terms of a single flow pattern - as opposed to assuming, e.g., that a wandering beam has deflected off the walls of a cavity in a way that can only be analysed after the fact.

The parabolic shape of the southern lobe and its very smooth shape led me to identify it with the leading bow shock. I was willing to consider, as others are not, that relativistic particles might be present in this region. Further evidence for this bow shock interpretation of the southern hot spot came from examination of the emissivity profile as a function of distance from the axis of the parabola. Fig. 3 is a plot of the observed brightness (and derived emissivity) as a function of the "local Mach number", i.e., the local advance speed of the shock assuming that it maintains its shape. Two important conclusions arose from this analysis:

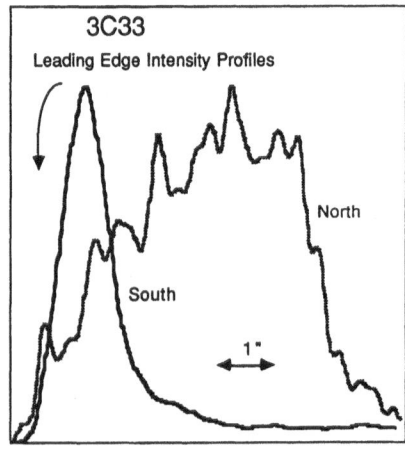

Figure 2. Brightness profiles along the source axis, from the leading edges (indicated) toward the nucleus.

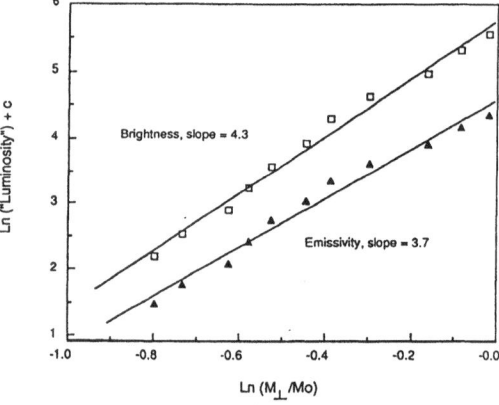

Figure 3. Observed brightness and derived emissivity along the leading edge of 3C33 south. The observed emissivity slope of 3.7 is close to the equipartition derived value of 3.5 (diagram from Rudnick, 1988).

1. At this resolution there is <u>no evidence for a distinct, isolated "hot spot"</u> separate from the overall structure. Instead, the brightness appears to be a smooth function of position along the leading edge.

2. The dependence of emissivity on the derived pressure jump across the bow shock matches that expected if the <u>equipartition (or minimum) relativistic energy densities scaled with the underlying fluid pressure</u>. This is only suggestive, but it surely is tantalizing!

I approached the analysis of the northern lobe hoping to find another potential bow shock, perhaps with a different opening angle to test the scaling relationship found in the south. Instead, I found a completely different structure which was not amenable to that type of analysis. The northern "mushroom cap" shape looks to me like the contact discontinuity seen in the numerical simulations. I have also identified the **H** feature as the final shock before the flow stagnation point (Fig. 1). There is a great deal of time variability in the simulated contact surface shapes, and no way to predict which should be seen.

Why should these two hot spots appear so different? Global parameter variations are not attractive, because of the similarities in size, distance, luminosity and alignment of the large-scale features. On the other hand, instabilities or small-scale inhomogeneities should not lead to such simple, regular shapes as observed at high resolution. Simple, but very different structures - that is the dilemma!

D. A Note on Particle Acceleration

As noted by several people at this meeting, most radio hot spot observations do not require localized particle acceleration. However, the optical synchrotron emission at the radio peak comes from electrons with lifetimes shorter than even the light travel time across the hot spot (Rudnick, Saslaw, Tyson and Crane 1981). A variety of mechanisms have been suggested and rough numbers worked out for converting bulk kinetic energy into relativistic particles at the jet terminus. But what theoretical constraints exist if instead we want to do this particle acceleration at the external bow shock, as I suggest above?

If the southern edge is a bow shock, then I have found that simple compression of a possible external population of relativistic particles and fields is not sufficient to produce the observed emissivity (Rudnick, 1988). Again, we could ask whether bow shock particle acceleration can explain these observations, or we could consider alternatives such as making the shock (synchrotron?) radiative.

IV. Final Remarks

On the *observational* side, I think it is important for us to find and study more examples of well-resolved hot spot regions which show "simple" structures, i.e., those for which we do not need to resort to beam or external inhomogeneities. On the *theoretical* side, we must explore the coupling between the fluid properties and the relativistic plasma, both inside and outside the contact discontinuity. Populations of relativistic particles outside the (current) contact discontinuity are found in diffuse relic lobes and, e.g., in diffuse galaxy cluster sources - they must not be ignored in the theoretical calculations.

REFERENCES

Rudnick, L., 1988, Astrophys. J. **325**, pp. 189-203.

Rudnick, L., Saslaw, W., Tyson, J.A. and Crane, P., 1981, Astrophys. J. **246**, pp. 647-652.

ACKNOWLEDGEMENTS: Martha Anderson has performed most of the analysis on 3C33 North. Jeff Pedelty assisted in the original observations and analysis. Partial support for this research came from the U.S. National Science Foundation, AST 83-15949. Support from the American Astronomical Society, the Max Planck Gesellschaft and the Office of International Education at Minnesota enabled me to participate in this meeting. I apologize to everyone for not putting proper references in this paper.

Larry Rudnick   Wayne Christiansen
      Phil Crane   Phil Hardee

Chris O'Dea with Connor

3C332: A SOURCE WITH AN "EXCEPTIONAL" MULTIPLE
HOT SPOT MORPHOLOGY

Wayne A. Christiansen
Department of Physics and Astronomy
University of North Carolina
Chapel Hill, NC 27514, USA

In the hope that exceptions may illuminate the rules, I would like to present a 6cm VLA image of the radio galaxy 3C332 which I believe shows that this source violates Robert's (Laing, of course) Rules of (jet-hot spot) Order. Referring to Figure 1, this source clearly has an interesting set of hot spots on the northeastern side. It is not at all clear to me which of these hot spots is the "primary" and whether the rest are indeed "secondary". Our map does not have high dynamic range but it is clear that the only jetlike feature on this image is pointing in the opposite direction, ie. away from the hot spot complex.

Also, in Figure 1, 6cm polarization vectors are superimposed on the total intensity contour map of 3C332. Here we do find some agreement with Robert's Rules (and Colin's as well) in the sense that the polarization vectors do indeed indicate that the magnetic field appears to be directed from one hot spot toward its adjacent neighbor.

Conclusions:

The single most important point concerning these observations of this particular source relates to the fact that we do indeed find (possibly two) hot spots which, based on their observed properties alone, would seem to qualify them as "primary" hot spots, but there is no evidence for a jet terminating in either one of them. Therefore, although the catagorization of primary or secondary hot spots is indeed useful it, perhaps, is not yet justified to take this morphological classification one step further by identifying "primary" hot spots as always marking the termini of active jets.

Figure 1

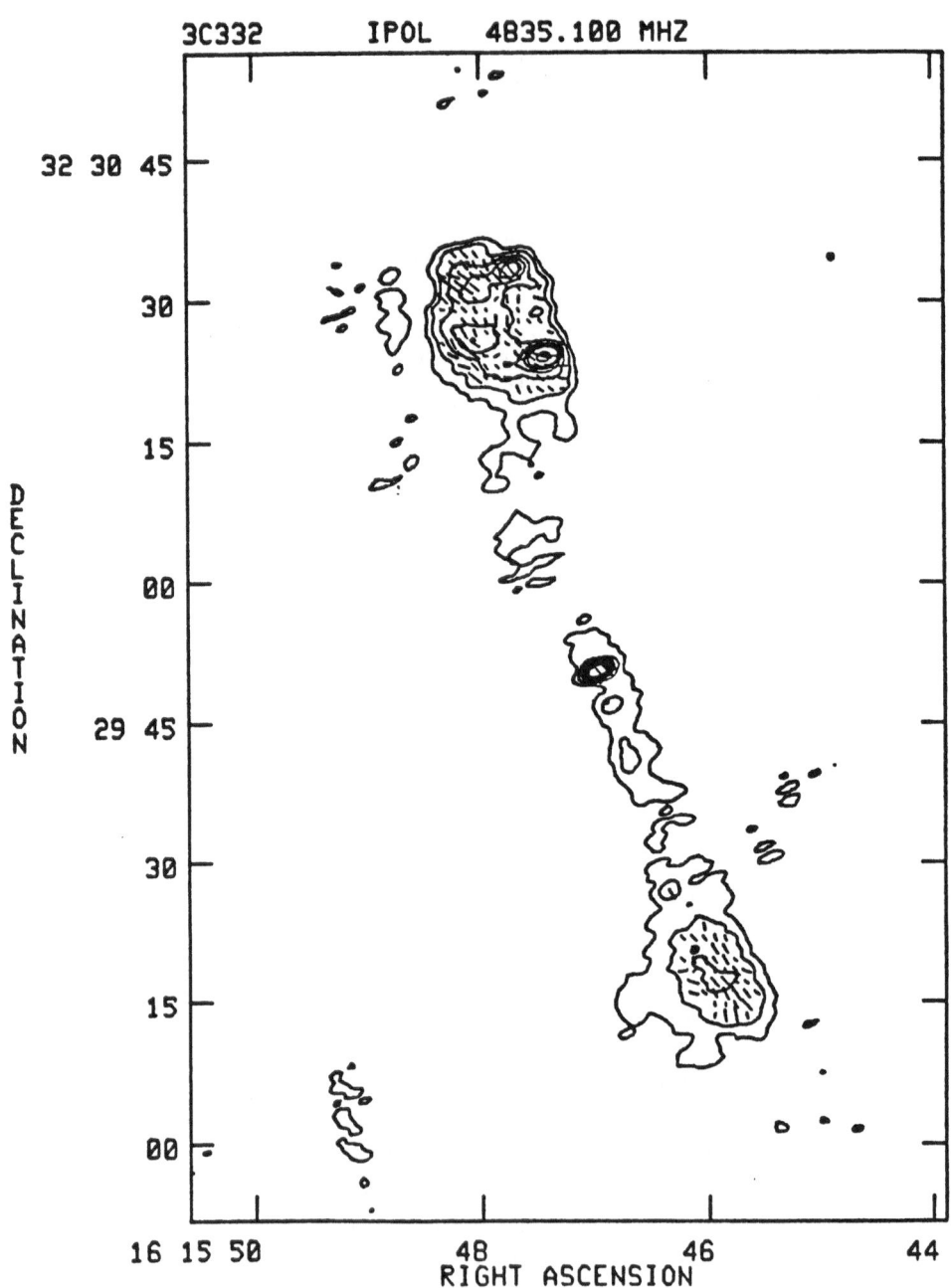

3C332   IPOL   4835.100 MHZ

PEAK FLUX =   1.5983E-02 JY/BEAM
LEVS =   0.6393E-03 * (-1.0,   1.0,   2.0,
    4.0,   6.0,   8.0,   10.0, 20.0, 40.0,
   60.0, 80.0, 100.0)

CONSTRAINTS ON THE HOTSPOT ADVANCE SPEED IN
THE CORE-DOMINATED QUASAR 1510-089

Christopher P. O'Dea
Netherlands Foundation for Radio Astronomy
Postbus 2, 7990 AA Dwingeloo, Netherlands

ABSTRACT

We present subarcsecond resolution VLA observations at wavelengths of 6 and 1.3 cm of the core-dominated quasar 1510-089. We find a counter component opposite to the main jet at a distance of ~0$\overset{\prime\prime}{.}$3 from the core. It is not clear yet whether this counter component is a knot near the base of a counter jet or a hotspot at the end of a counter jet. If the latter, and the two jets are ejected symmetrically from the nucleus, then the estimated hotspot advance speed is $\beta_{hs}\cos\theta \approx 0.93$ assuming that the difference in angular extents of the two sides of the source is due to the relative light travel times. This value seems high but is not currently ruled out.

I. OBSERVATIONS AND RESULTS

The quasar PKS 1510-089 (OR-017) at z=0.361 was observed as part of a project to image the extended structure around core-dominated quasars with subarcsecond resolution. Additional details and results are given by O'Dea, Barvainis, and Challis (1988; hereafter referred to as OBC). The source was observed in the A configuration of the VLA (Thompson et al. 1980). Observations at two adjacent frequencies were combined to obtain an effective bandwidth of 100 MHz at 4860 MHz (6.2 cm), 14940 MHz (2 cm) and 22460 MHz (1.3 cm) with resolutions of ~0$\overset{\prime\prime}{.}$45, ~0$\overset{\prime\prime}{.}$15, and ~0$\overset{\prime\prime}{.}$08, respectively.

At 20 cm, Perley (1982) detected a one-sided jet with length $\theta_1 \approx 8"$ at pa~160°. Our 6 cm image of the jet is shown in Figure 1. The projected length of the jet is ~40 kpc and is similar to that of 3C273. The VLBI jet has an orientation of ~170° and points towards the base of the southern jet (Padrielli et al. 1986). At both 2 and 1.3 cm (Figure 2) we detect a slightly resolved (deconvolved size at 1.3 cm is 0".1x0".07 [i.e.,~450x310pc]) secondary component with an integrated flux density at 1.3 cm of $S\approx26$ mJy and an integrated spectral index between 1.3 and 2 cm of $\alpha\approx-0.6$. The component is at $\theta_d\approx0".3$ at pa $\approx-28°$ which is approximately 180° from the inner part of the main jet (at pa $\approx155°$) and thus, it is likely that this component is part of a counterjet. At 2 cm the E vector of this component is roughly aligned with the direction to the core (OBC). Thus, if the rotation measure is small, the apparent projected magnetic field is perpendicular to the counterjet as found in many knots and hotspots (e.g., Dreher 1981; Bridle 1984; Bridle and Perley 1984).

II. DISCUSSION

It is not currently known whether this counter component represents a knot near the base of a possible counterjet or a hotspot at the end of the counterjet. However, neither OBC nor Perly (1982) have detected any structure beyond the component on the counterjet side. If the component is indeed a terminal hotspot then a constraint on the hotspot advance speed is obtained assuming that the difference in apparent lengths of the two sides of the source is due to the difference in light travel times (e.g., Longair and Riley 1979). If the axis of the source makes an angle Θ to our line of sight, then

$$\beta_{hs}\cos\Theta=\frac{R-1}{R+1} \qquad (1)$$

where R is the ratio of the lengths of the two sides of the source and β_{hs} is the advance speed of the hotspot through the intergalactic medium in units of the speed of light. We assume that β_{hs} is the same for both sides of the source. Under this hypothesis, the observed ratio of the lengths of the two sides gives an inferred hotspot advance speed of $\beta_{hs}\cos\Theta\approx0.93$. This estimate is independent of the distance to the source, but is subject to uncertainty caused by possible intrinsic asymmetry in the source. We note that Kundt and Gopal-Krishna (1986) have suggested $\beta_{hs}\approx0.7$ in 3C273.

Are these high values of hotspot advance speed plausible? The value we have estimated of $\beta_{hs}\gtrsim0.93$ is much larger than the value of $\beta_{hs}\lesssim0.3$ obtained from statistical analysis of samples of double sources (e.g., Ingham and Morrison 1975; Longair and Riley 1979; Banhatti 1980; Macklin 1981). However, the samples used were restricted to sources with fairly symmetric double structure and thus do not rule out the existence of a tail in the distribution to high values of β_{hs}.

A more serious problem was raised by Conway et al. (1981) who pointed out that large hotspot advance speeds require extremely low densities in the external medium.

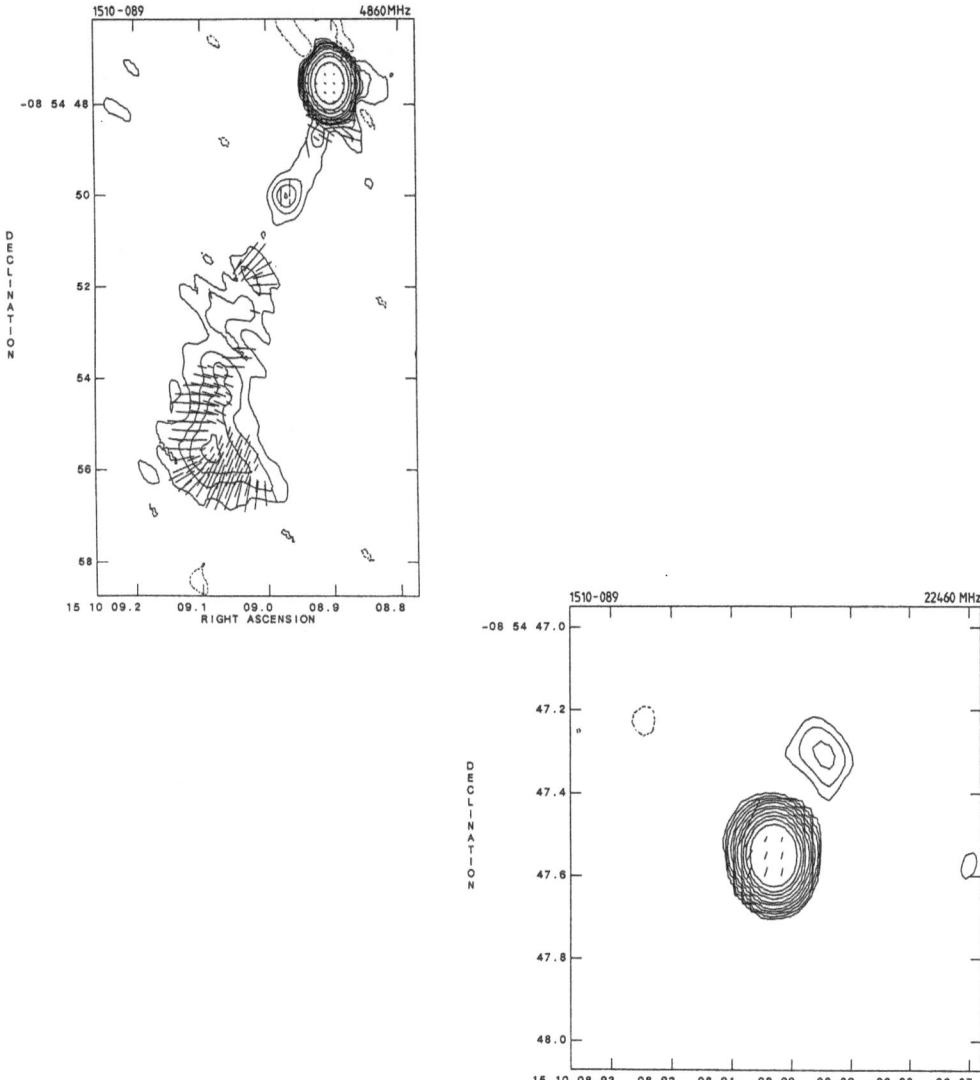

Figure 1 (Top). 1510-089. Contour plot of total intensity at 6 cm with E-vectors superposed whose length is proportional to the fractional polarization. 1 arcsecond = 67%. Contour levels are -1 (dashed), 1, 2, 3, 4, 5, 8, 12, 20, 50, 100, and 300 mJy per beam. the clean beam FWHM is 0".59x0".42 @ 3°.

Figure 2 (Bottom). 1510-089. Contour plot of total intensity at 1.3 cm with E-vectors superposed whose length is proportional to the fractional polarization. 1 arcsecond = 70%. Contour levels are -5 (dashed), 5, 8, 12, 20, 30, 50, 75, 125, 200, 300 and 500 mJy per beam. The clean beam FWHM is 0".10x0".08 @ 2°.

For 1510-089 the condition that the hotspot pressure P_{hs} is balanced by the external ram pressure $\rho_{ext}(\beta_{hs}c\gamma)^2$ where γ is the Lorentz factor gives an external number density of

$$n_{ext}=1.1\times10^{-5}cm^{-3}(\frac{P_{hs}}{10^{-8}dynes\ cm^{-2}})(\frac{1-\beta_{hs}^2}{\beta_{hs}^2}) \qquad (2)$$

We have calculated the minimum pressure in the hotspots (see OBC) ignoring possible relativistic boosting of the hotspot flux density sence the hotspot advance speed may be decoupled from the Doppler boosting (Lind and Blandford 1985; OBC). The hotspot could also be out of minimum pressure by a large factor. Equation (2) gives $n_{ext}\lesssim10^{-6}cm^{-3}$ for $\beta_{hs}\gtrsim0.93$ assuming $P_{hs}=10^{-8}$ dynes cm^{-2}. A density this low might be feasible if 1510-089 is extremely isolated and the densities of $<2\times10^{-8}cm^{-3}$ estimated for the hot intercluster medium (Forman, Jones, and Tucker 1984) are correct. However, 1510-089 lies in the direction of a group of galaxies (Yee, Green, and Stockman 1986). If it is established that 1510-089 is a member of this group then densities of $\sim10^{-4}cm^{-3}$ might be more realistic (e.g., Kriss et al. 1983). For a density of $10^{-4}cm^{-3}$ the hotspot advance speed is reduced to $\beta_{hs}\approx0.33$ if $P_{hs}=10^{-8}$ dynes cm^{-2}. A hotspot advance speed of $\beta_{hs}=0.93$ is consistent with an external density of $10^{-4}cm^{-3}$ only if $P_{hs}\gtrsim10^{-6}$ dynes cm^{-2}.

ACKNOWLEDGEMENTS

These results were obtained as part of a larger project with Rich Barvainis and Peter Challis. I thank them for their fruitful and generous collaboration as well as for their indulgence. I am also grateful to Tony Foley, Richard Strom, Stefi Baum, and Wolfgang Kundt for comments on the manuscript.

References

Banhatti, D.G. (1980). Astron. Astrophys. **84**, 112.
Bridle, A.H. (1984). A.J. **89**, 979.
Bridle, A.H. and Perley, R.A. (1984). Ann. Rev. Astron. Astrophys. **22**, 319.
Conway, R.G., Davis, R.J., Foley, A.R., and Ray, T.P. (1981). Nature **294**, 540.
Dreher, J.W. (1981). Astron. J. **86**, 833.

Forman, W., Jones, C., and Tucker, W. (1984). Astrophys. J. **277**, 19.

Ingham, W. and Morrison, P. (1975). Mon. Not. R. Astr. Soc. **173**, 569.

Kriss, G.A., Cioffi, D.F., and Canizares, C.R. (1983). Astrophys. J. 272, 439.

Kundt, W. and Gopal-Krishna (1986). J. Astrophys. Astr. **7**, 225.

Lind, K.R. and Blandford, R.D. (1985). Astrophys. J. **295**, 358.

Longair, M.S. and Riley, J.M. (1979). Mon. Not. R. Astr. Soc. **188**, 625.

Macklin, J.T. (1981). Mon. Not. R. Astr. Soc. **196**, 967.

O'Dea, C.P., Barvainis, R., and Challis, P.M. (1988). Astron. J. in Press.

Padrielli, L., Romney, J.D., Bartel, N., Fanti, R., Ficarra, A., Mantovani, F., Matveyenko, L., Nicholson, G.D., and Weilder, K.W. (1986). Astron. Astrophys. **165**, 53.

Perley, R.A. (1982). Astron. J. **87**, 859.

Thompson, A.R., Clark, B.G., Wade, C.M., and Napier, P.J. (1980). Astrophys. J. Suppl. **44**, 151.

Yee, H.K.C., Green, R.F., and Stockman, H.S. (1986). Astrophys. J. Suppl. **62**, 681.

Bob Fosbury

The extended structure of the radio galaxy PKS 0521–36: radio polarization and optical emission lines

W M Goss & **R D Ekers** *National Radio Astronomy Observatory, P.O. Box 0, Socorro, New Mexico 87801, USA*

R A E Fosbury[1] & **C N Tadhunter** *Space Telescope-European Coordinating Facility, European Southern Observatory, Karl-Schwarzschild-Straße-2, D-8046, Garching bei München, FRG*

I J Danziger *European Southern Observatory, Karl-Schwarzschild-Straße-2, D-8046, Garching bei München, FRG*

Abstract of paper to be submitted to Mon. Not. R. astr. Soc.

Summary. The radio galaxy/*BL Lac* object PKS 0521-36 ($z = 0.055$) has been mapped at 1.49, 4.86 and 14.97 *Ghz* with scaled arrays of the Very Large Array (VLA). The angular resolution is $\approx 4\,arcsec$. The distributions of spectral index and depolarization between 1.49 and 4.86 *Ghz* have been derived. Using the observations at all three frequencies, the rotation measure and projected magnetic field orientations have been derived at a resolution of $\approx 6\,arcsec$. The radio structure consists of a compact VLBI core, a resolved hot spot to the southeast and an extensive low brightness halo to the west. The radio structure associated with the optical jet has also been mapped with a 1 *arcsec* resolution at 4.86 *Ghz*. The magnetic field is parallel to the jet near the core and perpendicular at greater distances.

Long-slit spectroscopy and CCD images in Hα+[NII] have revealed extended low ionization emission emission to the east and southeast of the nucleus; this region is associated with an area of pronounced depolarization and large rotation measure.

---

[1]Affiliated to the Astrophysics Division, Space Science Department, European Space Agency

Frazer Owen
       Chris O'Dea

New VLA Results on M87

Frazer N. Owen

National Radio Astronomy Observatory *
Socorro, NM 87801

Philip E. Hardee

Department of Physics and Astronomy
University of Alabama, Tuscaloosa, AL 82071

T.J. Cornwell

National Radio Astronomy Observatory *
Socorro, NM 87801

Dean C. Hines

Department of Astronomy
University of Texas, Austin, TX 78712

and

Jean A. Eilek

Department of Physics
New Mexico Tech, Socorro, NM 87801

Abstract

New results obtained with the VLA at $\lambda 2$ and $\lambda 6$cm show a wide variety of interesting radio structure in the inner two kiloparsecs of M87.

1) The radio emission from the M87 jet is dominated by complex filaments and loops. The jet also appears limb brightened at several points along the structure. The magnetic field appears to thread these features. The complex structure certainly requires that the jet has a filling factor much less than one. The most consistent interpretation of the image is that most of the emission is coming from a narrow boundary region.

---

* The National Radio Astronomy Observatory is operated by Associated Universities, Inc., under contract with the National Science Foundation.

2)   The emission from the inner radio lobes also shows complex
     filamentation. The volume emissivity in the brightest filaments
     exceeds the average background by at least a factor of 100. In
     regions where the sensitivity is adequate the magnetic field
     also appears to thread features.

3)   The Faraday rotation over most of the radio lobe exceeds 1000
     rad m^{-2}. In one small region it reaches 8000 rad m^{-2}. Over most
     of the source, the rotation values are positive. The smallest
     values occur in the direction of the jet (~100 rad m^{-2}).

     If the optical synchrotron emission also occurs in a boundary
layer, the light travel times across the emitting region are smaller
than previously assumed and thus it is easier to understand how the
relativistic particles live long enough to radiate. The particles may
be accelerated in the boundary layer, or they may be accelerated in
the nucleus, they then propogate down the center of the jet in a low
field region and diffuse into a high field region in the boundary
layer.

     The minimum pressure derived from the synchrotron emission from
the jet exceeds the exterior pressure estimates from the X-ray images
by factors of 2 to 50 along all of the jet beyond knot D. This
suggests that the magnetic fields in the boundary layer are responsible for some of the confinement of the jet and/or some of the
emitting regions are not in pressure equilibrium with the external
medium. The overall morphology and structure of the jet are consistent with a dynamically important magnetic field.

     The filaments in the lobes have minimum pressures very similar
to the estimated X-ray pressure. A thermal cooling instability could
be responsible for their creation. Alternatively, magnetic instabilities could have produced these structures.

     The high values of the rotation measure suggest that a dynami-
cally important magnetic field exists at least in some of the inner
regions of M87. The generally positive values of the rotation measure
require a large scale field in the central region with a magnitude
of several microgauss. The low values of rotation in the direction of
the jet imply that the jet lies in front of the lobes and the Faraday
rotation screen.

I. Introduction

Since its discovery as a radio galaxy in the early 1950's and the first arguments connecting the radio emission with the optical jet (Baade and Minkowski, 1954), M87 has been the most important example for the study of non-thermal phenomena in extragalactic jets. In 1980 VLA observations began to reveal the close correspondence between the morphology at centimeter and optical wavelengths (Owen, Hardee and Bignell, 1980). In 1983, we reported the first $0''.1$ maps of the M87 jet with the VLA at 2cm (Biretta, Owen, and Hardee, 1983). These images, which have a dynamic range of about 5000 to 1, seemed most consistent with a shock at knot A being responsible for the brightening of the jet at that point. In this paper we report new observations at $\lambda 2$cm which have reached a dynamic range of 50,000 to 1 with the same resolutions as the 1983 images and $\lambda 6$cm images with $0''.4$ resolution and over 100,000 to 1 dynamic range. As we will discuss below, these results seem to us to require new interpretations.

II. Observations, Reductions, and Results

The observations reported in this paper were made in 1985 and 1986 in the VLA A array at $\lambda 2$cm and $\lambda 6$cm. Some data from 1983 C array observations was also included in the $\lambda 6$cm images. A combination of some improved observing techniques with the VLA, some special calibrations to correct the closure errors, self-calibration of the images, and finally maximum entropy deconvolution of the images have allowed us to reach very high dynamic ranges. The techniques used are discussed more fully in the individual papers on the observations and references therein (Owen, Hardee and Cornwell 1988; Hines, Owen, and Eilek, 1988).

The $\lambda 2$cm image has a beamsize of $0''.1$ and a dynamic range of about 50,000 to 1. The $\lambda 6$cm image has a beamsize of $0''.4$ and a dynamic range of about 150,000 to 1. Since the nucleus of M87 is so bright, these high quality results are necessary in order to see the details in the structure of the jet and lobes. Away from the nucleus, the dynamic range is never greater than 1000 to 1 at $\lambda 2$cm or 10,000 to 1 at $\lambda 6$cm and many individual features are only a factor of 5 to 10 above the effective noise. However, even these more modest ranges of intensity are impossible to display with a linear transfer function and thus we have used histogram equalized images in order to display a wider range of intensities in a single gray scale image. Thus the brightness levels in the grey scale images of the total intensity maps should be viewed qualitatively not quantitatively.

In figure 1 we show the λ2cm image of the jet at 0".1 resolution. Even in this histogram-equalized image, the brightest complex of knots, ABC are saturated. Other displays of this region are contained in Owen, Hardee and Cornwell (1988). The bright nucleus has been almost totally subtracted from this image. For scale the bright discontinuity in the jet at knot A occurs about 12 arcsec or 1 kpc from

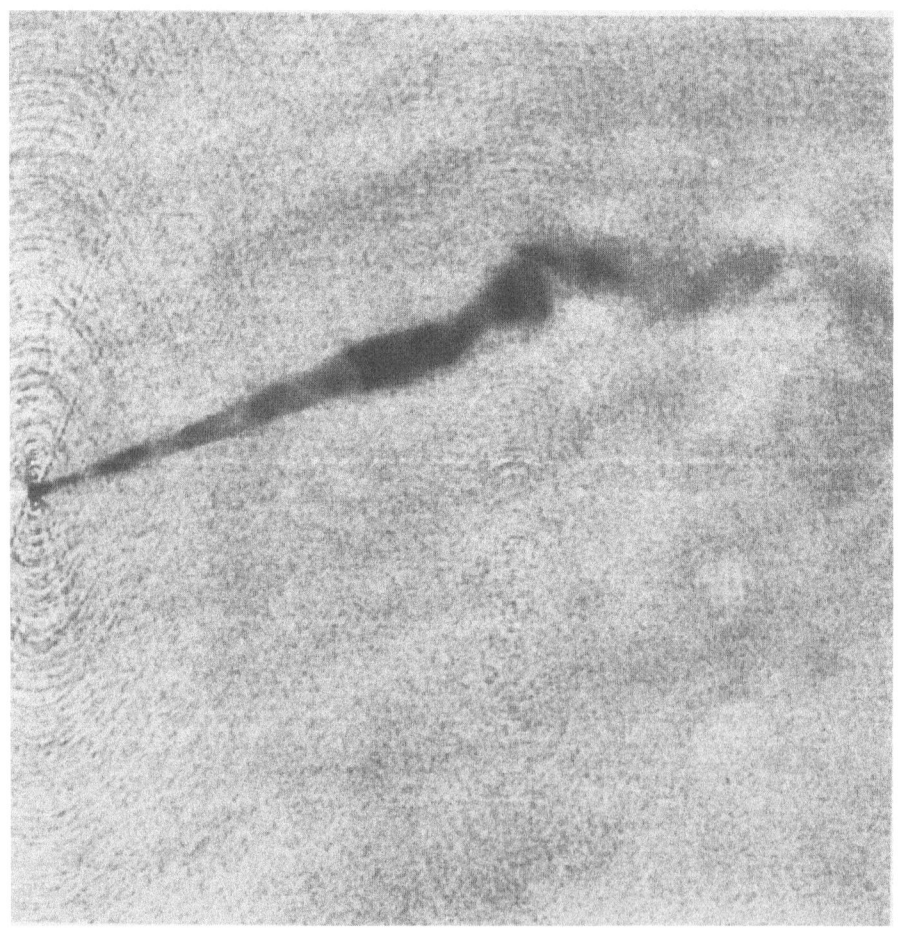

Fig. 1. VLA total intensity image of the M87 jet at λ2cm with 0".1 resolution.

the nucleus. Figure 2 contains the λ6cm image of the jet and lobes. This display is intended mainly to show the filaments and thus the nucleus and jet are mostly saturated. The beamsize is 0".4 and the distance to the edge of the lobes from the nucleus, east or west, is about 30 arcsec or 2.5 kpc. In figure 3 we show the pattern of Faraday rotation over the lobe as obtained from the six centimeter dataset. One can see the same outline of the emitting region as in figure 2 but a very different pattern in detail. The brightest regions have typical values of 2000 rad m^{-2}. The peak value in the figure reaches 8000 rad m^{-2}. The medium intensity regions extending over most of the display have values near 1000 rad m^{-2}. The dark areas on the jet have values of only about 100 rad m^{-2}. Only a few negative regions are found near the edges of the lobes in mostly low surface brightness regions.

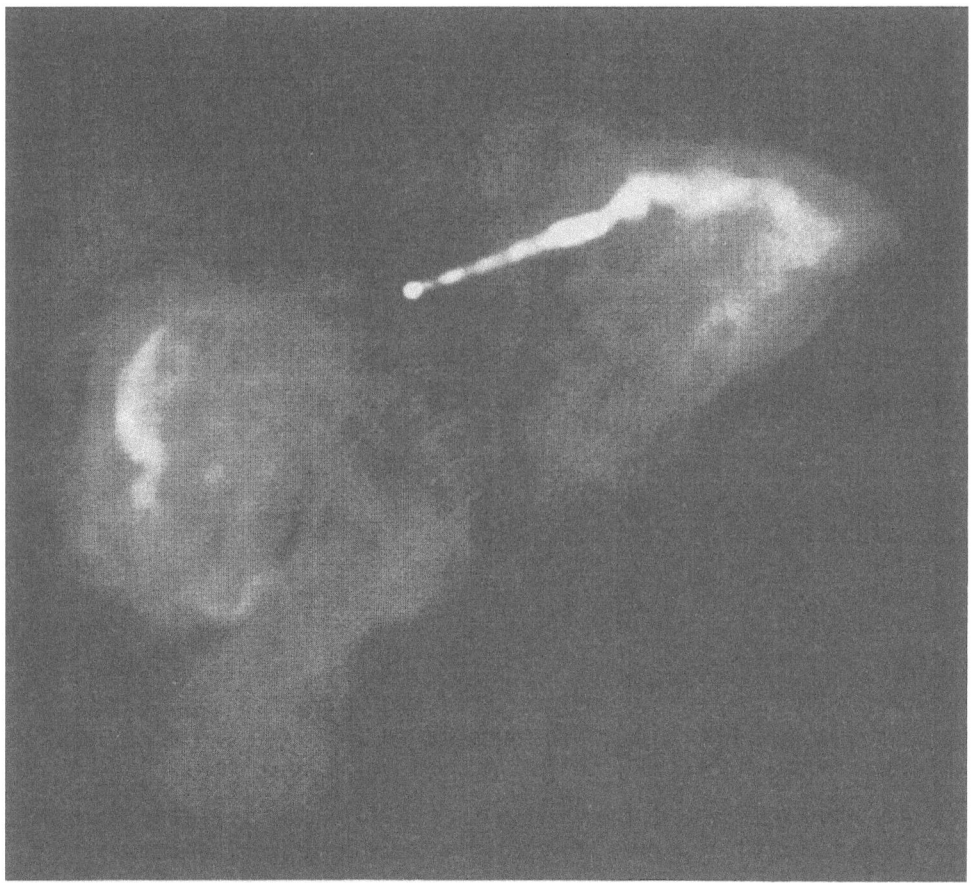

Fig. 2. VLA total intensity image of the M87 jet and lobes at λ6cm with 0".4 resolution.

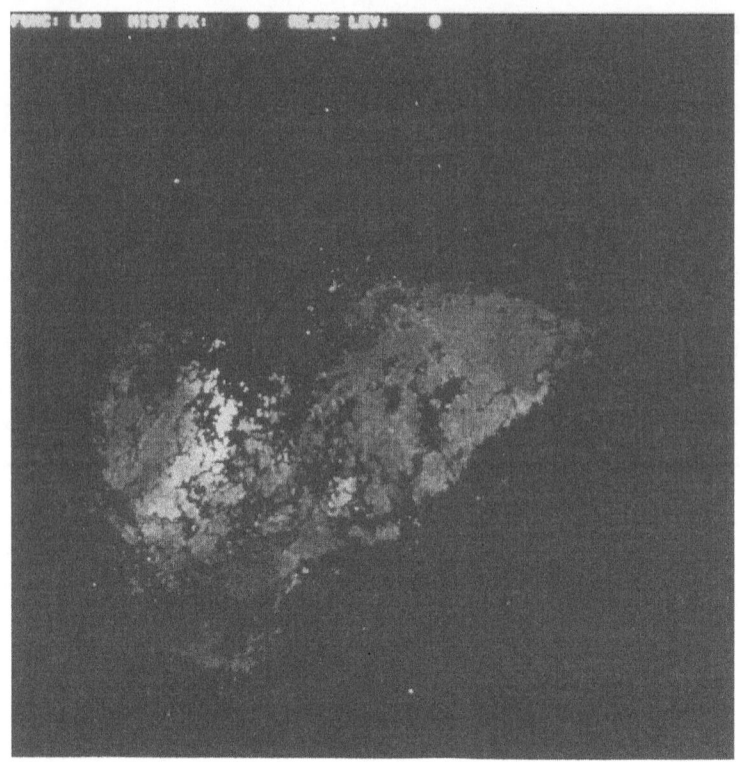

Fig. 3. VLA image of Faraday rotation in front of M87 with 0.4 resolution.

III. The Jet

The intensity maxima in the jet initially referred to as "knots" because of their optical appearance at ~ 1 arcsec resolution are now all resolved and show complex structure. For convenience we will continue to use the classical lettering system for the general regions on intensity maxima in the jet. These features are G, D, E, F, I, A, B, C which are located at about 1".3, 3".5, 6".0, 8".5, 11", 12", 14", 17" from the nucleus, respectively (see figure 1). The jet is limb brightened over much of its length. Extreme limb brightening can be seen on the intensity image in figure 1 between E and F. In addition to the limb brightening, the intensity image reveals the presence of one or more filaments which cut diagonally across the jet between D and F. From F through I the filaments appear to merge into sheets. In general there is only a small intensity decrease between the filaments and the sheets. The exception is a dark diagonal line across the jet between F and I.

Knot A also appears to consist of a filamentary feature which is more tightly wrapped in pitch angle around the jet than the features further upstream. Close examination of the images shows that the tightly wrapped feature begins just upstream of A between A and I. Very faint, tightly wrapped filaments appear to continue downstream of A into B but are confused by irregular bright features in B. A particularly interesting feature is the dark line that runs down the interior of the jet between A and C. Another interesting feature is the apparent ring of emission downstream of C.

The projected magnetic field, corrected for Faraday rotation, generally lies along the filaments. Even small bends of 10 or 20 degrees in the filaments usually have an associated change in the polarization direction. The percentage polarization is typically 30 percent.

The appearance of the filaments, the magnetic field direction, and the limb brightened or flat topped profiles seen in these data seem to require an unfilled emitting structure with a volume emissivity rising toward the outer boundary of the jet. The brightest regions in this emission pattern appear to be flux ropes wrapped around the outside of the jet. Sharp edges on the features at the beginning of knot A and the end of knot C are also most consistent with surface features. In this picture one would not expect to see limb brightnening all along the jet since the filaments are wrapping around the jet and also apparently changing their brightness along the filament.

The low emissivity in the core of the jet suggests one or more of the following possibilities: 1) only relatively weak magnetic fields, 2) a field aligned distribution of relativistic electrons (positrons) or 3) cold electrons (positrons) exist in the jet's core. On the other hand, the constant average spectral index along the jet, the similarity between radio and optical structure in the jet (Keel 1988) and the short lifetime to synchrotron losses in the optical imply that highly energetic electrons (positrons) must either be generated *in situ* within the boundary layer or be deposited in the layer from the jet's interior. For example, *in situ* acceleration by turbulent wave-particle interaction may be restricted to the surface

layer because turbulent energy input driven by Kelvin-Helmholtz instabilities is highest at the surface (Hardee 1983) and is likely to
result in a surface brightened jet (Eilek 1982). If the high energy
electrons are accelerated in the nucleus and deposited in the surface
layer after propagating down the jet's interior, then they must be
able to reach the end of the jet without significant synchrotron or
Compton losses. The timescale for Compton losses can be used to set a
lower limit to the velocity in the jet's interior. Taking a photon
field with energy density $u_{rad,-13} = u_{rad}/10^{-13}$ erg cm^{-3}, electrons
which radiate with a critical frequency $\nu_{c,14} = \nu_c/10^{14}$ Hz in a magnetic field, $H_{\mu G}$ in microgauss, this lower limit is

$$v_l \approx 20\ \nu_{c,14}^{1/2}\ u_{rad,-13}\ l_{j,kpc}\ H_{\mu G}^{-1/2}\ \text{km s}^{-1}$$

where $l_{j,kpc}$ is the jet's length in kpc. If we assume

1)   a critical frequency 3×10^{14} Hz at the spectral break in the
      optical,

2)   the radiation energy density to be that of the starlight at 2
      kpc from the center which is 4×10^{-12} erg cm^{-3} (Keel, private
      communication), an order of magnitude above the 3 K background,

3)   a typical minimum energy field of 300 μG in the surface fila-
      ments and

4)   a jet length of 3 kpc,

then $v_l \approx 240$ km s^{-1} in the non-radiating interior.

The magnetic field strength corresponding to similar synchrotron
loss times is 10μG. While the jet probably has a larger velocity than
this limit, this calculation shows that the velocity of the jet could
be much smaller than the speed of light without requiring *in situ*
particle acceleration.

We have calculated minimum pressures from the data shown in
figure 1 all along the jet. These can be compared with external
pressures in the same region estimated from the X-ray images obtained
by the *Einstein* observatory (Lea *et al.*, 1982; Schreier *et al.*, 1982;
Stewart *et al.*, 1984). Details of the calculations are discussed in
Owen, Hardee and Cornwell (1988). To summarize all of the minimum
pressures downstream of knot G are at least a factor of 2 above the
X-ray pressure estimates, even in between the "knots". Along most of
the jet, the minimum pressure exceeds the X-ray pressure estimates by
a factor of 4 to 50. While a variety of explanations are possible,

this result seems easiest to understand either if the emitting regions are transient features out of pressure equilibrium with the
external medium and/or if the jet is magnetically confined.

IV. The Lobes

In Figure 2 we show the λ6cm image of the radio lobes. One can
see the jet and its outer extensions blending into the more diffuse
lobe structure. A complex distribution of small scale structure,
including loops, linear features, and almost unresolved blobs can be
seen against a diffuse background. Minimum pressures in these
features range from about 10^{-9} to 3×10^{-10} dyn cm^{-2}. Typical minimum pressure fields are 60 to 120 μG. For these fields, particle
lifetimes at λ6cm to synchrotron losses are in the range $1-3\times10^6$
years. The diffuse background may be a uniform medium or it could be
a mass of fainter filaments blended together by our resolution.
Assuming it is uniform, the minimum pressure is about 10^{-10} dyn cm^2,
the minimum pressure field is 37 μG and the corresponding particle
lifetime is about 10^7 years.

The minimum pressures in the filaments are close to the
pressures exterior to the jet estimated from the X-ray observations
at 1 kpc from the nucleus. The minimum pressure in the diffuse regions is lower, although not far below the Stewart *et al.* (1984)
model at 2 kpc. Thus it seems possible that the filaments are in
pressure equilibrium with the thermal gas in the region. Therefore
either the pressure contributed by relativistic particles and
magnetic field is probably less than the thermal pressure or the
synchrotron emitting regions are far from the minimum pressure/
equipartition condition.

If the filaments are formed out of the diffuse region, there are
at least three mechanisms possible.

1) The filaments may have formed out of the diffuse region by
    evolving toward equipartition from a state very far from
    equipartition. This might occur by means of a synchrotron in-
    stability (e.g. Eilek and Caroff 1979). When explored in detail
    this model seems to require very special particle energy

distributions in order to make bright filaments instead of dark filaments.

2) If the diffuse region contains a mixture of relativistic particles, magnetic fields and thermal gas, the filaments may form by means of a thermal cooling instability. This requires that the lifetimes of the relativistic particles to synchrotron losses be longer than the thermal cooling time. This seems possible if the fields in the filaments are at or below the minimum pressure values.

3) The filaments could be magnetically confined themselves. The minimum pressures are above the X-ray pressures in the brightest filaments and, of course, could have even higher pressures. If the magnetic field is dynamically important throughout this region it is possible that magnetic tearing instabilities could be responsible for the filaments (e.g. Furth, Killeen, and Rosenbluth, 1963).
Further discussion of this problem is contained in Hines, Owen and Eilek (1988).

An additional clue which suggests the magnetic field may be important throughout this region is given by the high Faraday rotation measures found in the direction of the source as shown in figure 3. Assuming the range of X-ray properties suggested by the three papers discussed earlier, the highest Faraday rotation region (8000 rad m^{-2}) suggests a field of at least 100 to 200 μG. A net ordered field of a least a few μG must exist over the entire region in order to account for the net predominence of positive rotation measures. If the field is complex with many field reversals along the line of sight, the magnitude of the field would be much larger. Thus we may be seeing the tip of the iceberg.

In order to obtain such large Faraday rotation values, the region responsible for the Faraday rotation must be in front of the synchrotron emitting regions (i.e. a "Faraday screen"). If the two regions were mixed, the synchrotron source would be depolarized (e.g. Burn 1966). However, since most of the polarization seems to be associated with filaments, the rotation could be occuring in the diffuse emission region discussed earlier (the interfilament region). It seems unlikely that the Faraday screen is far out in the cluster since given the estimated density from the X-ray images further out in the cluster and the gradients in rotation measure observed over the source, the magnetic field pressure would need to exceed the thermal pressure considerably.

The small values of rotation measure in the direction of the jet also suggest that the Faraday screen is close to the center of the cluster. In fact, the much smaller values of rotation measure in the direction of the jet are most consistent with the jet being *in front* of the Faraday screen and the other filaments. Further analysis of these data may allow a true 3-dimensional picture of the source to be constructed.

V. Conclusions

In this short review we have discussed several new results from the recent VLA observations. A more complete discussion of each of these results occurs in the referenced papers which should be appearing about the time this book is published. The dominant conclusions are that

1) the jet and lobes are both unfilled, filamentary structures;
2) the magnetic field probably plays a more important role than previously suspected, possibly dominating the physics of the region we are observing and
3) the physics of the jet, especially the problems with particle lifetimes, are easier to understand if the jet emission from the jet is mostly in a boundary layer as suggested by these observations.

References

Baade, W. and Minkowski, R. (1954), *Ap.J.* **119**, 221.
Biretta, J.A., Owen, F.N. and Hardee, P.E. (1983),
        Ap.J. (Letters) **224**, L27.
Burn, B.J. (1966), *M.N.R.A.S.* **133**, 67.
Eilek, J.A. (1982), *Ap.J.* **254**, 472.
Eilek, J.A. and Caroff, L.J. (1979), *Ap.J.* **233**, 463.

Furth, H.P., Killeen, J. and Rosenbluth, M.N. (1963),
   Phys. Fluids **6**, 459.

Hardee, P.E. (1983), *Ap.J.* **269**, 94.

Hardee, P.E. (1987), *Ap.J.* **313**, 607.

Hines, D.C., Owen, F.N. and Eilek, J.A. (1988), in preparation.

Keel, W.C. (1988), *Ap.J.* **329**, 532.

Lea, S.M., Mushotsky, R. and Holt, S.S. (1982), *Ap.J.* **262**, 24.

Owen, F.N., Hardee, P.E., and Bignell, R.C. (1980),
   Ap.J. (Letters) **239**, L11.

Owen, F.N., Hardee, P.E., and Cornwell, T.J. (1988), in preparation.

Schreier, E.J., Gorenstein, P., and Feigelson, E.D. (1982),
   Ap.J. **261**, 42.

Stewart, G.C., Canizares, C.R., Fabian, A.C. and Nulsen, P.E.J.
   (1984), *Ap.J.* **278**, 536.

Max Camenzind   Martin Schlötelburg

Digital Photometry of the Jet in M 87[1]

I. Pérez-Fournon[a], L. Colina[b], J. I. González-Serrano[b], P. L. Biermann[c]
[a]Instituto de Astrofisica de Canaris, La Laguna, Tenerife, Spain
[b]Universitäts-Sternwarte, Göttingen, G.F.R.
[c]Max-Planck-Institut für Radioastronomie, Bonn, G.F.R.

The jet in M 87 was observed with a CCD in the filters r an g. Using accurate models for the underlying elliptical galaxy, the morphology and spectrum of the jet can be described with confidence. The data show, that the local optical spectral index is steeper than at radio wavelength, and steepens with distance from the core. This can be understood in terms of local acceleration of energetic particles in shocks (Biermann and Strittmatter 1987). The low upper limit to the flux-density of a counter-jet constrains bulk relativistic motion models. Details are given in Pérez-Fournon *et al.* (1988).

Imaging Polarimetry of the Jet in M 87[2]

M. Schlötelburg, K. Meisenheimer, and H.-J. Röser
Max-Planck-Institut für Astronomie, D-6900 Heidelberg 1, Federal Republic of Germany

Summary. Optical polarization along and across the jet of M 87 was mapped at 1".5 effective resolution in the R band. For the first time the optical polarization structure can be followed all the way from 2" to 26" from the core. Values of $24.2 \pm 0.3\%$ at knot A (the brightest knot), $22.1 \pm 1.0\%$ at knot D (the innermost knot) and a maximum of $41 \pm 3\%$ at a distance of 22" from the nucleus were obtained. On a global scale the magnetic field in the synchrotron region exhibits a highly ordered pattern: it is aligned with the jet except at knots A and C where it stands perpendicular to the jet axis. In addition, the data reveal positional mismatches of polarization and brightness. The polarization peaks corresponding to knots D and F are shifted by ≈ 1" downstream. At knot F and close to knot B a significant gradient of polarization is found across the jet.

[1]Astrophys. J., **329**: L81-L84 (1988)
[2]Astron. Astrophys. **202**: L23-L26 (1988)

Hermann-Josef Röser

Continuum observations of hot spots at wavelengths < 1 cm

Hermann-Josef Röser

Max-Planck-Institut für Astronomie
Königstuhl 17
D-6900 Heidelberg 1
Federal Republic of Germany

Abstract

Identifications of optical counterparts for radio hot spots in classical double radio sources (Fanaroff-Riley-class II) are reviewed. For those detected studies at infrared and mm wavelengths are combined with radio and optical data to derive the synchrotron continuum and verify the synchrotron origin of the radiation. The majority of optically detected hot spots show a high frequency cut-off indicating that the energy of the radiating relativistic particles is limited to values below some critical upper bound.

I) Introduction

Until recently hot spots in extragalactic double radio sources (FR class II) have been the sole domain of radio astronomers. Around 1980 with the advent of linear detectors with high quantum efficiency their optical colleagues saw a fair chance to take part in the game and several groups began to search extensively for optical radiation from the outskirts of radio galaxies. The motivation behind these efforts was discussed in a paper by Saslaw et al. (1978) presenting the physical background for several emission mechanisms expected in principle to operate in the outer parts of radio sources. Since then optical synchrotron radiation was shown to be the source for the detected continuum radiation (Meisenheimer, Röser 1986) and this review is confined to its discussion (line emission, which was also detected in the meantime, is discussed by Wil van Breugel in this volume). For synchrotron sources a major advantage of high frequency observations is the short lifetime of the radiating particles (about

100 years in the optical compared to 10,000 years in the radio range). Thus visible light directly points to the locations where these particles are accelerated. Furthermore the observations at high frequencies - mm, infrared and optical - together with radio data constrain the synchrotron continuum of hot spots over the widest range currently possible and allow the reconstruction of the spectrum of the underlying relativistic particles. This way we hope to learn more about the physical processes operating in giant radio sources.

II) Observational situation

In 1984 White & Birkinshaw published the first 89 GHz observations of Cygnus A, the source with by far the brightest radio hot spots. Together with measurements at lower frequencies they derived a spectral index, which would predict a hot spot flux in the optical R band of 15.7 mag for the brightest hot spot D. However, an optical counterpart of this strength can be borne out by the observations of Kronberg et al. (1977). We are thus forced to conclude, that a spectral turnover has to occur somewhere between the highest radio frequencies and optical wavelengths. As is well known from canonical synchrotron theory (Pacholczyk 1970) such continua could result from an upper bound in the energy distribution of the underlying particle population. Once the magnetic field is known (e.g. from equipartition considerations) the maximum particle energy could be determined from detailed observations of the synchrotron continuum in optically identified hot spots.

For the hot spots in Cygnus A optical identification of faint counterparts is severely hampered by the high density of foreground stars ($b^{II}=6°$). But the situation is not too different for high galactic latitude fields if one is pushing the detection limits down to R = 24 ^{mag}. Then the high surface density of faint galaxies is the main source of confusion (see figures A1b and A3b. The same is true for work in the infrared, where these galaxies are relatively bright). Thus in a study of optical radiation from radio hot spots positional coincidence between radio and optical signals can only be a first step. It has to be followed by detailed investigations of spectral shape and polarization properties to verify the synchrotron nature and thus the identity of the object under consideration with the radio hot spot.

In the following the observational status of non-radio continuum work on radio hot spots will be described. In contrast to the two previous topics the amount of material to be reviewed is quite limited so I hope to present a complete list of observations available to date in this area. As almost all work at high frequencies has been done in the optical, this review is guided by them. Objects are - in deviation from the talk given at Ringberg - arranged in order of their right ascension and divided into three categories. They should reflect the degree to which confusion with unrelated objects can be ruled out. First all detections are collected, for which more than just the astrometric information is available. For these photometry, polarimetry and/or spectroscopy put forth an association with the radio hot spot (group A). The second group (B) encompasses all those objects where an optical object has been detected on deep images at the radio position but no further information is yet available. Therefore the connection with the radio hot spot remains to be established. Finding charts are given in Appendix A for both groups in cases where they are not available in the literature. The final and last group (C) lists objects for which either optical identification has been unsuccessfully attempted, or where reported identifications were subsequently shown to be inaccurate. A summary of the essential points and a brief outlook to future observations conclude this review.

III) Observations of individual objects

As pointed out above confusion with faint unrelated objects is a major problem in optical hot spot identification. Thus before any physical conclusions can be drawn from the observations one has to decide upon criteria which classify an object as the optical counterpart. Certainly one single measurement in the radio and one in the optical are not enough to derive a "synchrotron powerlaw". But if - as in the case of 3C 303 - a single powerlaw over the whole optical and radio range is found there is hardly a way out. The ultimate test in any case would be to compare the optical or infrared polarization with the radio data. This, however, has only been done for very few sources. In the case of extremely steep optical spectra as found for several hot spots, a measurement in the infrared K band is also a crucial descriminator against field galaxies, which should level off in their continua at these wavelengths and not connect smoothly to the mm and radio flux of the hot spot.

The classification presented below follows these arguments to subdivide objects for which high frequency measurements are reported. The majority of the objects are taken from the work by Crane et al. (1983) and from observations which our group at the MPIA (Hiltner, Meisenheimer and myself) has collected since 1984 on Calar Alto and La Silla. Most of the hot spots in the work by Crane et al. are comparatively weak at 5 GHz, whereas we have set up a larger optical survey based upon the 3C-166-sample (Jenkins et al. 1974), selecting mainly the brighter ones. Nevertheless the collection of objects studied thus far is still rather heterogeneous and statistical conclusions are premature.

Group A objects:

0040+517 (3C 20): Our identification of the western hot spot in 3C 20 is based on two deep, photometrically calibrated CCD images in the R and one in the B band (figure A1a). The continuum shape of this source is further constrainted by a 230 GHz point obtained with the IRAM 30m telescope on Pico Veleta (Meisenheimer et al. 1988). The reader is referred to Hiltner et al. (1989) and the former paper for detailed references to the radio data, especially the morphology of this hot spot. The optical radiation, which is unresolved at 1″3 seeing, originates from the compact component B (Laing 1981). Combining all available fluxes and an interstellar extinction value of $A_V = 1.09$ ($b^{II} = -11°$) a sharp continuum turnover from the power-law with $\alpha = -0.53$ ($S_\nu \sim \nu^\alpha$) is found at far IR wavelengths (fig. 1a). This is perfectly fitted with a synchrotron spectrum as described by Meisenheimer in this volume. The major support for a synchrotron origin comes, however, from polarization measurements (Hiltner et al. 1989). The degree of linear polarization in the optical of $(48.8 \pm 6.7)\%$ at position angle $(68 \pm 4)°$ is - within the errors - in agreement with the radio data (Laing, priv.comm.).

0106+130 (3C 33): Following the initial discovery of an optical counterpart to the southern hot spot by Simkin (1978), this object was studied by several groups and is the best observed object to date (see figure A1b for a deep finding chart). Dreher, Simkin (1986) used new VLA and CCD R-band observations for an astrometric analysis and report exact coincidence between the emission peaks at 6cm and in the R band. They claim that on the other hand the blue emission (based on Arp's 200 inch III-aJ plate) is not congruent with the hot spot's radio emission. However, on our recent CCD images, which have a much higher S/N ratio than Arp's plate, we do not see such a morphological difference between the B, R or I images (see figure 1). These data

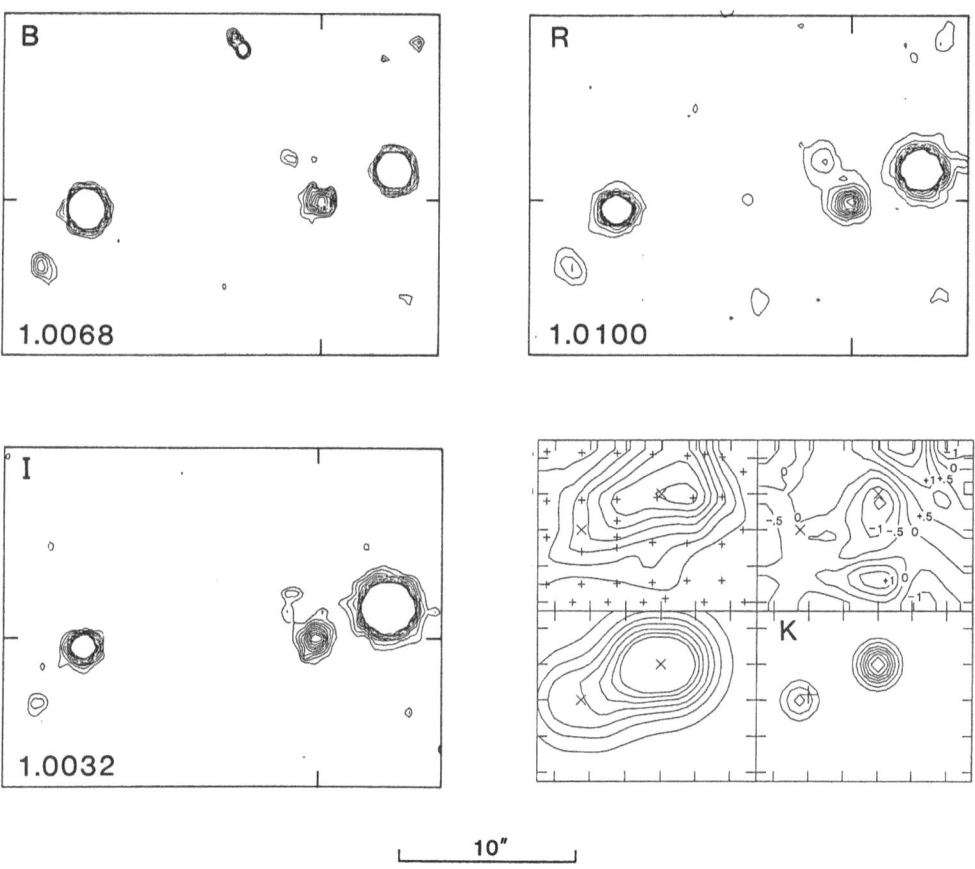

Fig. 1. Logarithmic isophote plots of the hot spot 3C 33 (south) at various wavebands. The optical data are from observations in the prime focus of the 3.5 m telescope on Calar Alto. The spacing factor of the isohotes is given in each plot.
The K-Band image with linear isophotes is a reconstruction from observations with UKIRT. The upper left panel gives the raw data with the locations of the individual integrations indicated by crosses. The lower left is the model describing these data, the rms deviations to the raw data are plotted at upper right. Finally the lower right gives the two point sources as reconstructed (details see Meisenheimer et al. 1988).

also clearly show that the source is extended (≈ 1" intrinsic extension). The association of this object with the radio source was unequivocally proven via polarization measurements in the R band by Meisenheimer, Röser (1986). Their linear polarization of (29.2 ± 2.4)% at position angle (54 ± 4)° is well in agreement with the radio data by Rudnick et al. (1981). The flux of this hot spot has subsequently been measured by Simkin (1986), Meisenheimer, Röser (1986), Crane et al. (1987), our group and by Crane, Stockton (this volume) at various high frequencies. These data are all combined in figure 2. The first infrared (see figure 1) and mm results for radio hot spots have been obtained for this source (details are given in Meisenheimer et al. 1988). There is now a general consensus that the spectrum steepens considerably at wavelength shortward of 1 cm and again the shape is very well represented by a synchrotron cutoff spectrum (figure 3a).

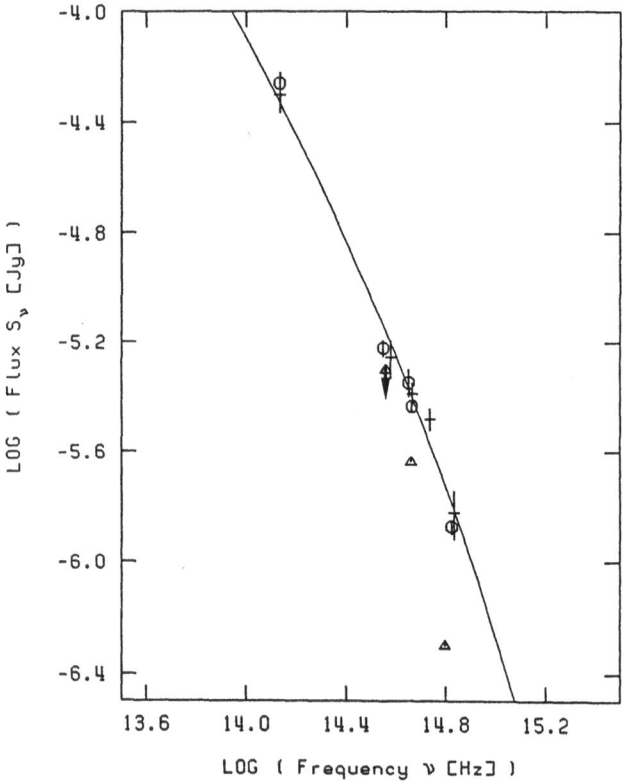

Fig. 2. Multifrequency photometry of the hot spot 3C 33 (south) by various authors: Simkin 1986 (Δ), Meisenheimer, Röser 1986 (O), Crane et al. 1987 (+), Crane, Stockton, this volume (+) and Meisenheimer et al. 1988 (O). The continuum is a fit with a synchrotron cutoff spectrum as in figure 3. Data have been corrected for interstellar extinction (except for Simkins's data, whose discrepancy remains unexplained).

One complication arises from the complex structure of this hot spot: the contrast between the brightness peak and its surroundings is unusually low (Rudnick 1988, see figure 3 in Perley's contribution). The hot spot is bound at its leading edge by a parabolic arc whose surface brightness (at 5 GHz) reaches its maximum in the apex. So it is not immediately clear, which part of the radio flux has to be taken to construct the overall hot spot spectrum. Under the assumption that the optical radiation is associated only with the most compact feature, Meisenheimer et al. (1988) modelled the source by a superposition of a broad (FWHM = 2".9) and a narrow (FWHM = 0".95) Gaussian peak. The latter is identified with the optical hot spot. Thus a rather flat hot spot continuum with a slope of -0.59 ± 0.08 is obtained between the radio and mm range. This is the fit shown in figure 3a.

0415+379 (3C 111): This object was again first identified in our optical survey on several CCD images in the R band (fig. A1c). The observed R flux of 1.3 ± 0.1 μJy - to be corrected for interstellar extinction (A_V = 1.86) - together with a K band and a 1.3 mm measurement (Meisenheimer et al. 1988) are once more leading to a turnover in the near infrared range (figure 3a). No optical polarization information is available yet so the association with the radio source rests solely on the very steep optical-to-IR continuum, which is too steep to be caused by confusion with an unrelated faint galaxy.

0518-458 (Pictor A): Pictor A is among the brightest radio sources in the southern sky and was included in the southern extension of our optical hot spot survey for that reason. Published radio maps are of low resolution, but Perley et al. (1989) have observed it with the VLA in all its hybrid configuration and present high resolution data at 2, 6 and 20 cm. Figures 4a, and b are included here from their work to aid discussion of high frequency data to be presented in this review.

An unusually bright optical counterpart (B $\approx$ 19.5 ^{mag}) was identified with the western hot spot by Röser and Meisenheimer (1987, figures 4d and A1d) on the basis of its high optical linear polarization. Follow-up observations show the polarization to be (45.6 ± 0.7)% at position angle (110.5 ± .5)°. Combining the broadband optical and infrared (J,H,K) measurements (Meisenheimer et al. 1988), a

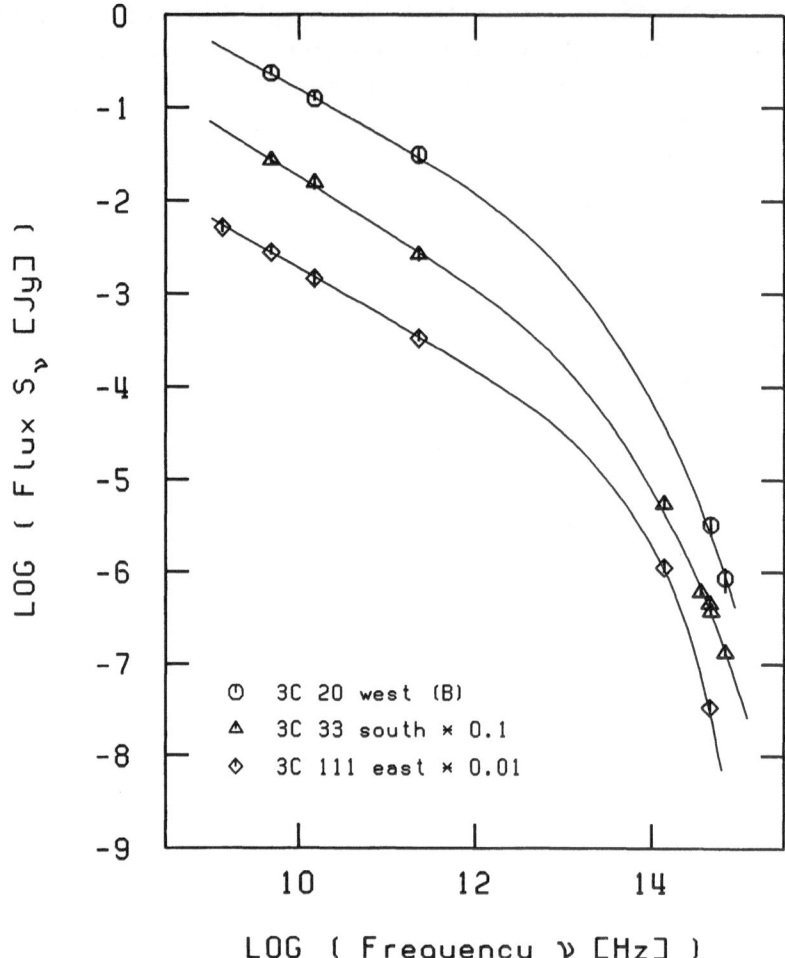

Fig. 3. Synchrotron continua of radio hot spots.
The fits are models according to Meisenheimer et al. (1988) (see also Meisenheimer, this volume).
a)   Continua without observable low frequency break have been found in 3C 20
(west), 3C 33 (south) and 3C 111 (east).

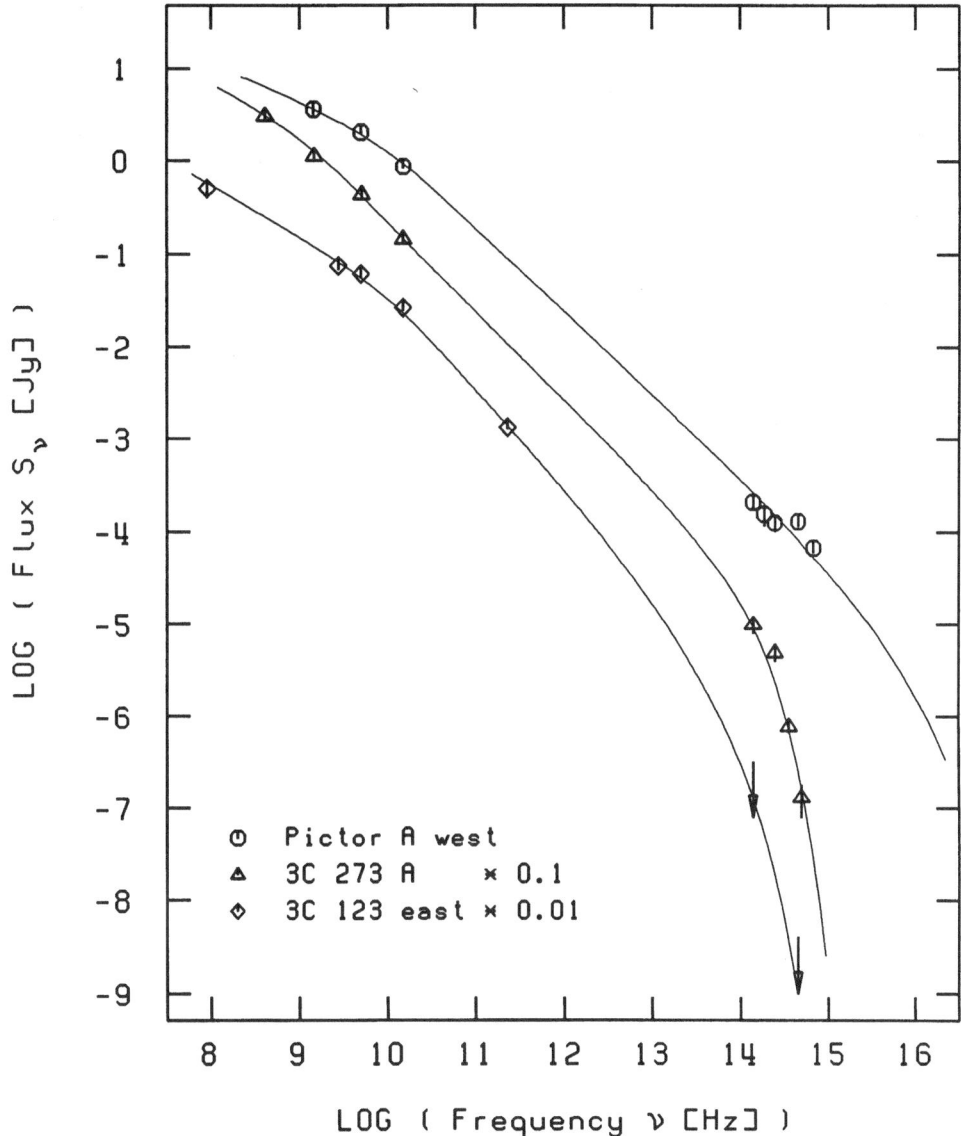

Figure 3

b) The hot spots of Pictor A (west), 3C 273A and 3C 123 (east) do show a low frequency break around or below about 10^{10} Hz.

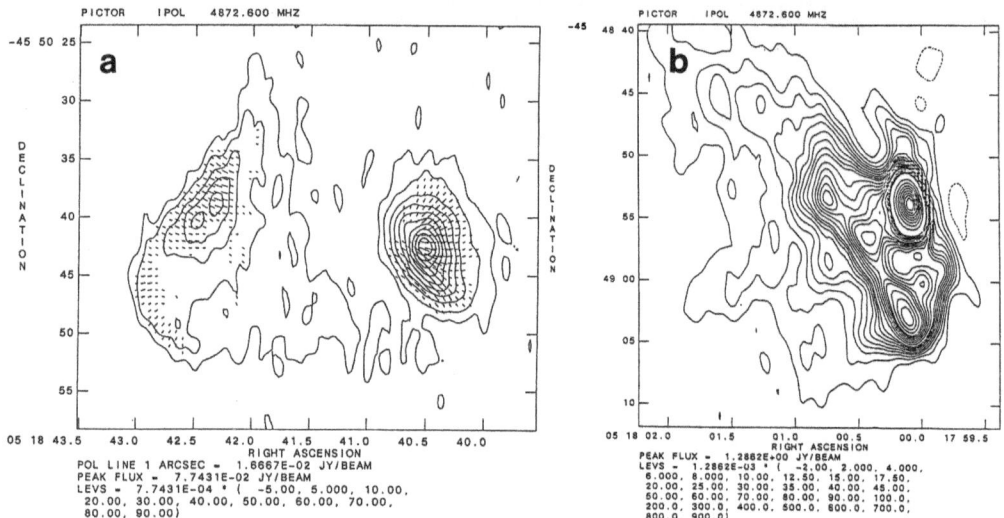

Fig. 4. Radio and optical maps of the hot spots in Pictor A (see Perley's figure 2 for a radio map of the whole source!):
a)   Eastern hot spot at 6cm, observed with the VLA, including polarization E-vectors. The optical counterpart in figure 4c is coinciding with the object at RA = $5^h18^m40.5^s$, however only with the part where the magnetic field is along the source axis. Note that such a magnetic field configuration is quite unusual for a hot spot.
b)   Western hot spot at 6cm. The bright component is unresolved here. Compare this with the optical image in figure 4d!

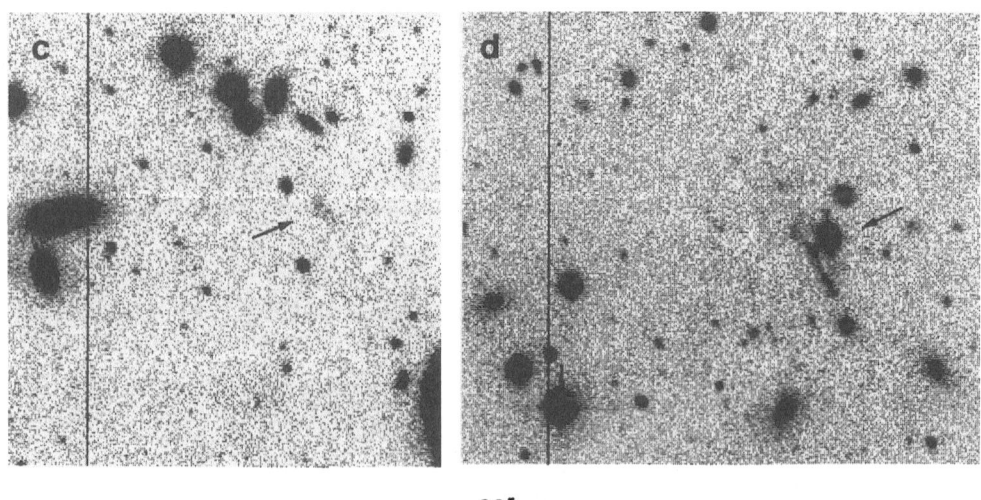

20"

c)   Optical (R band) image of the field around the eastern hot spot. The candidate optical object is marked by an arrow. Note the morphological similarity to the radio part.
d)   Western hot spot at 1" seeing (also R band) to be compared with figure 4b. Both optical frames are from CCD data taken with EFOSC.

straight powerlaw continuum is found between radio and optical wavelengths (figure 3b). This may even extend to X-ray frequencies, as this hot spot seems to be present on an EINSTEIN observatory IPC frame.[1] Low resolution spectra (0.8 nm FWHM) are featureless as expected for pure synchrotron emission.

The first observations also showed extended structure surrounding the hot spot (figure 4d). Although detailed photometry or polarimetry of the filament are not yet available, its synchrotron origin is evident - the morphology in the radio is exactly the same as that in the optical range. This was shown by smoothing our high resolution VLA map to the resolution of the optical R image. The isophotes turned out to be completely identical.

Following the identification of Pictor A (west) and the availability of high resolution VLA maps a search for a counterpart to the eastern hot spot also proved successful (figure 4c). As there is no further information yet available, its synchrotron origin is not established (group B hot spot !). Should it turn out, however, that the object detected at the hot spot position is indeed the optical counterpart, one would have direct evidence that a jet must be operating on both sides of the source : optically radiating synchrotron particles are tracers for continuing energy input due to their short lifetime.

1226+023 (3C 273): This quasar is not a FR class II object and is included here as the "hot spot" at the outer end is exhibiting a synchrotron spectrum very similar to the other hot spots described here. It was identified in the optical by Röser and Meisenheimer (1986), whose results were roughly confirmed by Keel (1988). The synchrotron nature of the radiation is only supported by the spectral shape (see figure 3b). Polarization measurements are very difficult to obtain due to contamination by the optical knot in the jet only 1.''5 away.

1251+278 (3CC 277.3, Coma A): The knots K1 and K2 in Coma A (Bridle

---

[1] Our new optical measurements in the B, R, and I band have largely confirmed the published data. The I point is even more above the straight powerlaw! As there is nothing obviously wrong with the J, H, K data, we may see a deviation from the pure synchrotron continuum. Repetition of the infrared photometry and multiband polarimetry will clarify this point.

et al. 1981) are not hot spots in the usual sense as they are located in the middle of the lobe. Miley et al. (1981) studied both continuum and line emission from this source. They identified a blue compact object with radio knot K1 whose continuum would have a spectral index of - 0.56 ± 0.03 between the radio and the B band. A linear polarization of (14 ± 3)% at position angle (71 ± 6)° was reported by them for K1 (K2 is much weaker than K1 in the optical). They and subsequently van Breugel et al. (1985) argued that this knot may provide the ionizing radiation for the line emitting region (see van Breugel, this volume). Keel (1988) also obtained multiband photometry of both knots K1 plus K2. He got a considerably lower flux than Miley et al. (1981) (see figure 5). Based on the relative spectral shapes in the radio and optical for both knots K1 and K2, a common origin of radio and optical radiation is not obvious to me and needs further investigation.

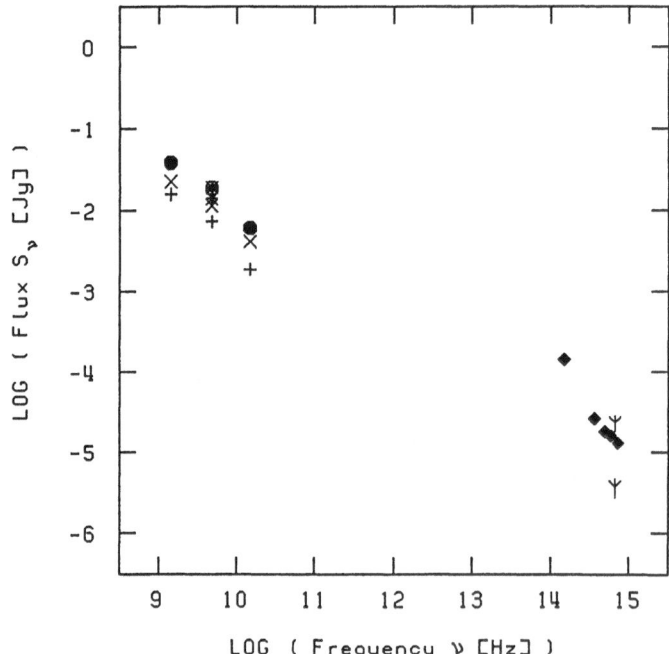

Fig. 5. Data collected from the literature for Coma A (3C 277.3) are displayed for knots K1 and K2 separately as well as for their sum (filled symbols). Data are from Miley et al. 1981 (Y), Bridle et al. 1981 (*), van Breugel et al. 1985 (+,x) and Keel 1988 (◊).

<u>1441+522 (3C 303)</u>: A radio map of this source is given in figure 6 by
Perley in this volume. Lelièvre, Wlérick (1975) and independently
Kronberg (1976) have found a fuzzy object with UV excess at the location of the western hot spot in 3C 303. In fact this is the first
optical hot spot identification altogether. In a subsequent paper

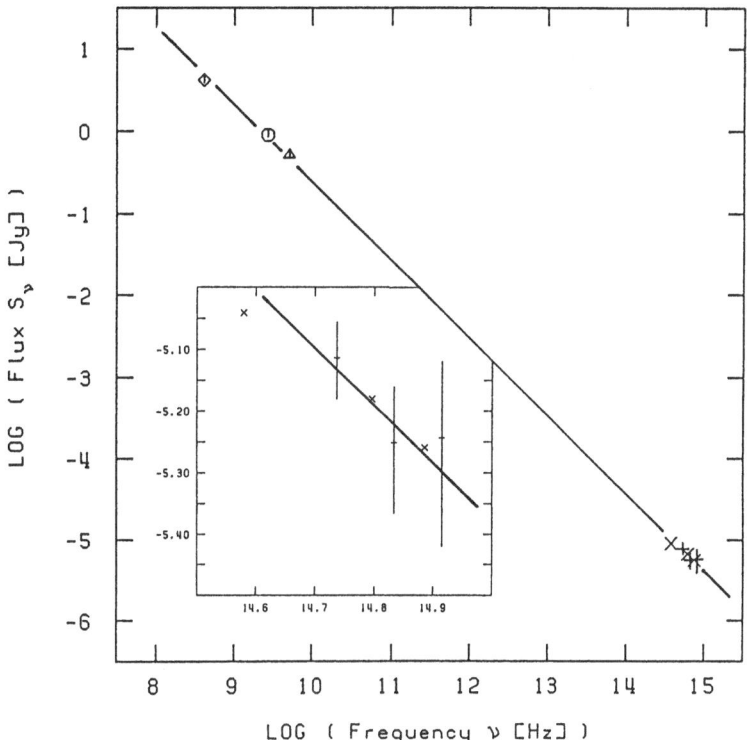

Fig. 6. The western hot spot in 3C 303 does not show a high frequency cutoff.
Data are assembled from Lonsdale et al 1983 (◇), Laing 1981 (O), Pooley and
Henbest 1974 (Δ), Lelièvre and Wlérick 1975 (+) and Keel 1988 (x). The least
square fitted powerlaw with index -0.95 gives a prefect description of the data.

Kronberg et al. (1977) presented a detailed radio and optical study
of the whole complex field including a spectrum of the hot spot counterpart, a featureless blue continuum. Lonsdale et al. (1983) and
Keel (1988) both argue that this hot spot shows a straight powerlaw
from the radio to the UV, Keel presenting new optical photometry. I

have plotted all available photometric data in figure 6. One power-
law of spectral index α = - 0.95 represents perfectly all measurements. So this hot spot - as Pictor A (west) - does not show a cutoff
longward of the UV. Two-dimensional polarimetry is urgently needed to
finally verify the synchrotron nature of this object.

2152-69 (PKS): High linear polarization and a continuum steeply
rising towards higher frequencies (!) are reported by S. di Serego
Alighieri below.

Group B

In addition to objects described above, there are quite a few
more hot spots observed in the optical, for which only the detection
is reported but no further information is available. These objects
are listed below together with the relevant observational information
currently available.

0518-458 (Pictor A east):   see above

1030+585 (3C 244.1): Crane et al. (1983) have studied a large number
of hot spots in the optical with the KPNO 4m telescope and various
detectors. Their candidate object for the northern hot spot of
3C 244.1 (V ≈ 22 ^{mag}) is coincident with the radio source to within
the astrometric error. We have looked at this field on a deeper R
band image and found that object "b" of Jenkins et al. (1977) is
obviously the brightest galaxy in a cluster (figure A2a). A comparison of our frame with that of Crane et al. is difficult, partly
because different filters were used and partly because they reproduced their image at extremely high contrast. We also see a candidate
at the position of the northern hot spot. But as the field is covered
with cluster galaxies, a chance superposition is quite likely. This
problem can only be solved by better astrometry and polarimetry.

1056+432 (3C 247): Laing et al. (1978) detected optical objects at
both hot spot positions during identification work of 3C sources, the
brighter (western) was confirmed by Crane et al. (1983). We found no
high polarization (3.9 ± 4.8 %) for this candidate but verified also
the eastern one. As the field is again quite crowded, careful astrometry and better polarimetry are needed.

1142+318 (3C 265): Identified by Saslaw et al. (1978), this object lies along a chain of galaxies in a rather crowded field. Positional coincidence is not perfect ($\Delta\delta \approx 2''$) judged on an overlay of a new high resolution VLA map (Laing, priv. comm.) aligned to the optical core position.

1319+428 (3C 285): Tyson et al. (1977) and Saslaw et al. (1978) identified a $V \approx 23^{mag}$ object with the peak in the eastern radio lobe as observed by Hargrave. Positional agreement is to within 1".

1609+660 (3C 330): Crane et al. (1983) report a $V = 24^{mag}$ object 3" south of the western radio peak. We confirm this object on a deeper prime focus CCD image. In addition we find a series of fainter objects to the north of this, the closest being about 1" from the hot spot position (figure A2b). As the astrometry was done from an overlay copied from the figure in Crane et al. at the scale of our image, the error probably is of the same order. The field is rather crowded, so even with this newly detected object, a chance superposition is highly probable.

1704+608 (3C 351): This quasar has a very strong multiple northern hot spot and the field was first studied optically by Kronberg et al. (1980). No optical object was detected down to J = 22 and F = 21 ^{mag}. The field was observed again by us under excellent conditions with the 3.5m telescope on Calar Alto. A compact object coinciding with the southern, compact and brighter hot spot as well as a diffuse, fainter object at the position of the northern component was found. Again a rich cluster of galaxies in the field is a major source of confusion (figure A2c).

1845+797 (3C 390.3): Again a radio map of this source is given in Perley's paper (figure 7). Saslaw et al. (1978) found a faint object of surface luminosity B $\approx$ 25 ^{mag} arcsec^{-2} coinciding to within 1" with the compact knot in the northern lobe on two 4 m plates.

1957+405 (3C 405, Cygnus A): Due to its low galactic latitude of $b^{II} = 6°$, this object is extremely difficult for optical identification work (see above). Kronberg et al. (1977) took deep prime focus images with the Palomar 5 m telescope and found no coincidences except for the most compact hot spot B, which may be identified with a faint stellar object of J $\approx$ 22 ^{mag}. Detailed photometry and polarimetry of this candidate is required until one can decide upon its connection with the radio source. For a detailed discussion of the radio hot spots of Cygnus A see the contribution by Carilli in this volume.

<u>2135-147</u>: Hawkins (1978) found a stellar object at the position of the radio peak in this QSO. Its magnitude is B = 21.6 ^{mag} and it is of neutral colour (B-R = 1.5 ^{mag}).

Group C

In the following hot spots are listed, for which optical identification has been attempted but no optical object was found. I know that this list is quite incomplete but nevertheless should be useful to aid future work in this field. Some information on the optical field around the hot spots may also be taken from deep identifications of the central sources for 3C objects. The following authors plot the positions of the hot spots in their finding charts: Kristian et al. 1974 and 1978, Longair, Gunn 1975, Laing et al. 1978, Riley et al. 1980, Gunn et al. 1981.

Objects for which there were reports in the literature on optical identifications but which were subsequently shown not to be the hot spot counterpart are also included in this group.

<u>0307+169 (3C 79)</u>: Crane et al. (1983), no object on prime focus plates down to V = 24.5 ^{mag}.

<u>0702+749 (3C 173.1)</u>: same as 3C 79

<u>0917+458 (3C 219)</u>: Crane et al. (1983) found an object of V = 24 ^{mag} at the edge of the northern lobe, several arcsec away from the hot spot.

<u>0936+361 (3C 223)</u>: same as 3C 79

<u>1108+359 (3C 252)</u>: Crane et al. (1983) see a faint wisp coinciding with an extension of the outermost radio contour in the eastern lobe on prime focus and CCD images. Its distance from the hot spot is $\approx$ 5".

<u>1158+318 (3C 268.2)</u>: An object with V $\approx$ 23 ^{mag} lies in the northern

lobe about 6" south of the peak in the 5 GHz emission (Crane et al., 1983).

1522+546 (3C 319): Crane et al (1983) report a resolved object under- neath the northern hot spot. From its morphology on deep CCD images, we classify this as a galaxy (figure A3a).

1529+242 (3C 321): Crane et al. (1983) found a stellar object at the position of the eastern hot spot. We find no high polarization for this object and spectroscopic observations show it to be a late type star (Crane, priv. comm.).

1549+628 (3C 325): Crane et al. (1983) see no candidate on their CCD images, neither do we (R > 23.5 ^{mag}).

1627+444 (3C 337): The object reported by Crane et al. (1983) is coinciding with a secondary radio maximum within the western lobe almost halfway between the nucleus and the hot spot.

1832+474 (3C 381): Crane et al. (1983) see no object close to the western hot spot on their CCD image (R > 23.5 ^{mag}).

During our optical hot spot survey we have thus far searched in the fields of more than 50 sources for optical emission from the extended radio component. The majority of the data has been inspected only visually on the TV. In most cases, where an object was found close to the hot spot position, the data have been completely analy- zed and the object was reobserved. This resulted in the positive identifications of 3C 20 and 3C 111 as described above. Although this way the selection of objects is not statistically controlled, it is evident that only a small fraction of all FR-II sources do have hot spots with continua extending to optical frequencies.

IV) Summary

What have we learned from the observations just described? First, the existence of optical synchrotron radiation asscociated with radio hot spots in classical double (FR-II) radio sources is now **_firmly established_** (3C 20, 3C 33, (3C 111), Pictor A and

107

(3C 303)). However, the fraction of sources detectable in the optical is certainly small, statistically useful numbers cannot yet be given. Hopefully we will find a much larger fraction of hot spots if we observe at infrared or even sub-millimetre wavelengths. This hope is derived from the finding that most of the sources detected so far show an extreme steepening of their continua between the infrared and the optical (see figures 3a,b). From the present point of view we would like to know the real distribution of the turnover frequencies: are they scattered all the way from highest radio to optical frequencies, or is there a preferred value sharply peaked around the current detection level of a few times 10^{14} Hz as suggested by Biermann and Strittmatter (1987)? In figure 3b, 3C 123 is given as an example of a non-detection. Despite its very bright radio emission, we could only derive strong optical and infrared limits. But the turnover may well be as low as 7×10^{11} Hz, much less than the plotted case of 6×10^{13} Hz!

One immediate consequence of optical synchrotron radiation is the requirement for **in situ acceleration** of particles due to the short lifetime of the optically radiating electrons. They cannot be transported out from the nucleus in a jet as they would lose all their energy due to synchrotron radiation in the jet's magnetic field and also due to Compton losses in the cosmic background radiation field (see Meisenheimer et al. 1988). Where this acceleration is taking place in detail is subject to further, more accurate astrometric measurements. A positional error of 0."1 from CCD frames relative to a compact nucleus should not be too difficult. For 3C 20 e.g. this would then enable a more specific location of the optical synchrotron source within hot spot B. Accurate astrometry at an even higher level will also soon be available for Pictor A, where we have recently acquired lots of short exposures under subarcsecond seeing conditions. For 3C 33 the situation is more complicated and a high frequency radio map is urgently needed to make progress with this source.

Optical synchrotron light is not confined to a tiny unresolvable region. The hot spot in 3C 33 is **clearly resolved** (intrinsic extend about 1.5 kpc, $H_0 = 50$ km/sec/Mpc) and the filament in Pictor A west (figure 4d) stretches over almost 15 kpc in length ! So the particle accelerators must really have huge dimensions.

Another interesting fact is the existence of a **maximum energy** in the particle population manifested in the existence of **high frequency cutoffs** in the spectra of most of the sources. We are studying the range where acceleration gains are balanced by synchrotron losses, which prohibit acceleration to higher Lorentz factors. The physical background, which in the end sets these limits, remains to be studied in detail. Here the very rare hot spots without a cutoff like Pictor A west and 3C 303 may provide important clues.

So further observations of hot spots at high frequencies are mandatory in order to fully exploit this tool in source diagnostics for radio galaxies. Very large optical/infrared telescopes, infrared arrays, ISO, mm telescopes, ROSAT and the HST will certainly bring new and vital flavour to the subject. But "old-fashioned" ground based observations will certainly continue to provide important data, whose capability remains to be exhausted.

Acknowledgements: I would like to thank Klaus Meisenheimer for a careful reading of the manuscript.

References

Biermann, P.L., Strittmatter, P.A.: 1987, Astrophys.J. **322**, 643

Bridle, A.H., Fomalont, E.B., Palimaka, J.J., Willis, A.G.: 1981, Astrophys.J. **248**, 499

Crane, P., Tyson, J.A., Saslaw, W.C.: 1983, Astrophys.J. **265**, 681

Crane, P., Stockton, A., Saslaw, W.C.: 1987, Astron.Astrophys. **183**,16

Dreher, J.W., Simkin, S.M.: 1986, A.J. **91**, 58

Gunn, J.E., Hoessel, J.G., Westphal, J.A., Perryman, M.A., Longair, M.S.: 1981, Month.Not.R.A.S. **194**, 111

Hawkins, M.R.S.: 1978, Month.Not.R.A.S. **185**, 23P

Hiltner, P.R., Meisenheimer, K., Röser, H.-J., Laing, R.A.: 1989, in preparation

Jenkins, C.J., Pooley, G.G., Riley, J.M.: 1977, Mem.R.astr.Soc. **84**,61

Keel, W.C.: 1988, Astrophys.J. **329**, 532

Kristian, J., Sandage, A., Katem, B.: 1974, Astrophys.J. **191**, 43
                                    1978, Astrophys.J. **219**, 803

Kronberg, P.P.: 1976, Astrophys.J. (Letters) **203**, L47

Kronberg, P., van den Bergh, S., Button, S.: 1977, A.J. **82**, 315
                  Erratum : A.J. **82**, 1039

Kronberg, P.P., Burbidge, E.M., Smith, H.E., Strom, R.G.: 1977,

Astrophys.J. **218**, 8

Kronberg, P.P., Clarke, J.N., van den Bergh, S.: 1980, A.J. **85**, 973

Laing, R.A., Longair, M.S., Riley, J.M., Kibblewhite, E.J., Gunn,
J.E.: 1978, Month.Not.R.A.S. **183**, 547

Laing, R.A.: 1981, Month.Not.R.A.S. **195**, 261

Lelièvre, G., Wlérick, G.: 1975, Astron.Astrophys. **42**, 293

Longair, M.S., Gunn, J.E.: 1975, Month.Not.R.A.S. **170**, 121

Lonsdale, C.J., Hartley-Davies, R., Morison, I.: 1983,
Month.Not.R.A.S. **202**, 1P

Meisenheimer, K., Röser, H.-J.: 1986, Nature **319**, 459

Meisenheimer, K., Röser, H.-J., Hiltner, P., Yates, M.G., Longair,
M.S., Chini, R., Perley, R.A.: 1988, Astron.Astrophys, in press

Miley, G.K., Heckman, T.M., Butcher, H.R., van Breugel, W.J.M.: 1981,
Astrophys.J. (Lett.) **247**, L5

Pacholczyck, A.G.: 1970, Radio Astrophysics (San Francisco, Freeman)

Perley, R.A., Röser, H.-J., Meisenheimer, K.: 1989, in preparation

Riley, J.M., Longair, M.S., Gunn, J.E.: 1980,
Month.Not.R.A.S. **192**, 233

Röser, H.-J., Meisenheimer, K.: 1986, Astron.Astrophys. **154**, 15

Röser, H.-J., Meisenheimer, K.: 1987, Astrophys.J. **314**, 70

Rudnick, L., Saslaw, W.C., Crane, P., Tyson, J.A.: 1981,
Astrophys.J. **246**, 647

Rudnick, L.: 1988, Astrophys.J. **325**, 189

Saslaw, W.,C., Tyson, J.A., Crane, P.: 1978, Astrophys.J. **222**, 435

Simkin, S.M.: 1978, Astrophys.J. (Letters) **222**, L55

Simkin, S.M.: 1986, Astrophys.J. **309**, 100

Tyson, J.A., Crane, P., Saslaw, W.C.: 1977, Astron.Astrophys. **59**, L15

van Breugel, W.J.M., Miley, G., Heckman, T., Butcher, H., Bridle, A.:
1985, Astrophys.J. **290**, 496

White, M., Birkinshaw, M.: 1984, Astrophys.J. **281**, 135

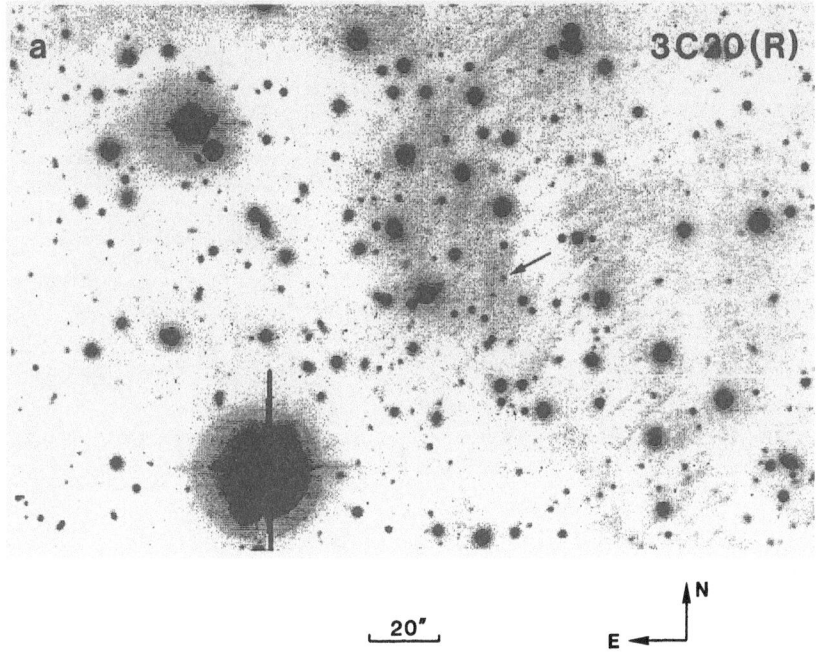

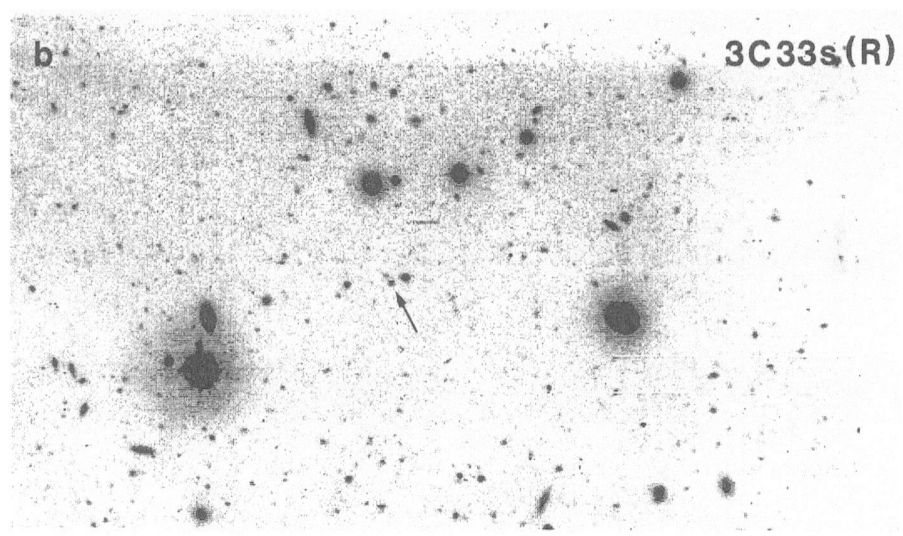

A1) Finding charts for optically identified hot spots. Execpt for Pictor A, which was observed with EFOSC at the ESO 3.6m telescope, all images where taken in October 1985 with the 3.5m telescope on Calar Alto. The hot spots are marked.

    a) Field of 3C 20. There is no counterpart to the eastern hot spot detected so far. The "Filaments" across this image are reflections due to a very bright star just outside the field to the southwest.

    b) Field of 3C 33 south (R).

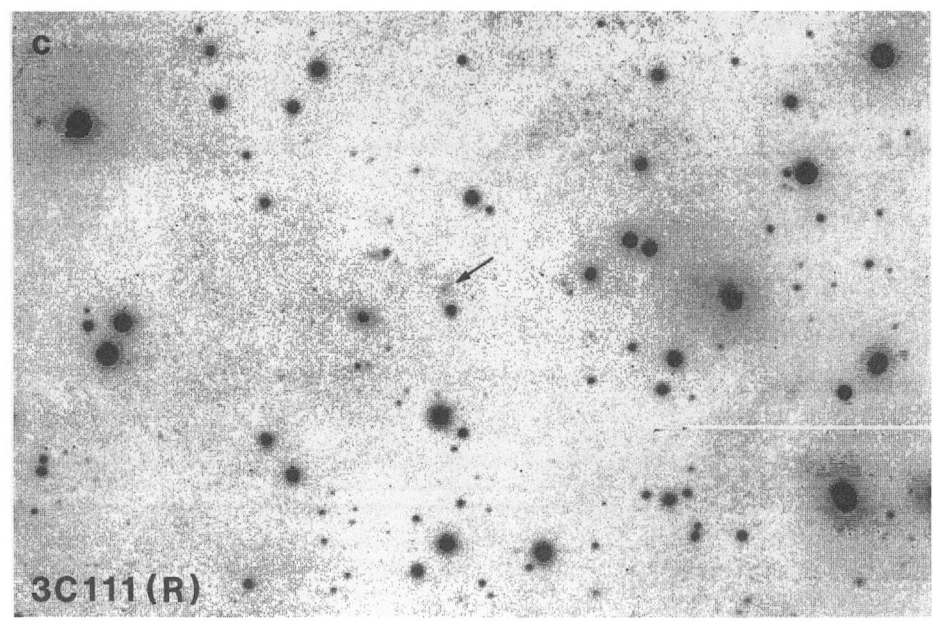

3C111(R)

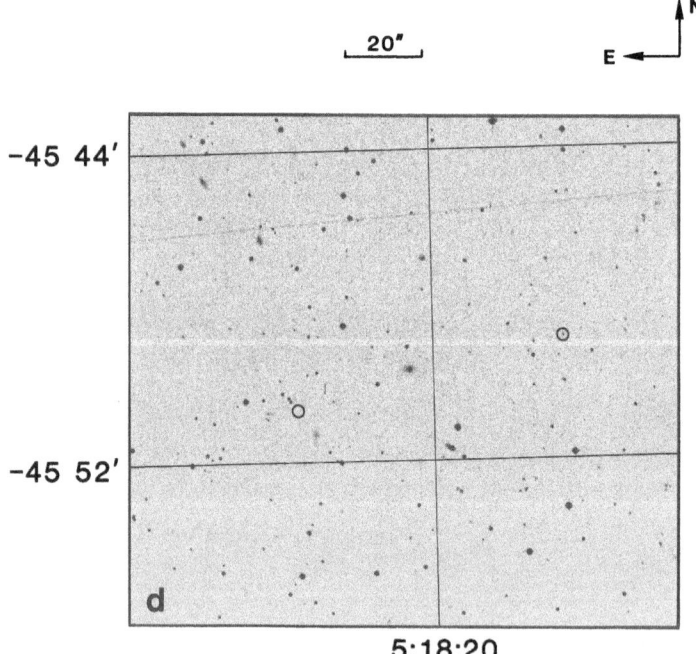

20"

N
E

-45 44'

-45 52'

5:18:20

A1)

c) Field of the eastern part of 3C 111 (R).

d) Field of Pictor A from the ESO Quick Blue Survey plate.
   Locations of the hot spots are marked. Please refer to figure 4 for
   close-ups of the two hot spots.

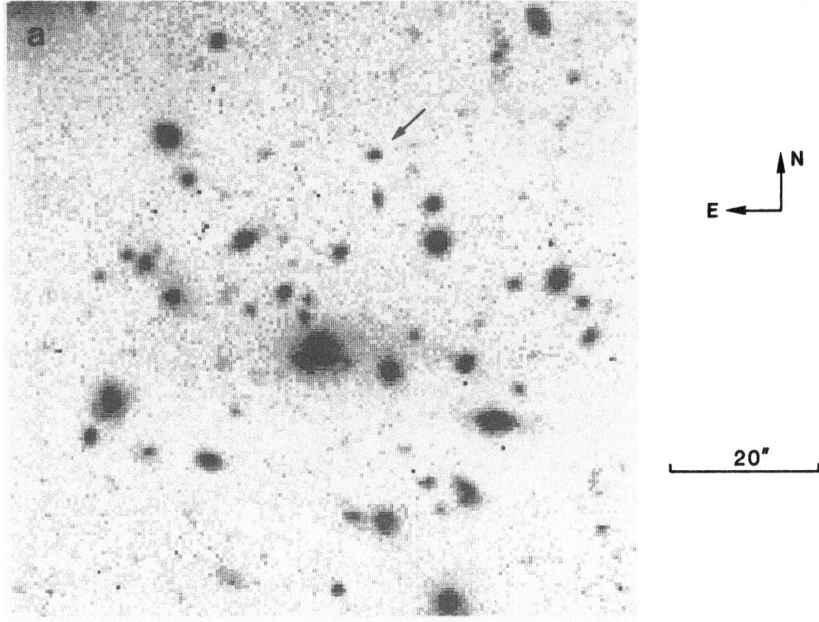

N

E

20"

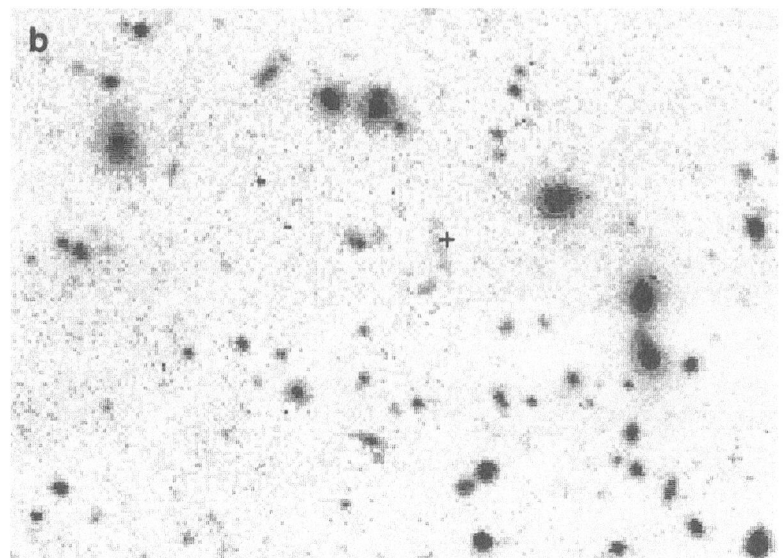

A2) Fields of optical hot spot candidates. Images have been taken with the 3.5m
     telescope on Calar Alto.
     a) 3C 244.1: Although Crane et al. (1983) give a finding chart for this
        source, this image is included to show the cluster of galaxies mentioned
        in the text. The arrow indicates the object regarded as the optical hot
        spot counterpart by Crane et al.
     b) 3C 330: The cross indicates the hot spot position as derived from Crane
        et al. (1983). Obviously they have detected the southernmost (brightest)
        part of the elongated structure close to the center. Note the presence
        of faint galaxies all over the field.

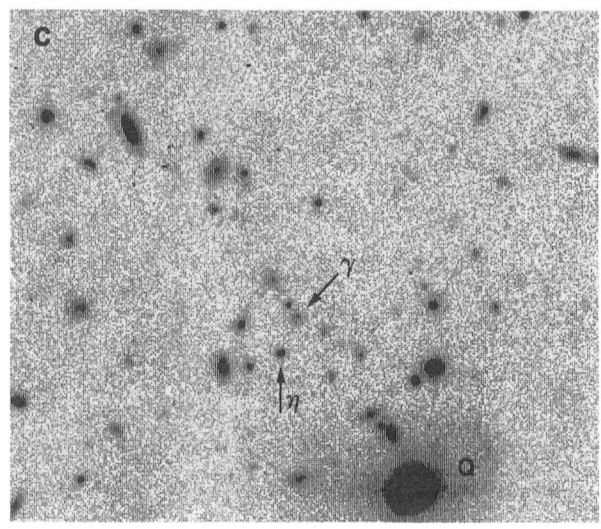

A2)

c) 3C 351: The quasar (Q) and the two candidate hot spot counterparts (γ, η) are superimposed on a cluster of galaxies.

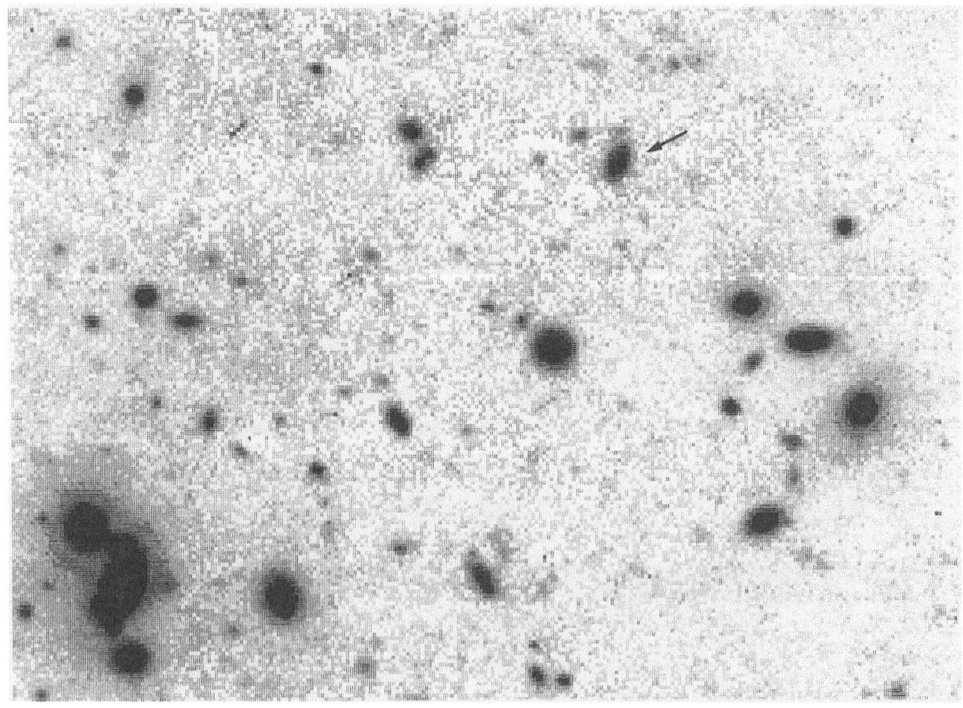

A3)   Update of the finding chart for the group C hot spot in 3C 319. Data have been taken in the R band at the 3.5m telescope on Calar Alto.

MULTIFREQUENCY FLUX DETERMINATION IN THE HOT SPOT OF 3C33 SOUTH

Philippe Crane
European Southern Observatory
Garching, West Germany

and

Alan Stockton
Institute for Astronomy
University of Hawaii
Honolulu, Hawaii, U.S.A.

Abstract

New optical (I,B) and near infrared (K) measurements of the hot spot in 3C33 South have been obtained. A reanalysis of the radio fluxes at 20 cm and 6 cm provides better comparison with the optical fluxes. The simplest interpretation of the data gives a radio spectral index of $\alpha = -0.89$ with a break in the spectrum at $\log \nu \sim 13.8$. At optical frequencies, the spectral index is $\alpha = -1.97$.

In a continuing effort to understand the mechanism that accelerates electrons to sufficient energies so that they produce optical synchrotron radiation, we have reobserved the south radio lobe of the double radio source 3C33. These observations add to and complement our previous work (Crane, Stockton and Saslaw, 1987). In three observing sessions, from August to November 1987, we obtained images of 3C33 south in Hα, I, R, B, and K bands. In this note we report an analysis of the I, B, and K images for the "hot-spot" (C1 of Crane et al., 1987). Additionally, we have remeasured the radio fluxes at 20 cm and 6 cm from the original VLA data of Rudnick et al. 1981. The Hα images have not yet been analyzed, and the R images merely confirm our previous work (Crane et al. 1987).

The new data presented here for the I and B band images were obtained at the University of Hawaii's 88 inch telescope using the Galileo / Institute for Astronomy CCD system and a TI 800×800 CCD chip. The data consisted of 8-10 minute exposures with the B filter and 8-10.3 minute exposures with the I filter. The images were reduced to standard magnitudes with reference to the stars F11 and 92-288 using magnitudes from Landolt (1983). The magnitudes were then converted to fluxes using the prescription of Bessel (1979).

The K-band images (see Figure 1) were obtained at the UKIRT telescope on Mauna Kea using the IRCAM. These consisted of 18-6 minute exposures readout every 45 sec. They were reduced to standard fluxes using the stars HD3029 and GℓL105.5.

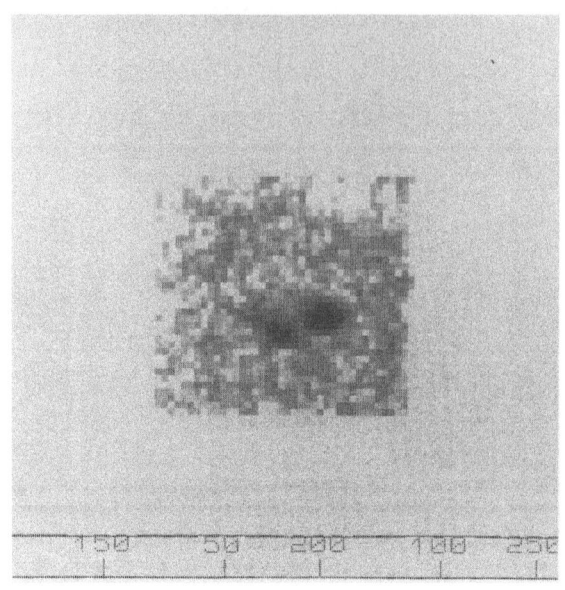

Figure 1: Reproduction of the summed K-band images of 3C33 South. North is up, east is left. Each pixel is 0.62 arcseconds. The bright object is a star. The object SE of the star is at the peak of the radio hot spot.

The VLA data from Rudnick et al. (1981) at 20 cm and 6 cm was smoothed to a common resolution of 0.3 arcsec. Fluxes were then determined within a 2".8 diameter aperture which corresponded to the same aperture as used for the optical and near infrared data.

A summary of the flux measurements is presented in Table 1. Also included in the table are results from radio measurements at 2 cm (Dreher, 1981) and at 1.3 mm (Röser, 1987).

Table 1: Summary of Flux Measurements

Band	log ν (Hz)	Flux (Jy)	Reference
B	14.8337	$1.3\pm0.3\times10^{-6}$	This work
V	14.7368	$3.0\pm0.3\times10^{-6}$	Crane et al. 1987
R	14.6655	$3.8\pm0.4\times10^{-6}$	Crane et al. 1987
I	14.5740	$5.3\pm0.9\times10^{-6}$	This work
K	14.1347	$49.2\pm7.0\times10^{-6}$	This work
1.3 mm	11.3010	$33\pm3\times10^{-3}$	Röser, 1987
2 cm	10.1761	0.44 ± 0.03	Dreher, 1981
6 cm	9.6989	0.90 ± 0.08	This work
20 cm	2.1761	2.50 ± 0.25	This work

It is gratifying to report that the fluxes reported here and those measured independently by Meisenheimer et al. (1988) (see also Röser in these proceedings) agree rather well.

Figure 2 shows a plot of these data. The data have been fit to two independent power laws, one for λ > 1.0 mm and one for λ < 1.0 mm. In the absence of other compelling evidence, this seems to be the simplest interpretation of the data. Additionally, this would be very close to what would be expected from the continuous injection of electrons with a power law energy distribution (Pacholczyk, 1970).

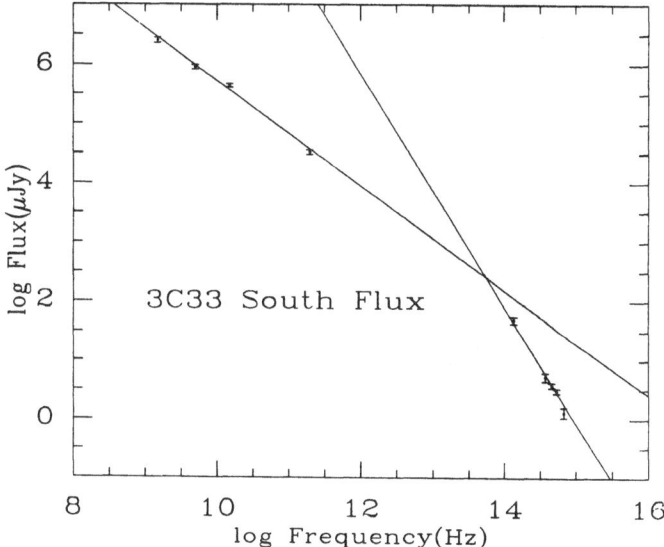

Figure 2: Plot of the fluxes in Table 1. The solid lines are fit to the data. For log ν < 13, log F = 14.58 - 0.885 log ν and for log ν log ν > 13, log F = 29.49 - 1.970 log ν.

The suggestion (Crane et al., 1987) that the optical spectral index is similar to the radio spectral index seems to be ruled out by these more complete observations. There seems to be little doubt that the spectrum of 3C33 South hot spot steepens considerably at wavelengths shortward of about 5 microns (log ν = 13.8). The challenge is to determine the unique mechanism which is responsible for this spectral behavior.

References

Bessel, M.S., 1979, Publ. Astron. Soc. Pacific **91**, 589.
Crane, P., Stockton, A., and Saslaw, W.C., 1987, Astron. Astrophys. **183**, 16-20.
Dreher, J.W., 1981, Astronomical Journal **86**, 833-847.
Meisenheimer, K., et al., 1988, preprint.
Pacholczyk, A.G., 1970, Radio Astrophysics (San Francisco; W.H. Freeman & Co.).
Rudnick, L., and Saslaw, W.C., Tyson, J.A., and Crane, P., 1981, Astrophys. J. **246**, 647-652.

Phil Crane   Larry Rudnick
                         Phil Hardee

Sperello di Serego Alighieri

A blue and polarized source along the radio axis of PKS 2152-69*

S. di Serego Alighieri, R.A.E. Fosbury and C.N. Tadhunter
ST-ECF, Garching bei München, FRG

Abstract. We report observations of the optical energy distribution and linear polarization of the continuum source in the high ionization cloud discovered by Tadhunter et al. (1987, *Nature*, **325**, 504) at 10 arcsec to the North East of the nucleus of the radio galaxy PKS 2152-69. After subtraction of the smooth contribution from the outskirts of the galaxy, the spectrum is steeply rising in the near ultraviolet with $f_\nu \propto \nu^{3\pm1}$ and the degree of linear polarization in the B band is $8.6 \pm 1.5\%$ in position angle $124° \pm 5°$, which is close to the perpendicular to the nucleus-cloud direction (P.A. = 45°). The degree of polarization of the continuum radiation could be higher, since about 27% of the flux in the B band is due to emission lines. These measurements refer to the whole of the cloud ($\sim 5 \times 4$ arcsec, B = 20.6). Although the cloud is resolved, no changes in polarization are detected across it.

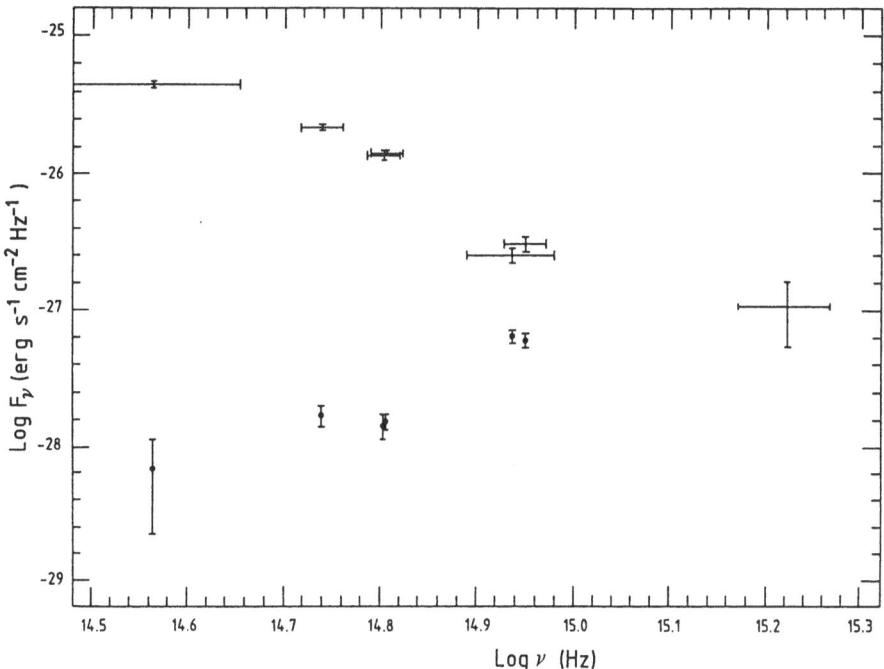

Figure 1. The continuum flux densities for the nucleus and the cloud (•) of PKS 2152–69. They where obtained by imaging the galaxy through medium-broad band filters avoiding the brightest emission lines The horizontal bars show the width (FWHM) of the filters, while the vertical ones represent the photometric errors. The flux density of the cloud has been corrected for the underlying galaxy contribution and for the emission lines. The rightmost point is a measurement of the nuclear flux taken with IUE.

A full account of this work is given by S. di Serego Alighieri et al., 1988, *Nature*, **334**, 591.

Wil van Breugel

Extended Optical Line Emission in Radio Galaxies

Wil J.M. van Breugel

Radio Astronomy Laboratory,
UC Berkeley, CA 94720, USA

Abstract

Optical and radio observations of nearby and distant powerful radio galaxies with extended optical emission-line regions are reviewed. Comparisons of optical emission-line images with radio continuum maps, supplemented with detailed optical spectroscopy, show that the interaction of radio jets and hot spots with ambient gas can have dramatic consequences for both the radio source and its surrounding medium. I will discuss evidence for the deflection and decollimation of (low power) jets, gas entrainment by jets and hot spots, Faraday depolarization/rotation near regions of dense ionized gas, enhanced star formation along radio source trajectories, and collimated ionizing radiation from radio galaxy nuclei.

1. Introduction

Narrow-band images in bright emission-lines (Hα, [OIII]λ5007, [OII]λ3727, Lyα), when combined with detailed radio maps, have shown that many radio galaxies have associated extended emission-line regions (EELR's; e.g. Fosbury 1985; van Breugel 1986). Long-slit spectroscopic observations can provide spatial information about the physical conditions of this ionized gas. The morphological and kinematic data of EELR's associated with radio galaxies in especially dense environments strongly suggest that gas entrainment (heating and acceleration) occurs by their radio jets and hot spots. The dense material may be cooling gas in a cluster, a gaseous disk intrinsic to the parent galaxy or acquired through merging or, at high redshifts, a proto-galactic medium. The source of ionization of the gas is often

121

unclear and may be related to the jet/gas interactions, photo-ionization by the active galaxy nucleus or other processes which may be related to the presence of the radio source. Observations of EELR's in radio galaxies might thus provide useful constraints for models of radio jets and hot spots, their galaxy nuclei and their environments.

In this review I will discuss the current status of the radio and optical observations of EELR's in radio galaxies, including recent results on emission-line surveys of optically unbiased samples of nearby and distant sources, detailed radio/optical studies of selected objects, their radio/optical morphological correlations, and the kinematics, ionization and origin of the EELR's. I will assume throughout a value for the Hubble constant of $H_0 = 75$ kms^{-1} Mpc^{-1} and the deceleration parameter $q_0 = 0$, and have scaled measurements from other authors accordingly. References to most specific objects are given in Table 1 and will generally not be repeated.

2. Nearby Radio Galaxies

Surveys

To study the general occurence of EELR's in nearby radio galaxies Baum and collaborators have made an imaging emission-line surve of a flux limited sample of nearby and relatively powerful radio galaxies (Baum 1987; Baum *et al.* 1988). 87% of the sources have redshifts z < 0.2, and the remainder has 0.2 < z < 0.5. The basic sample was selected from the Parkes and Molonglo catalogs, with declinations between -30° and +20°. Galactic latitudes -10° < b_{II} < +10° were excluded. To this were added some other nearby (3CR) radio galaxies with no prior knowledge of their optical properties. This optically unbiased, representative sample consisted of 38 objects. Approximately ~70% of the sources in this sample are of intermediate radio luminosity, with 5×10^{41} erg s^{-1} < L_R < 5×10^{43} erg s^{-1}. In ~25% of the sources L_R is above, and in ~5% below these limits. The main results from this work can be summarized as follows:
1) Spatially extended gas was detected in 85% of the sources.
2) The emission-line regions have linear sizes ranging from 1-100 kpc, with a median size of ~10 kpc.

3) The extent and luminosity of the emission-line gas is an order of magnitude larger than in normal (radio quiet) elliptical galaxies of similar optical magnitudes.

4) In general the ionized gas is located within a broad cone along the radio source axes.

5) In most cases estimates for the nuclear UV/X-ray continua suggest that the EELR's of the radio galaxies might be photoionized by their nuclei.

6) The total emission-line and radio luminosities are comparable and correlate over four decades in luminosity.

7) The emission-line regions span a range of linear scale sizes and exhibit a variety of morphologies. One can make the following distinctions, using a slightly modified version of Baum's classification scheme.

Unresolved emission-line regions

These emission-line regions are found in a variety of sources, including a wide angle tail source (3C 89), a binary system with multiple jets (3C 75) and a classical double source (3C 390.3) (6/38; 16%).

Small emission-line regions

These are small, near-nuclear EELR's with linear sizes < 10 kpc and < 1/2 times the size of their associated radio galaxies (14/38; 37%). Often the EELR's resemble small disks, rings or spheroids (3C 33, 3C 403), but they can also be irregular with filamentary extensions (3C 223; 3C 405 [Cygnus A]), or associated with a dustlane (3C 272.1 [M 84]). Because their angular sizes are generally small, morphological radio/optical correlations are difficult to determine. The EELR and wellknown dustlane in M 84 are perpendicular to the radio source axis, as is often found in radio galaxies with dustlanes (Kotanyi and Ekers 1979). In general, however, the small disk/ring/ spheroidal EELR's appear only marginally correlated with their radio source axes.

Large emission line regions

These are EELR's with linear scale sizes > 10 kpc, but still < 1/2 times the size of their associated radio galaxies (11/38; 29%). They are usually, but not always, found along the radio axes and their morphologies are generally complex. Three subclasses may be distinguished.

i) *S-shaped* regions, in some cases connected to nearby companion galaxies. These are probably tidal features and overall may show little morphological relationship to their radio sources (PKS 0349-278).

ii) Emission-line regions along the radio axes with *transverse* filamentary structure (PKS 0634-206). Other objects (not part of this sample) which show this phenomenon are 3C 321, 4C 29.30 and NGC 541 (see next section). All these objects have gaps in their emission line regions along nucleus/hot spot or jet directions. Deep broadband images of these galaxies often show (stellar) continuum loops or shells, suggestive of tidal interactions or merging (Toomre and Toomre 1972, Quinn 1984). Presumably the gaps in the filaments have been caused by the jets and hot spots passing through this gas.

iii) Filaments *parallel* to the radio axes, sometimes near the edges of diffuse radio lobes (3C 98 [Fig. 1], 3C 227). The filaments, in particular those along the edges, might be caused by thermal instabilities, shockheating or other processes which may occur at the interfaces between radio plasma and ambient gas, for example when the backflows of radio lobes re-enter their galaxian environments (3C 98), or when the radio sources are embedded in a dense cooling X-ray halo (3C 274). At least in some cases the parallel filaments may also be tidal features. An example, not a member of Baum's sample, is PKS 0511-484. This source has a long emission-line filament along the N radio axis, at the end of which there is a 'stripped' galaxy. The kinematics of the filament seem to support a tidal interaction interpretation.

Co-spatial emission-line regions

These are emission-line regions which have sizes > 1/2 times the size of their associated radio galaxies (7/38; 18%). They range in size from < 1-50 kpc and overlap with one or both radio lobes. The radio sources may be of relatively low luminosity, with compact (PKS 1345+125) or diffuse morphologies (3C 317, PKS 0745-191), or are

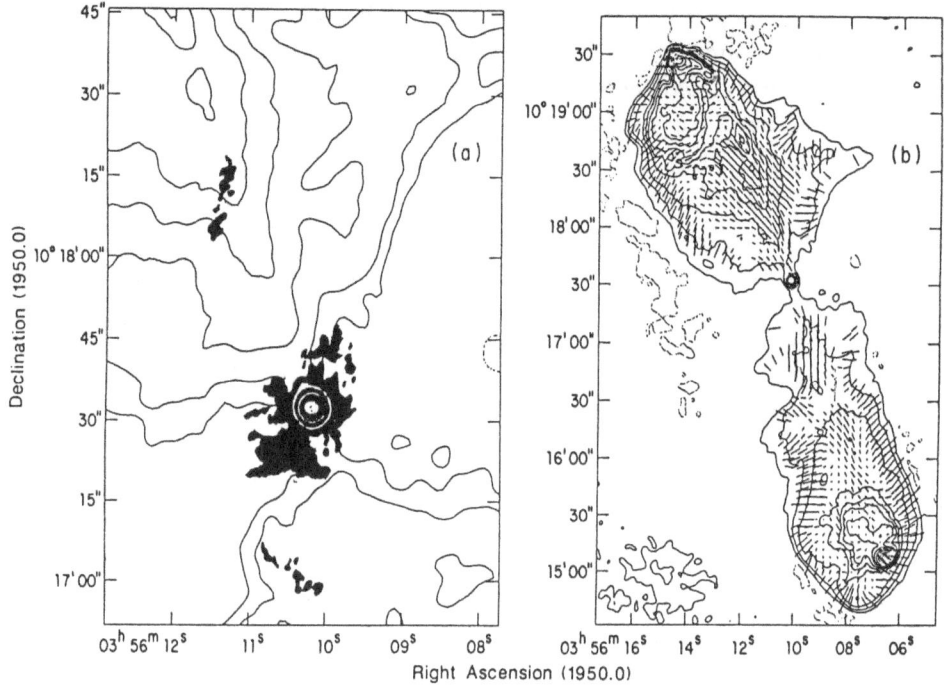

Figure 1: Radio image (contours and polarization vectors, Fig. 1*b*) of 3C 98 observed at 6 cm, with 3.8″ resolution (Baum *et al.* 1988). In Figure 1*a* the central region of 3C 98 is shown superimposed on an optical image of the extended Hα + N[II] line emission (dark regions). Note the filaments along the radio axis.

powerful and have classic double structures. Some of the latter (<u>4/38; 11%</u>) have EELR's associated with their hot spots (3C 63, 3C 196.1, 3C 275 and 3C 295). These sources are all relatively small, with a median size of ~30 kpc.

The parent galaxies of these objects tend to be peculiar and may have double nuclei (PKS 1345+125), tidal (?) tails (3C 63) or are members of rich clusters (3C 317, 3C 295). The radio sources in this class seem to be embedded in especially dense environments, and therefore are ideal objects for investigating their interaction with ambient gas. This class of sources will be discussed in more detail below.

Selected Sources

There are now a large number of radio galaxies known which have bright and very extended emission line regions. The best studied cases are listed in Table 1, together with some general information. The table shows that EELR's can be found in many types of radio galaxies, with a large range in radio power and linear sizes: Seyferts, normal and peculiar elliptical galaxies, steep spectrum quasars and young galaxies at high redshift. The galaxies are ordered in radio luminosity, which is not very different from an ordering in redshift in this sample. Most of these sources are embedded in dense ambient gas such as gaseous disks, cooling halos, merging systems and proto-galaxies, and they exhibit the most spectacular EELR's and detailed radio/optical correlations. The radio and optical observations of these sources show good morphological, kinematic and spectroscopic evidence that the radio emitting plasma and emission-line gas are interacting. These interactions can significantly affect the radio morphologies and can modify the physical conditions of the surrounding media.

REFERENCES TO TABLE 1

3C 326.1	McCarthy *et al.* 1987*b*	PKS 2152-69	Tadhunter *et al.* 1987
3C 294	Spinrad *et al.* 1988		Schilizzi and McAdam 1975
3C 295	Baum *et al.* 1988	3C 305	Heckman *et al.* 1982
	Henry and Henricksen 1986	4C 26.42	van Breugel *et al.* 1984
3C 405	Perley *et al.* 1984	M 87	Ford and Butcher 1979
	Dreher *et al.* 1987		Keel 1984
	Pierce and Stockton 1986		Baum *et al.* 1988
3C 368	Djorgovski *et al.* 1987	4C 29.30	van Breugel *et al.* 1986
3C 441	van Breugel and McCarthy 1987	PKS 2158-380	Fosbury *et al.* 1982
3C 337	Pedelty *et al.* 1988		Hansen *et al.* 1987
PKS 0812+020	Wyckoff *et al.* 1983	NGC 7385	Simkin *et al.* 1983
	Wehinger *et al.* 1984		Hardee *et al.* 1980
	Rudnick 1984	Centaurus A	e.g. Graham and Price 1981
PKS 0511-484	Smith and Robertson 1985	NGC 541	van Breugel *et al.* 1985*b*
3C 435A	van Breugel and McCarthy 1987		Brodie *et al.* 1985
3C 171	Heckman *et al.* 1984		Hansen *et al.* 1987
PKS 0521-365	Keel 1986	I0421+040	Beichman *et al.* 1985
3C 321	van Breugel *et al.* 1988	NGC 1068	e.g. Wilson and Ulvestad 1987
3C 277.3	Miley *et al.* 1981	NGC 5929	Whittle *et al.* 1986
	van Breugel *et al.* 1985*a*	M 51	Ford *et al.* 1985
PKS 0634-206	Baum *et al.* 1988		Cecil 1987
PKS 0349-278	Danziger *et al.* 1984		
	Baum *et al.* 1988		
	Hansen *et al.* *1987*		

TABLE 1

SELECTED RADIO GALAXIES WITH EXTENDED OPTICAL LINE EMISSION

1	2	3	4	5	6	7	8	9	10
3C, other	IAU	Type	z	$\log P_{21}$ erg s^{-1} Hz^{-1}	α	$\log L_{EL}$ erg s^{-1}	d_R kpc	d_{EL} kpc	Comments
3C326.1	1553+202	?	1.825	36.0	1.15	44.2 Ly-α	57	68	Ly-α near hotspot
3C294	1404+344	?	1.78	35.7	1.14	–	135	101	Very extended Ly-α
3C295	1409+524	E	0.4614	35.3	0.90	42.1 [OIII]	26	~ 15	Distant 'cooling halo' system
3C405	1957+405	E	0.0565	35.2	1.09	42.0 [OIII]	125	7	'Rosetta stone' of classical dbls.
3C368	1802+110	?	1.132	35.1	1.30	–	60	57	Asymmetric velocities
3C441	2203+294	E	0.707	34.8	0.78	–	216	140	LE beyond hotspot, one-sided
3C337	1627+444	E	0.635	34.8	0.91	–	261	85	LE near hotspot, one-sided
PKS	0812+020	Q	0.406	34.2	0.84	–	115	53	Gxy near hotspot, one-sided
PKS	0511−484	E	0.3063	34.1	0.8	–	850	209	LE nucleus-hotspot
3C435A	2126+073	E	0.471	33.9	~ 1	–	68	146	LE beyond both hotspots
3C171	0651+542	N	0.2384	33.9	0.94	42.8 [OIII]	34	34	Asymmetric velocities
PKS	0521−365	BL	0.055	33.2	(0.7)	39.7 Hβ	30	15	Opt. cont. knot; LE opposite, one-sided
3C321	1529+242	E	0.096	33.0	0.87	42.5 Hα	497	52	Optical EL/cont. knot
3C277.3	1251+278	E	0.0857	32.9	0.7	41.8 Hα	73	73	Optical EL/cont. knot
PKS	0634−206	E	0.056	32.9	0.8	41.7 Hα	870	37	'Dark' jet; EL loop/shell
PKS	0349−278	E	0.066	32.8	0.8	41.9 [OIII]	280	44	LE misaligned with radio axis
PKS	2152−69	E	0.0282	32.8	0.7	–	123	86	Opt. EL/cont. knot, one-sided
3C305	1448+634	E?	0.0417	32.2	0.88	41.8 Hα	3.7	5.3	SSSS with 'plumed' hotspots
4C26.42	1346+268	E	0.0630	32.1	1.0	41.8 Hα	14	14	LE along edges, 'cooling halo' Gxy
M 87	1228+126	E	0.0043	32.1	0.8	40.5 Hα	69	16	Opt. cont. jet; 'cooling halo' Gxy
4C29.30	0836+299	E	0.0643	32.0	0.65	42.2 Hα	67	41	LE misaligned with radio jet
PKS	2158−380	E	0.0333	31.7	0.72	41.9 Hα	74	17	LE misaligned with radio axis
NGC 7385	2247+111	E	0.0259	31.7	0.8	39.0 Hβ	412	12	Faint opt. EL/cont. knot, one-sided
Cen A	1322−427	E	5 Mpc	31.7	–	–	~ 10^3	45	Young stars along jet direction
NGC 541	0123−016A	E	0.0181	31.0	0.75	40.3 Hα	102	19	Starburst at end of jet
IRAS	0421+040	Sy	0.046	30.5	0.6	–	34	~ 25	EL filaments along lobes
NGC 1068	0240−002	Sy	0.0034	30.3	0.8	–	1	1	Sy II with edge brightened lobe
NGC 5929	1524+418	Sy	0.0088	29.2	0.8	40.3 [OIII]	0.2	0.6	Sy II with small double
M 51	1327+474	Sy	10 Mpc	28.0	0.8	39.1 Hβ	1	≳ 1	Shockheated 'bubbles'

Notes to Table 1

Col.1: name

Col.2: IAU designation

Col.3: galaxy type (E = Elliptical; Q = Quasar; N = N-type; BL = BL Lac-type; Sy = Seyfert)

Col.4: redshift

Col.5: radio power at 1400 MHz, assuming $H_0 = 75$ km s^{-1} Mpc^{-1} and $q_0 = 0$. The 3CR 1400 MHz flux densities were taken from Kühr $et\ al.$ 1979 (preprint nr. 55 of MPIfR), or Kellermann, Pauliny-Toth and Williams 1969, Ap.J. 157, 1. Other flux densities were from the literature. The 1400 flux density for Cen A was derived from the 406 MHz measurement by Cooper, Price and Cole (1965, Aust. J. 18, 589), asuming $\alpha = 0.8$.

Col.6: spectral index ($S_\nu = \nu^{-\alpha}$) near 1400 MHz, of total fluxdensity

Col.7: emission-line luminosity, Hα = Hα + [NII] $\lambda\lambda 6548,\ 6584$, [OIII] = [OIII] $\lambda 5007$, [OII] = [OII] $\lambda 3727$ doublet. The emission line fluxes of PKS 0349-278, 3C 98, PKS 0634-206, 3C 171, 3C 274, 3C 305, 3C 405 are taken from Baum $et\ al.$ (1987).

Col.8: linear extent between radio hotspots, or total linear extent (if only one hotspot) (Note: 3C 171 and 3C 305 have large radio 'plumes' perpendicular to the core-hotspot directions).

Col.9: linear extent emission-line gas (if two-sided), or maximum distance emission-line gas from nucleus (if one-sided; see comments).

Col.10: LE = Line Emission; EL = Emission-lines; SSSS = Small Steep Spectrum Source

The main conclusion one can draw from these observations is that the effects of the interaction of radio galaxies with dense ambient material are varied. While a number of relatively general properties seem to emerge, summarized below, they are not always present and do not occur in equal detail in all of the sources. It is clear that properties of EELR's depend on a complex mixture of environment, radio power (jet), epoch (redshift) and evolution.

Emission-line 'screens'

In general the brightest EELR's are found adjacent to the brightest radio components, such as knots in jets and hot spots, and sometimes fainter filaments are also seen along the edges of more diffuse lobes (Fig. 2). This suggests that the enhanced emission-line brightness is probably a line of sight effect, with longer lines of sight intersecting the line emitting screen at the boundaries of a source. In 3C 277.3 very faint line emission was also observed across the lobes, in agreement with such an interpretation. With the improved sensitivity and angular resolution of future optical (ground based and satellite) telescopes one would expect this to be quite common for radio sources embedded in dense gas. The screens are probably caused by the interaction of the radio sources with dense ambient bas, but the exact mechanism is yet unclear. For example, the expanding lobes might drive radiative shocks into pre-existing clouds or trigger thermal instabilities, both of which would lead to enhancend cooling and denser gas near the edges. Other processes might occur however and a better understanding requires finding what the source of ionization is and, perhaps, the origin of the gas.

Depolarization and Faraday rotation

Regions of optical line emission and radio polarization are usually anti-correlated (Fig. 2). In addition to depolarization, in some radio sources large rotation measures (RM, defined as $[\phi_{\lambda_2} - \phi_{\lambda_1}]/[(\lambda_2)^2 - (\lambda_1)^2]$, with $\lambda_2 > \lambda_1$) have been, or could be, measured. Most of these sources have also associated emission line gas (i.e. 3C 218, 3C 295 and M 87; Kato *et al.* 1987, Baum *et al.* 1988). The notable exception is Cygnus A (Dreher *et al.* 1987) but in this case optical observations are hampered by the large Galactic extinction in the direction of this low latitude object.

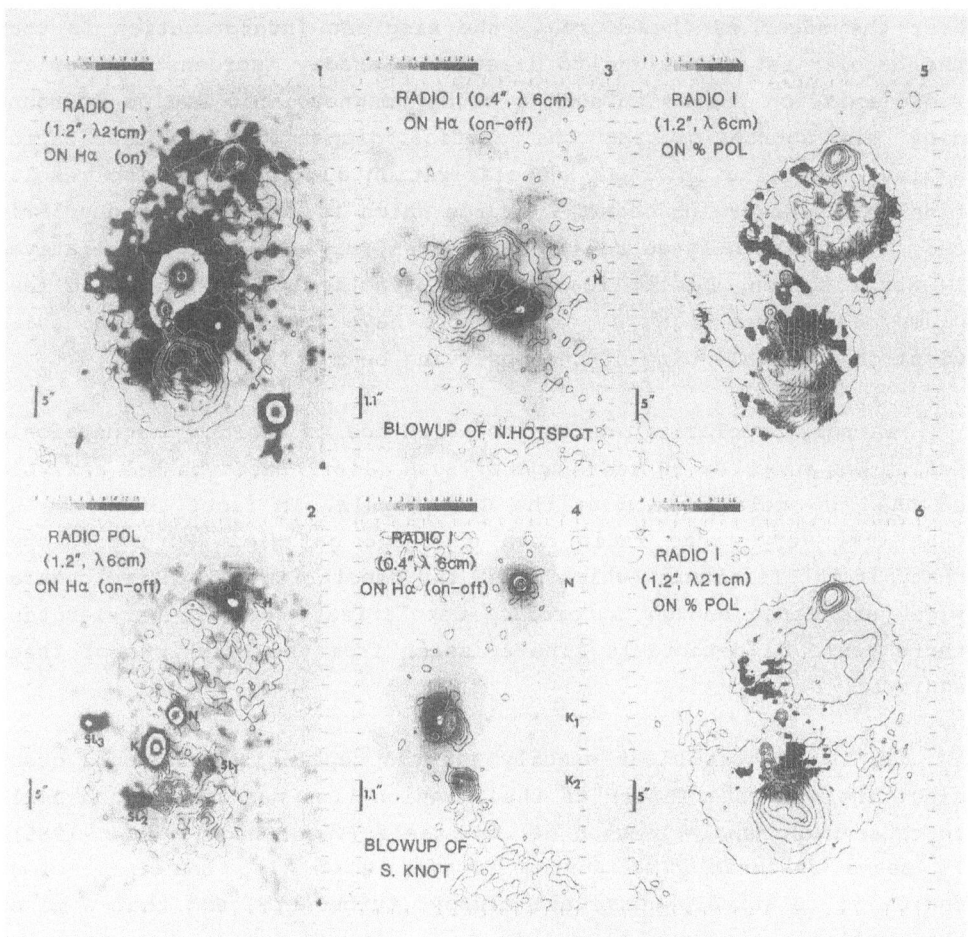

Figure 2: Various superpositions of radio and optical images (taken in sub-arcsecond seeing) to illustrate the morphological relationships of the radio continuum and the line-emitting gas of 3C 277.3 (van Breugel *et al.* 1985a). *Panel 1*, Gray scales represent the Hα + N[II] emission smoothed to ~ 1″.5 resolution with *no* continuum subtracted. The contours represent the total intensity at 21 cm with 1″.2 resolution. Note the extended line emission near the northern hotspot and along the NE edge of the more diffuse lobe. *Panel 2*, Gray scales representing the *pure* Hα + N[II] emission with the radio *polarized* intensity (6 cm, 1″.2) superimposed. Note the anti-correlation between optical line and polarized radio emission. *Panel 3*, Gray scales represent the *pure* Hα + N[II] emission smoothed to ~ 8″, near the northern hotspot. Contours represent the total intensity at 6 cm with ~ 0.4″ resolution. Note the displacement of the brightest radio and optical emission. *Panel 4*, As in panel 3 for the nucleus (N) and the knots (K_1 and K_2) in the jet. Note the displacement between radio and optical knots and the elongation of the optical emission in the direction of the jet. *Panel 5*, Gray scales represent the percentage polarization (increasing with darkness) at 6 cm with ~ 1.2″ resolution. The contours show the total intensity. Note the depolarized regions near the northern hotspot, along the knots and jet, and orthogonal to the jet in the center of the southern lobe. *Panel 6*, As in panel 5 for 21 cm with ~ 1.2″ resolution.

Since the line emission appears to be preferentially located near the edges of the sources, the simplest interpretation is that the depolarization is due to irregular Faraday 'screens': polarized radio emission passes through a clumpy magnetoionic medium surrounding the source causing the spatial dispersion of polarization angles, resulting in 'beam' depolarization due to the finite resolution of the observing beams. A source which is not entirely depolarized, but exhibits large rotation measures, may still have a relatively dense screen, but at least a good fraction of this must be less clumpy (and thus might be expected to have less optical line emission) on the scale size of the observing beam.

Faraday depolarization and rotation due to a clumpy magnetoionic medium was proposed 20 years ago by Burn (1966) based on his analysis of the (de-)polarization of the Crab nebula. In fact, using the at that time very crude radio data on radio galaxies, Burn predicted that 'if the filaments [which cause the depolarization in radio galaxies] are dense enough to produce depolarization at cm wavelengths, there could be detectable line emission from the outskirts of these sources'!

It is not yet clear exactly how the depolarization would occur since the covering factor of the emission-line gas may be too small to cause much depolarization all by itself (van Breugel *et al.* 1986). It seems reasonable to assume however that also the *inter*-cloud medium is relatively dense and clumpy (turbulent), and that some of the depolarization may be due to this tenuous but larger volume of magnetoionic gas. At short enough radio wavelengths this medium would be Faraday transparent and large Faraday rotation may be observed (with RM $\sim \lambda^2$). In some distant radio galaxies the high redshifts make the restframe wavelengths sufficiently small that large (restframe) RM values could be measured (Section 3).

Jet deflection

There are now several radio jets with good morphological and kinematic evidence for associated emission line regions and it is possible to take a first look for some trends. These sources are, in order of increasing radio luminosity: Cen A, NGC 541, 4C 29.30, 3C 277.3, 3C 321, and PKS 0812+02. The published data for PKS 0812+02 are very incomplete. PKS 2152-69 almost certainly belongs to this list (between 3C 321 and 3C 277.3) but no radio data is yet available. 3C 277.3, 3C 321 (and PKS 2152-69), in addition to EELR's also have radio/optical continuum knots (see Röser, these Proceedings). The low luminosity sources Cen A, in sofar observable beyond its dustlane, and NGC 541 (Fig. 3) do not have such knots but instead have extended regions of starformation.

It appears that with increasing radio power (~ jet kinetic energy?) 1) the jets are more collimated, 2) the ionization state (mechanism) of the EELR's seems to increase (change), and, possibly,

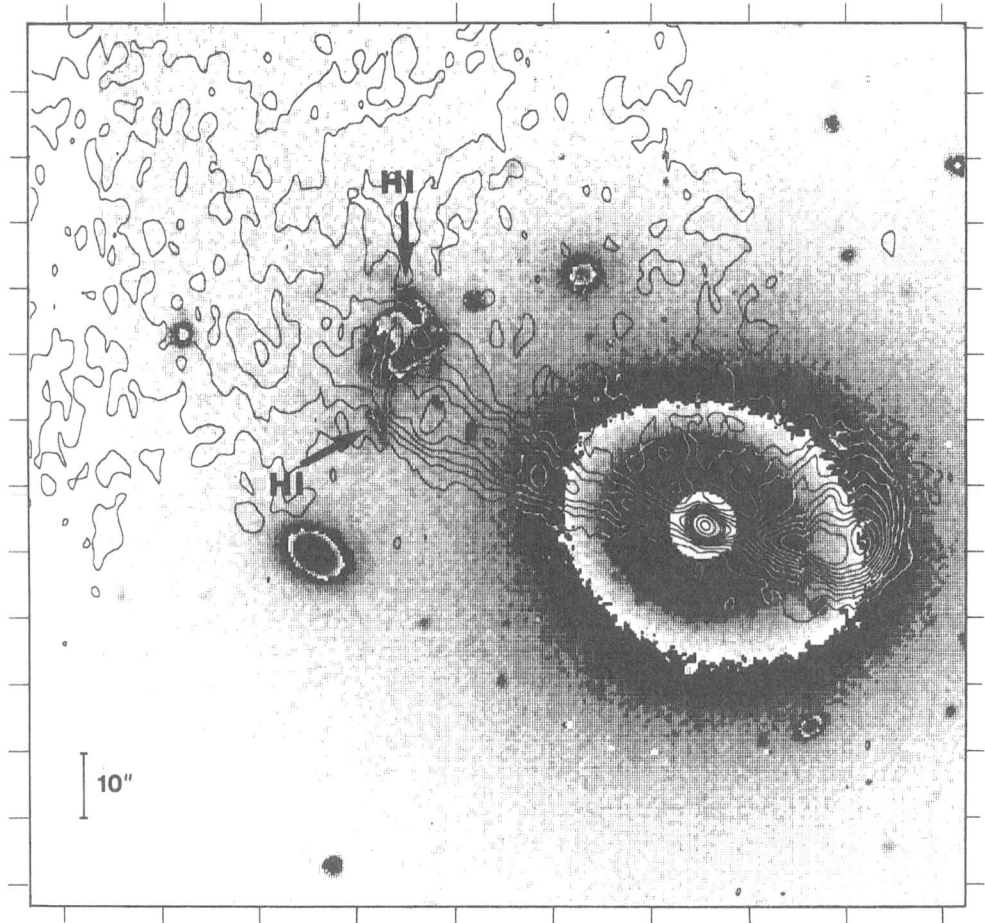

Figure 3: Overlay of a 21-cm radio continuum map (~ 3″ resolution) on an optical narrow-band image of Minkowski's Object (M.O.) that includes the Hα emission-line (van Breugel *et al.* 1985*b*). Neutral hydrogen (HI) observations resulted in the two detections as indicated by arrows (van Breugel *et al.* 1987). Note the filamentary structure of M.O. downstream and sharp boundaries upstream from the jet, the gap in M.O. near the center of the jet and the decollimation of the jet downstream from M.O.

3) the bright radio/optical continuum knots are closer to the galaxy nuclei. Of course these trends must be regarded with extreme caution because of the few sources and many possible selection effects involved. However if true, this seems to indicate that low power jets are more easily deflected and may have insufficient momentum to develop strong shocks (knots) near locations where they encounter dense interstellar material. On the other hand, their more gentle flows may trigger the large scale collapse of clouds, leading to star formation.

In the more powerful sources jet/gas collisions may cause very strong shocks and the optical continua associated with these might ionize this gas (Section 5). This would at least seem the best explanation for the knots in 3C 277.3 and PKS 2152-69, but the situation in 3C 321 is entirely unclear and other explanations may be required.

Hot spots

EELR's near hot spots are relatively rare in nearby radio galaxies. In addition to those in the 'representative' sample of Baum *et al.*, they have also been found in 3C 277.3 and 3C 171 as well as in some small steep spectrum sources (3C 305) and Seyfert II's (NGC 5929), when images with sufficient angular resolution are available. The line emission is usually brightest at the leading edge of the hot spots (3C 305) or slightly downstream from this (3C 277.3, 3C 171, Seyfert II's). The small distances between these emission line regions and the hot spots, assuming they are not far from their bow-shocks, imply short cooling lengths and hence relatively dense pre-shocked gas. In most cases, 3C 277.3 being one of the exceptions, independent evidence exists which indicates the likely presence of such dense gas (gaseous disk, dust, cooling halo, proto-galaxy). Together with kinematic data, and many assumptions, EELR's associated with hot spots may be used to obtain further insight in the advance speeds of hot spots and gas entrainment (Section 4).

Several of these radio galaxies have rather flat hot spots with extensive plumes at nearly right angles to the radio source axes. Examples are 3C 305, 3C 171, and possibly the N hot spot of 3C 295. An extreme example may be 3C 293 (Bridle *et al.* 1981; van Breugel *et* al. 1984). These morphologies may be caused by the relatively large ram pressure (ρv^2) due to excessively dense gas (ρ large), which results in slowly advancing hot spots with little backflow and plumes nearly perpendicular to the radio axes.

3. Distant Radio Galaxies

To study the radio/optical correlations of powerful radio galaxies at high redshifts (z > 0.5), McCarthy and collaborators have obtained optical narrow- and broad-band images of most of the 3CR radio galaxies in this redshift range. These observations show similar radio/optical correlations as have been found for the nearby sources, except that they are more extreme: the EELR's are aligned with the radio galaxies and sometimes even extended beyond the radio hot spots, also the galaxy *continua* are elongated in the same directions, some of the sources show very large rotation measures near hot spots with extended line emission, and the emission-line velocities and line widths are considerably larger.

Line emission beyond radio hot spots

The majority of (~60%), powerful (L_R > 5 x 10^{43} erg s^{-1}) radio galaxies at moderate to high redshifts (z ~ 0.5-1.8) have extended (> 10 kpc) emission line regions, nearly all of which are parallel to the radio axes (McCarthy *et al.* 1987a; McCarthy 1988). Many sources have EELR's associated with their hot spots: all those in Table 1 with z > 0.4, as well as several others (McCarthy *et al.* 1987a). In some of these the hot spots also have very weak optical/UV continua associated with them although not always in a one-to-one correspondance to the radio emission. Whether this is due to synchrotron, stellar continuum or another emission mechanism is at present not clear and may differ for different objects.

In some distant radio galaxies (3C 441, 3C 435A; Fig. 4) the line emission extends more than ~ 30 kpc *beyond* the radio hot spots (van Breugel and McCarthy 1987). This indicates that the source of ionization, at least in these objects, is not directly related to the interaction of the radio galaxies with their environment. The

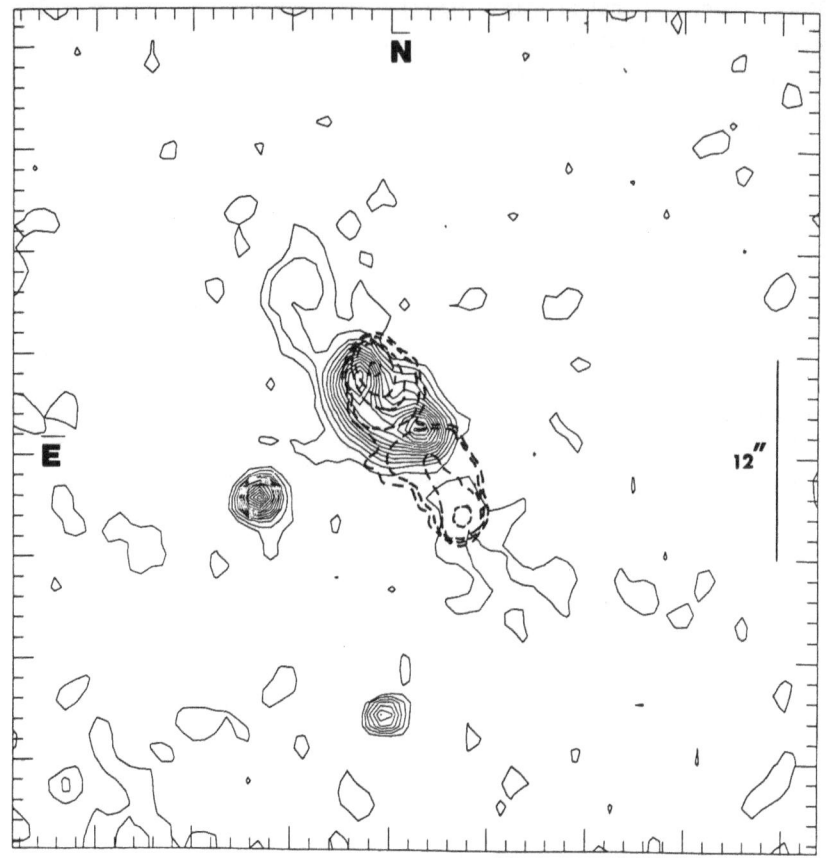

Figure 4: Contour map of the extended optical line emission ([OII], continuum subtracted) associated with 3C 435A. Overlaid (dashed lines) is a 21 cm map with 1.2″ resolution (van Breugel and McCarthy 1987). Note the extended optical line emission beyond the radio hotspots.

simplest explanation for these observations would be that these, and perhaps all, radio galaxy nuclei have an anisotropic source of (photo-) ionization, which may be associated with the jets which power the radio lobes (Section 5).

Radio source and galaxy alignments

Broad-band images show that at large redshifts many galaxies are elongated and exhibit multimodal structure (Lilly and Longair 1984; Spinrad 1987; Djorgovski *et al.* 1987; Le Fèvre *et al.* 1988). The long axes of these galaxies are generally aligned with those of their associated radio sources, both at optical (McCarthy *et al.* 1987a [Fig. 5]; Chambers *et al.* 1987) and, at least in one case, at infrared wavelengths (Chambers *et al.* 1988). It shows that their morphologies are fundamentally related. This considerably complicates galaxy evolution and cosmology models, regardless of the exact origin of this effect, since distant galaxies can sofar only be found through their associated radio sources.

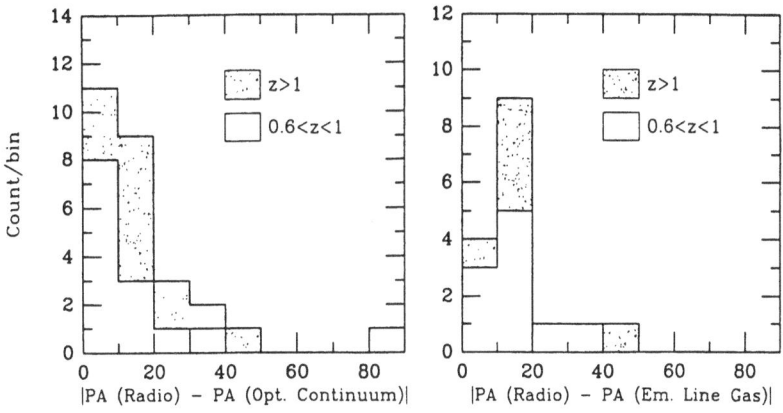

Figure 5: Histogram of the difference between the radio and optical (continuum Fig. 5a, emission-line Fig. 5b) position angles for distant 3CR radio galaxies (McCarthy *et al.*, 1987a).

Most of these radio galaxies have blue optical and IR colors, suggesting that they are undergoing large scale bursts of star formation (e.g. Lilly and Longair 1984; Spinrad and Djorgovski 1987). A straight forward interpretation of the radio/optical alignments is then that these star bursts have been triggered by the bowshocks and backflows associated with the radio hot spots (McCarthy *et al.* 1987a). Other explanations such as gravitational amplification (Le Fèvre *et al.* 1988), fuelling along the radio axes (Djorgovski 1988a) and inverse Compton scattering (Chambers *et al.* 1988) appear more contrived with the currently available data. Induced starformation implies that an active nucleus develops very early in the formation of the parent galaxy, perhaps even before the bulk of the star formation is completed. A possible example of such a "proto-galaxy" may be 3C 326.1 (McCarthy *et al.* 1987b). Theoretical models of galaxy formation along these lines have been proposed by Silk and Norman (1981) and by Silk and Szalay (1987).

Large rotation measures

Multifrequency radio observations of two of the most distant 3CR radio galaxies (3C 326.1, z = 1.825; and 3C 256, z = 1.819) have shown that their hot spots have a low degree of polarization (a few %) and large rotation measures. In 3C 326.1 the *minimum observed*, absolute rotation measures are ~ 302 rad m^{-2} for the W hot spot, and ~ 0 rad m^{-2} for the E hot spot. In 3C 256 polarized emission was detected at both wavelengths only from the NW hot spot, which has a minimum observed, absolute RM of 329 rad m^{-2}. The rest frame rotation measures (RM$_{rf}$) are a factor $(1+z)^2$ larger than this. This results in *minimum* RM$_{rf}$ values of 2500 rad m^{-2} for the N and NW hot spots in 3C 326.1 and 3C 256, respectively.

The optical narrow-band (Lyα) observations of 3C 326.1 by McCarthy *et al.* have shown that the hot spot with the largest RM$_{rf}$ coincides with the brightest emission-line gas (Fig. 6). A similar radio/optical comparison for 3C 256 shows that its NW hot spot also appears embedded in optical emission-line gas. It suggests that the Faraday rotation (and depolarization) occurs in magnetoionic gas associated with these emission-line regions. By analogy to the nearby radio galaxies this gas may form an irregular Faraday screen surrounding the radio hot spots.

Gas densities in the interstellar media of young galaxies such as 3C 326.1 and 3C 256 may be relatively large. Also large magnetic fields might exist, generated through dynamo action (Pudritz and Silk 1988). The entrainment and compression of such a medium by the bowshock of a powerful source like 3C 326.1 could conceivably result in a dense magnetoionic screen around the hot spot and in the back flowing radio lobes. As an example : for an intercloud density n$_e$ ~ 0.1 cm^{-3} (~ 10 times that of our Galaxy), and a screen thickness

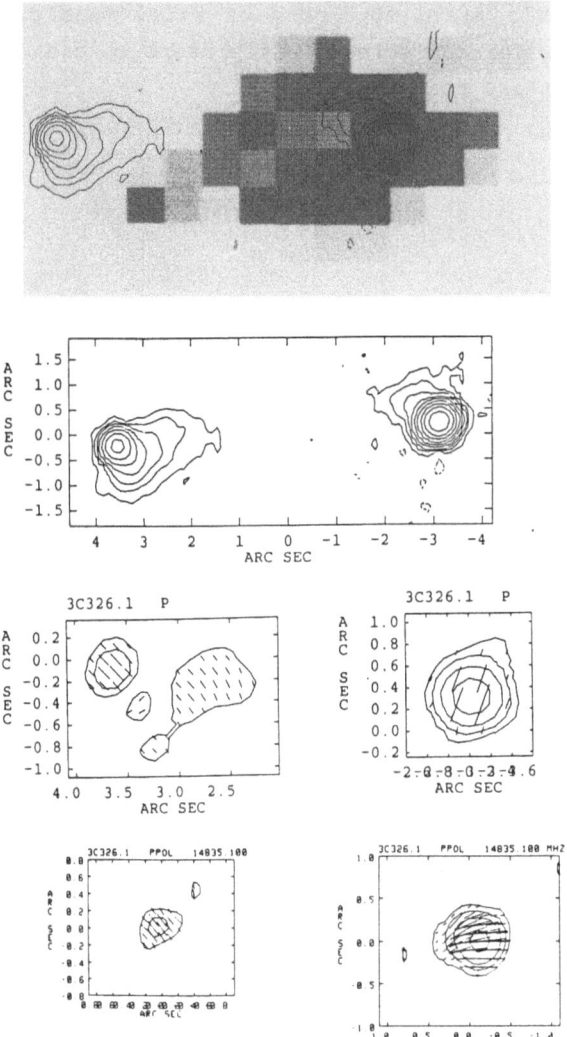

Figure 6: Contour maps of 3C 326.1 at 6 cm and 2 cm, all with 0.4″ resolution. From top to bottom the figures show (a) a 6 cm total intensity map with a Ly-α image superimposed (McCarthy *et al.* 1987*b*), (b) the same 6 cm map, with sky coordinates and a better display of the W hotspot, (c) 6 cm polarized intensity + position angles (E and W hotspots) and, (d) 2 cm polarized intensity + position angles (E and W hotspots). Note that the Ly-α emission is associated with the brightest (western) of the two radio hotspots, which also shows the largest Faraday rotation.

137

of ~ 1 kpc, magnetic field strengths of a few tens of μG would be required within the Faraday screens. At present we can only speculate whether such values are reasonable.

4. Kinematics

Longslit spectroscopic observations with relatively high spectral resolution (~ 2-5 Å) yield information on the kinematics of the EELR's, which can be compared with radio properties. Two classes can be distinguished: the EELR's which *are*, and those which are *not* associated with bright, high pressure radio components such as knots in jets and hot spots. The kinematics of the gas in these two cases may be dominated by the radio sources and the environment of the galaxies respectively.

Environment: rotation and quiescent gas

In general emission-line gas which is not closely associated with bright radio features does not appear much disturbed, even though it may be located along the radio axes. Radio galaxies where the kinematics is dominated by the rotation of a gaseous disk in the parent galaxy, with possible distortions due to tidal interaction or merging, are for example PKS 0349-278, PKS 0511-484, PKS 2158-380, and possibly in the quasar 3C 275.1 (Hintzen and Stocke 1986).

The gas velocities of EELR's of radio galaxies in cooling halos tend to be small and erratic (Heckman *et al.* 1988). Except near bright radio features, where the emission-lines are broad, these gas velocities are not much influenced by the radio source and may just reflect motion of the cooling gas in the halo.

Gas entrainment

In radio galaxies where the ionized gas *is* associated with bright radio components there is often good evidence that the kinematics is disturbed by the radio source. EELR's in the nearby radio galaxies 3C 277.3 and 4C 29.30 show large velocity gradients and line widths of several hundred km s^{-1} near knots and hot spots. The total velocity gradient along the jet in 4C 29.30 is even larger ($\sim$ 700 km s^{-1}). The filament along the radio jet in 3C 321, downstream from the bright knot, has a $\sim$ 1000 km s^{-1} velocity shear. In PKS 2152-69 the asymmetric line profiles in its high ionization knot indicate an even larger velocity change of $\sim$ 2000 km s^{-1} along the radio source axis.

In small steep spectrum sources such as 3C 305 and the lower luminosity Seyfert II galaxies (e.g. M 51, NGC 1068, NGC 5929), the kinematics (and ionization) appears to be bimodal. The velocities and line widths are large in regions which overlap with the radio source, and the high excitation gas is found predominantly along the radio axes. The lower excitation gas, often seen through the galaxies, and the high excitation gas beyond the radio hot spots, have relatively small line widths and seem to partake in the overall rotation of the galaxy disks (Heckman *et al.* 1982; Unger *et al.* 1987). The (extra-nuclear) [OIII]λ5007 emission-line profile near the brightest lobe of 3C 305 shows an extremely large "blue wing" of $\sim$ 2000 km s^{-1}, with its centroid offset by $\sim$ -400 km s^{-1} (Heckman *et al.* 1982). Blueward line asymmetries have also been found in Seyfert II galaxies (Heckman *et al.* 1981, Whittle 1985a).

Many of the more powerful (L_R > 5 x 10^{43} erg s^{-1}), high redshift radio galaxies with extended emission-line regions exhibit large velocity gradients and line widths of up to $\sim$ 2000 km s^{-1} (e.g. McCarthy *et al.* 1987c; Spinrad *et al.* 1988). Considering the relatively low linear resolution (1" $\sim$ 10 kpc at redshifts of $\sim$1.8) even larger localized velocities are likely. The velocities and line widths are usually largest near the hot spots.

In at least two of these radio galaxies (3C 171 [Fig. 7], 3C 368) the velocity field along the radio axes is very asymmetric: large gradients (600-1000 km s^{-1}) are seen on one side, and much smaller or negligible velocity changes are found near the opposite lobes. On the other hand, the line widths are large throughout. Furthermore, recent observations of a number of very distant radio galaxies, including 3C 326.1 (Djorgovski 1988b) and 3C 294 (Spinrad *et al.* 1988), show blueward asymmetric Lyα profiles near radio hot spots. In general it seems that the $P_R \sim \delta v_{EL}$ relation which has been found for small steep spectrum sources and Seyferts (e.g. Heckman *et al.* 1981; Whittle 1985b), may perhaps be extended to include also

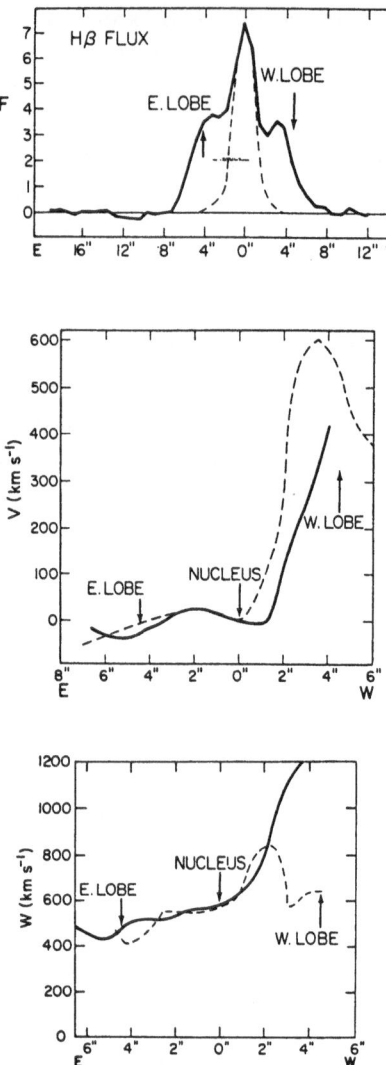

Figure 7: Brightness and velocity profiles of emission line gas associated with the powerful radio galaxy 3C 171 (z = 0.2384; Heckman *et al.* 1984). From top to bottom the figures show (a) the [O III] λ5007 flux along the major (solid line) and minor (dashed line) radio axes, (b) the gas velocity along the major axis using two different long-slit spectrographs (the dashed line representing the most sensitive data, obtained during the best seeing but with the lowest spectral resolution), and (c) the FWHM line widths along the same position. Note the large velocity asymmetry between the E and W radio lobes, and the large line widths (up to 1200 km s^{-1}).

the more powerful and larger sources in sofar these are also embedded in their emission-line regions.

These kinematic properties of EELR's near bright radio components are inconsistent with rotation, infall or other large scale ordered velocity fields unperturbed by the radio source. Instead they strongly support the morphological evidence for interaction between radio sources and dense ambient material, and suggest that gas entrainment occurs. Entrainment also provides a natural explanation for the possible $P_R \sim \delta v_{EL}$ correlation if one assumes that the radio luminosity is primarily dependent on the bulk kinetic energy of the jets ($L_R = 0.5 \epsilon \rho_J v_J^3 A_J$, where ϵ is an efficiency factor, and ρ_J, v_J and A_J the jet density [gm cm^{-3}], velocity and diameter respectively) and that the observed emission line velocities and line widths are at least some (indirect) measure of these bulk velocities. In principle kinematic data with enough spatial resolution, as one might hope to obtain with the Hubble Space Telescope, could be used to measure the velocity profiles of entrained gas transverse to the jet directions in objects such as 3C 277.3. This might help constrain models of the kinematics of the (unseen) jet material.

Models

Kinematic data, together with various assumptions pertaining to the (radiative) bowshocks associated with the hot spots, the physical conditions of the pre-shocked ambient medium and ram pressure confinement may be used to obtain estimates of the advance speeds of hot spots, and possibly of the jet velocities and densities.

Such models, with various degrees of detail, have been applied to the radio galaxies 3C 277.3 and 3C 305, and several Seyfert galaxies (NGC 1068, M 51, NGC 5929). In 3C 277.3 no prior knowledge existed with respect to its ambient medium. In fact one of the more important conclusions from this analysis was that, despite its normal elliptical appearance, 3C 277.3 must possess an inhomogeneous halo with relatively dense ($n_e \gtrsim 0.4$ cm^{-3}) clouds at distances $\gtrsim 50$ kpc from the galaxy nucleus, with a total mass of the cloud system of 10^7-10^8 M$_\odot$. The filling factor of this gas is probably rather small (10^{-6}-10^{-3}) and its effect on the morphology and advance speed of the radio hot spot is probably negligible.

This is unlike in 3C 305, where the filling factors are considerably larger (10^{-2}–10^{-1}). Assuming an *average* ambient gas density of ~ 3 cm^{-3} and temperature ~ 10^6 K, for which values there may be some indirect evidence, Heckman *et al.* find that the hot spots can be ram pressure confined at velocities as low as ~300 km s^{-1}. Such low advance speeds of the hot spots would yield a natural explanation for their flat, plumed morphologies.

A slightly different analysis was made by Wilson and Ulvestad (1987) for NGC 1068 to explain its limb brightened, arrow shaped hot spot. These authors propose that a radiative bowshock not only shocks and compresses the ISM clouds but also the magnetic fields and cosmic rays which are present in the galaxy disk. This would explain simultaneously the limb brightening of the radio lobe and the line emitting gas close to it. For cloud densities of 1-400 cm^{-3} they arrive at bowshock velocities of 76-770 km s^{-1}, a mass swept up by the shock of 1x10^8 - 2.5x10^5 M$_\odot$, a jet velocity of 4,000 - 15,000 km s^{-1}, and a jet density of 0.1 - 0.003 cm^{-3}.

From the above one might conclude that the larger emission line velocities found in the very powerful distant radio galaxies (~ 1000 km s^{-1} or more) would simply be evidence for larger bowshock and entrainment velocities in these sources. While this may be the case, the exact mechanism of how to accelerate the clouds to such high speeds appears problematic. For shock velocities above a few hundred km s^{-1} the cooling time may become too long (McKee and Hollenbach 1980), and the gas might therefore not emit optical line radiation. (Note: The presence of EELR's in radio sources is not necessarily evidence for radiative shocks, even if such shocks occur [Section 5]). More sophisticated models of radiative bowshocks might be needed.

The blue-asymmetric emission-line profiles in 3C 305 and Seyfert II's have been interpreted as further evidence for outflowing gas, accelerated by the expanding radio sources, with dust absorbing the redshifted component. In these objects the radio sources are entirely embedded in their parent galaxies, and the presence of dust is clearly established from spectroscopic (line ratios) and imaging (dust-lanes) observations.

In the very distant galaxies the presence of dust is unclear and, considering the often large luminosities and extents of their associated Lyα regions, may well be very small (e.g. Spinrad 1987).

The blue asymmetries in these objects may thus require a different explanation. Neufeld and McKee (1988) have proposed a 'Fermi acceleration'-like model in which the line asymmetries are caused by multiple scatterings of Lyα photons across shocks in partially ionized clouds. The shocks are thought to be driven into the clouds by the passage of the radio hot spot triggering star formation and creating the (local) source of Lyα photons. In this model the bulk velocity of the emission line gas is not necessarily assumed to be due to entrainment by the radio source.

Kinematics and jet sidedness

In principle the kinematics of emission-line gas associated with radio jets/hot spots might be used to determine whether the outflowing gas is at the backside of the source, or at the side directed towards the observer. In particular one might investigate whether EELR's associated with bright or one-sided jets are systematically blue shifted, and gas in their opposite lobes red shifted. If true this might provide further support for Doppler beaming models for jets.

A preliminary inspection of objects with good radio and optical data yields encouraging, though inconclusive results. The sample is still small and there are many systematic effects such as rotation, tidal interaction, backflow in radio lobes, insufficient spectral and spatial resolution, and sensitivity to isolate regions with high velocity gas which can dilute any existing correlation.

In Table 1 the following sources have bright radio jets: 3C 294, 3C 441, PKS 0812+02, 3C 321 (NW), 3C 277.3, 4C 29.30, PKS 0521-36, 3C 305 (NE), NGC 541, M 87, and Cen A. In 3C 294 the kinematics is different for different atomic species, while in Cen A, NGC 541, and possibly 3C 441 the gas velocities are small enough that they may be dominated by local conditions such as rotation, infall, tidal interaction *etc*. In 3C 441, 3C 405, PKS 0521-36 and M 87 there is no evidence for a direct association between the extended emission line gas and the radio jets. In the remaining sources there is such evidence however, and in all cases the gas at the side of the brightest jet is blueshifted relative to the parent galaxy. The total velocity shears along the jets are -700 km s^{-1} (PKS 0812+02), -1000 km s^{-1} (3C 321, filament), -200 km s^{-1} (3C 277.3), - 730 km s^{-1} (4C 29.30), and

-400/-2000 km s^{-1} (3C 305, NE). The last three objects also exhibit extended line emission near hot spots opposite of the jets, and the gas in these regions is redshifted.

If this trend would persist in a larger sample, it would support models in which the radio jet brightness is Doppler boosted in the direction of the observer. Thus one would predict for example that a radio map of PKS 2152-69 would show a radio jet at the same side as its high ionization emission-line cloud with its blue asymmetric emission line profile, but also that there should be radio galaxies with bright extra-nuclear *red*-shifted emission-line knots and *no*, or only very faint associated radio jets since the optical line emission is unbeamed, while the radio emission would be attenuated because it would be beamed away from the observer.

5. Ionization

In addition to measuring the kinematics of EELR's, spectroscopic observations can also be used to determine, at least in principle, many important physical parameters of the gas such as gas densities, temperatures, abundances, filling factors etc. Using close pairs of emission lines to minimize the effects of extinction, phenomenologi-cal diagrams have been constructed which show that different sources of ionization can be separated such as photoionization by hot stars or powerlaws, and collisional ionization by shocks (e.g. Baldwin *et al.* 1981). Do we really need a source of ionization for the EELR's in radio galaxies? Both the luminosities and spectra indicate that this is indeed the case.

When hot gas cools from ~10^6 to ~10^3 K the emergent Hβ luminos-ity is $L_{H\beta} = 10^{37} \dot{M} H_{rec}$ erg s^{-1} (see for example Johnstone *et al.* 1987). Here $\dot{M}$ is the amount of cooling gas in $M_\odot$ yr^{-1}, and H_{rec} is the number of times each hydrogen is ionized and recombines. In radio galaxies typical values for $L_{H\beta}$ are ~ 10^{40} - 10^{43} erg s^{-1}. In cooling halos in clusters $\dot{M}$ can be estimated from X-ray data and the observed $L_{H\beta}$ is orders of magnitude larger than predicted if no source of ionization is present ($H_{rec} = 1$; e.g. Johnstone *et al.* 1987; Heckman *et al.* 1988). In the EELR's of other radio galaxies $\dot{M}$ is not known but it can hardly be expected to be larger than in the objects which have X-ray emitting halos. Since the recombination time scales t_{rec} ~

$10^5 \, n_e^{-1}$ yrs are short (10^2-10^3 yrs in 3C 277.3 for example), and EELR's in radio galaxies are quite common, the cooling gas must be re-ionized continuously (H_{rec} large). These sources of ionization could be for example: shocks, hot stars, non-thermal (synchrotron jet) or thermal (accretion disk) UV emission from the galaxy nucleus, cosmic rays, X-ray heating etc. Some of these are discussed below, a more extensive discussion can be found for example in Robinson *et al.* 1987, Baum (1987) and Heckman *et al.* (1988).

Shocks

Since shocks occur in a wide variety of astrophysical conditions, and almost certainly in jets and hot spots of radio galaxies, they could be a natural source of ionization. Shockheated gas emits mostly low ionization lines at a relatively high temperature. Unfortunately temperature estimates are difficult to obtain, since for most practical purposes they depend upon measurements of the very weak [OIII]λ4363 emission line, so that the primary diagnostic is the presence of a low ionization spectrum. Low ionization EELR's associated with radio galaxies are usually found in cooling cluster halos. The observed emission line filaments have been interpreted as being caused by re-pressurizing shocks in the thermally unstable gas (Cowie *et al.* 1980).

However, the efficiency (H_{rec}) of collisional shockheating appears to be much too low and alternative models need to be considered (e.g. Johnstone *et al.* 1987; Robinson *et al.* 1987; Heckman *et al.* 1988). The main result of the (radiative) shocks may be not so much that they can explain the line emission, but that they compress the gas so that it shows up better through another source of ionization.

Star formation

Young hot stars may be available as a source of ionization in EELR's if the spectra resemble those of HII regions and their optical continua blue and extended. In good signal-to-noise data Balmer absorption lines should also be seen. Evidence for starformation along radio jets is known to exist in the nearby radio galaxies Centaurus A (eg. Graham and Price 1981) and PKS 0123-016A (Minkowski's Object near NGC 541), and has been suggested to possibly occur in NGC 7385

(Hardee *et al.* 1980) and along the jets in 3C 120 (Wlérick *et al.* 1985) and Fornax A (Graham 1987). In particular in Minkowski's object the evidence is strong, and is based on both its HII region spectrum and extended blue continuum. Large scale starformation is also thought to occur in the very distant radio galaxies. Here the evidence is mostly based upon their extended continua and blue colors. Because of their large redshifts most of the important emission lines can not be observed and consequently the spectroscopic evidence for starformation is not strong. Several of the high redshift radio galaxies do have relatively low ionization spectra which would be consistent with starformation (i.e. 3C 326.1), but sources with higher excitation spectra also exist (Spinrad 1986) and other sources of ionization (galaxy nuclei?) might be present as well.

Powerlaw photoionization

Other possible sources of photoionization that are commonly considered for radio galaxies are: nonthermal (synchrotron) UV emission associated with radio knots, hot spots and radio galaxy nuclei, or thermal emission from very hot (accretion?) disks. The optical continuum knots in 3C 277.3 and PKS 2152-69 are polarized suggesting that they might be due to synchrotron emission. Using the radio/optical spectral index to extrapolate the continuum into the UV suggests that the continuum would be strong enough to ionize the line emitting gas near the knots in 3C 277.3. In PKS 2152-69 no such information is available yet, but the presence of strong local ionization gradients which correlate with the optical continuum of the high ionization cloud is good evidence that also here the gas is ionized locally (Tadhunter, private communication).

Some radio hot spots are also known to exhibit optical synchrotron emission (3C 33, Pictor A; Röser, these Proceedings). None of these show convincing evidence for associated EELR'S so that no statement can be made whether they could be sources of local photoionization. Apparently not much dense, cool gas surrounds these hot spots, which is not too surprising since they are at large distances from their parent galaxy nuclei and well away from their interstellar media. Several of the more distant radio galaxies do have associated optical continua *and* line emission (Section 3), but further observations and analysis are required to explore their relationships.

The most common sources of photoionization are probably the radio galaxy nuclei themselves. Simple models, with basically only one parameter $U(r) \sim q(r)/r^2 n_e(r)$ (where $q(r)$ is the photon flux and $n_e(r)$ the thermal electron density), can reproduce the spectra quite well over a large range of emission-line ratio's (e.g. Robinson *et al.* 1987 [Fig. 8]; Tadhunter *et al.* 1987; McCarthy *et al.* 1987). $q(r)$ can for example be estimated from measurements of the optical con-

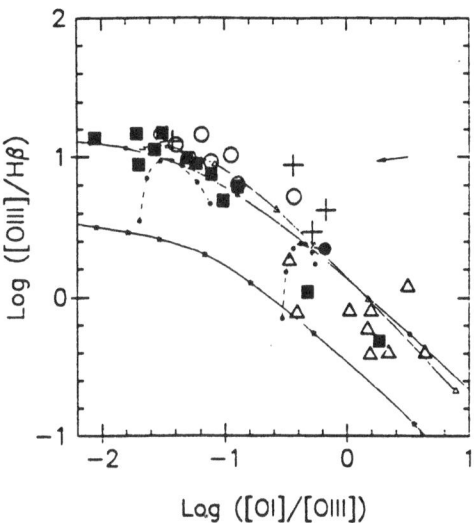

Figure 8: The [O III] $\lambda5007/H\beta$ ratios of extended (black squares) and nuclear (open circles) narrow emission line regions and cooling filaments (open triangles) plotted against [O I] $\lambda6300/$[O III] $\lambda5007$ (Robinson *et al.* 1987). Crosses indicate data for which no reddening correction (general direction indicated by arrow) could be obtained. Solid lines indicate loci of different photoionization model calculations, best fits are for powerlaw and high temperature blackbody continua. Dashed lines indicate the effects of abundance variations.

tinua (extrapolated to the UV), while $n_e(r)$ can be derived when the S[II]$\lambda\lambda6716,6731$ line ratio's are known. Both powerlaw and high temperature blackbody (~10^5 K) are acceptable as sources of ionization (Robinson *et al.* 1987).

Two complications can arise when applying these models. In cooling halos the radial variations of the predicted and observed line ratios are not consistent (Heckman *et al.* 1988). Instead and *in situ* source of ionization seems required which may, directly or indirectly, be related to the presence of the radio source. Another problem is that in some sources the *observed* nuclear optical continua are insufficient to provide the (UV) photons for ionization (e.g. Robinson *et al.* 1987; McCarthy *et al.* 1987c). A possible explanation for this would be that the optical continuum in these sources is partially obscured by a dusty torus or disk close to the nucleus, or that it is partially beamed along the radio axes by relativistic Doppler effects. Such a collimated source of ionization would be attractive in that it would simultaneously explain several observations: (i) the alignment of EELR's with radio major axes (ii) the presence of EELR's beyond radio hot spots in some sources and (iii) other wellknown, aspect dependant effects in other types of active galaxies (Seyferts, BL Lacs and quasars).

6. Conclusions

A general conclusion one can draw from the observations is that extended emission-line regions in radio galaxies are common and that they are related to the mechanism that produces the radio sources. In a statistical sense the emission-line gas is generally found along the radio axes but detailed radio/optical correlations are not always obvious. This is because there are three major ingredients which seem to be involved in producing EELR's in radio galaxies, and one major problem in interpreting the data: 1) The condition of the ambient medium (pre-existing dense and cool gas), 2) Physical properties of the radio source (power, collimation) and its interaction with the ambient gas 3) Optical (UV) properties of the AGN producing the radio source (collimated ionizing radiation?), 4) The detailed interpretation of the optical spectra.

To more fully exploit the presence of EELR's in radio galaxies for a better understanding of radio galaxies and their environment one needs: 1) Further detailed and more sensitive radio/optical observations of powerful (distant) radio galaxies. 2) Observations of (steep spectrum) quasars, which also seem to show EELR's along their radio axes (Baum 1987; Stockton and MacKenty 1987) to determine what their relationship is to the radio galaxies. 3) Hydrodynamic/MHD computer simulations of hot spots (bowshocks) propagating through a clumpy medium, with the inclusion of radiative shocks driven into the clouds. Such models would be extremely valuable in helping to understand the kinematics of the ionized gas in the lobes of powerful doubles. 4) Further investigations of ionization models, such as the importance of X-ray heating and cosmic rays, magnetic field reconnection etc. Major problems to be addressed are for example the extremely high excitation observed in PKS 2152-69, the astounding identical emission-line ratios in 3C 321, and the source of ionization in 'cooling halos'.

I wish to thank the observing staffs at Lick Observatory, Kitt Peak National Observatory and the VLA for their exellent support. Most of the results on extended emission-line regions described here are from work by Fosbury, Tadhunter and collaborators (on Southern Hemisphere objects), Baum, Heckman and collaborators (on nearby 'Equatorial' radio galaxies) and McCarthy, Spinrad and collaborators (on distant 3CR radio galaxies). I am grateful for the many discussions we had and their supply of pre-publication material. The work described here was supported in part through NSF grants AST 84-16177, AST 85-13416 and from the California Space Institute. I also greatly appreciate the travel support provided by the Max Planck Institut für Astronomie and the organizers of the meeting on 'Hot spots'.

References

Baldwin, J. A., Phillips, M. M., and Terlevich, R. 1981,
    Pub. A. S. P. **95**, 5.
Baum, S. A. 1987, Ph.D. Thesis, Univ. of Maryland.
Baum, S. A., Heckman, T. M., Bridle, A. H., van Breugel, W. J. M.,
    and Miley, G. K. 1988, *Astrophys. J. Suppl. Series* **68**.
Beichman, C. *et al.* 1985, *Ap. J.* **293**, 148.
Bridle, A.H., Fomalont, E.B., and Cornwell, T.J. 1981, *A.J.* **86**, 1294.
Brodie, J.P., Bowyer, S., and McCarthy, P. 1985, *Ap.J.Lett.* **293**, L59.
Burn, B. J. 1966, *M.N.R.A.S.* **133**, 67.
Cecil, G. 1988, preprint.
Chambers, K. C., Miley, G. K., and van Breugel, W. J. M. 1987,
    Nature **329**, 604.
Chambers, K.C., Miley, G.K., and Joyce, R.R. 1988, *Ap.J.Lett.*,
    in press.

Cowie, L. L., Fabian, A. C., and Nulsen, P. E. J. 1980, *M.N.R.A.S.* **191**, 399.

Danziger, I. J., Fosbury, R. A. E., Goss, W. M., Bland, J., and Boksenberg, A. 1984, *M.N.R.A.S.* **208**, 589.

Djorgovski, S., Spinrad, H., Pedelty, J., Rudnick, L., and Stockton, A. 1987, *A. J.* **91**, 1267.

Djorgovski, S. 1988a, in *Starbursts and Galaxy Evolution*, Th. Montmerle (ed.), Paris: Editions Frontières, in press.

Djorgovski, S. 1988b, in *Towards Understanding Galaxies at Large Redshifts*, A. Renzini and R. Kron (eds.). Dordrecht: Reidel.

Dreher, J.W., Carilli, C.L., and Perley, R.A., 1987, *Ap. J.* **316**, 611.

Ford, H. C., and Butcher, H. R. 1979, *Ap. J. Suppl.* **41**, 147.

Ford, H. C., Crane, P. C., Jacoby, G. H., Lawrie, D. G., and van der Hulst, J. M. 1985, *Ap. J.* **293**, 132.

Fosbury, R. A. E. *et al.* 1982, *M.N.R.A.S.* **201**, 991.

Fosbury, R. A. E. 1985, in *Structure and Evolution of Active Galactic Nuclei*, Giuricin *et al.* (editors) pg. 297, Dordrecht: Reidel.

Graham, J. A., and Price, R. M. 1981, *Ap. J.* **247**, 813.

Graham, J. A. 1987, *BAAS Abstract, 171st AAS meeting*, in press.

Hansen, L., Norgaard-Nielsen, H.U., and Jörgensen, H.E. 1985, *Astron. Astrophys. Suppl.* **71**, 465.

Hardee, P. E., Eilek, J. A., and Owen, F. N. 1980, *Ap. J.* **242**, 502.

Heckman, T. M., Miley, G. K., van Breugel, W. J. M., Butcher, H. R. 1981, *Ap.J.* **247**, 403.

Heckman, T. M., Miley, G. K., Balick, B., van Breugel, W. J. M., and Butcher, H. R. 1982, *Ap. J.* **262**, 529.

Heckman, T. M., van Breugel, W. J. M., and Miley, G. K. 1984, *Ap. J.* **286**, 509.

Heckman, T. M., Baum, S., van Breugel, W. J. M., and McCarthy, P. J., 1988, *Ap. J.*, submitted.

Henry, J. P., and Henricksen, M. J. 1986, *Ap. J.* **301**, 689.

Hintzen, P., Stocke, J. 1986, *Ap. J.* **308**, 540.

Johnstone, R. M., Fabian, A. C., and Nulsen, P. E. J. 1987, *M.N.R.A.S.* **224**, 75.

Kato, T., Tabara, H., Inoue, M., and Aizu, K. 1987, *Nature* **329**, 223.

Keel, W. C. 1984, *Ap. J.* **279**, 550.

Keel, W. C. 1986, *Ap. J.* **302**, 296.

Kotanyi, C. G., and Ekers, R. D., 1979, *Astron. Astroph.* **73**, L1.

Le Fèvre, O., Hammer, F., Nottale, L., and Mazure, A., and Christian, C. 1988, *Ap. J. Lett.* **324**, L1.

Lilly, S., and Longair, M. 1984, *M.N.R.A.S.* **211**, 833.

McCarthy, P., van Breugel, W., Spinrad, H., and Djorgovski, S. 1987a, *Ap. J. Lett.* **321**, L29.

McCarthy, P., Spinrad, H., Djorgovski, S., Strauss, M. A., van Breugel, W. J. M., and Liebert, J. 1987b, *Ap. J.* **319**, L39.

McCarthy, P., Spinrad, H., van Breugel, W. J. M., Djorgovski, S., Strauss, M. A., Dickinson, M. 1987c, In *Cooling Flows in Clusters and Galaxies*, ed. A. C. Fabian; Cambridge, England.

McCarthy, P. 1988, Ph.D. Thesis, Univ. of California, in preparation.

McKee, C. F., and Hollenbach, D. J. 1980, *Ann.Rev.Astr.Ap.* **18**, 219.

Miley, G. K., Heckman, T. M., Butcher, H. R., and van Breugel, W. J. M. 1981, *Ap. J. Lett.* **247**, L5.

Neufeld, D. A., and McKee, C. F. 1988, Ap. J. Lett. **331**, L87.

Pedelty, J. *et al.* 1988, preprint.

Perley, R.A., Dreher, J.W., and Cowan, J. 1984, *Ap.J.Lett.* **285**, L35.

Pierce, M. J., and Stockton, A. 1986, *Ap. J.* **305**, 204.

Pudritz, R., and Silk, J. 1988, preprint.

Quinn, P. J., 1984, *Ap. J.* **279**, 596.

Robinson, A., Binette, L., Fosbury, R. A. E., and Tadhunter, C. N. 1987, *M.N.R.A.S.* **227**, 97.

Rudnick, L. 1984, in NRAO *Workshop on Physics of Energy Transport in Extragalactic Radio Sources*, pg. 114.

Schilizzi, R. T., and McAdam, W. B. 1975, *Mem.R.A.S.* **79**, 1.

Silk, J., and Norman, C. 1981, *Ap. J.* **247**, 59.

Silk, J., and Szalay, A. 1987, *Ap. J.* in press.

Simkin, S. M., Bicknell, G. V., and Bosma, A., 1984, *Ap. J.* **277**, 513.

Smith, R. M., Robertson, J. G. 1985, *M.N.R.A.S.* **212**, 809.

Spinrad, H. 1986, *P.A.S.P.* **98**, 269.

Spinrad, H., and Djorgovski, S. 1987, *Observational Cosmology*, proceedings of IAU Symposium 124, ed. G. Burbidge. Dordrecht: Reidel, in press.

Spinrad, H. 1987, in *High Redshift and Primeval Galaxies*, J. Bergeron, D. Kunth, B. Roccavolmerange and J. Tran Thanh Van (eds.), Paris: Editions Frontières.

Spinrad, H., McCarthy, P. J., van Breugel, W. J. M., Liebert, J., Dickinson, M., and Djorgovski, S. 1988, in preparation.

Stockton, A., and MacKenty, J. W. 1987, *Ap. J.* **316**, 584.

Tadhunter, C. N., Fosbury, R. A. E., Binette, L., Danziger, I. J., and Robinson, A. 1987, *Nature* **325**, 504.

Tadhunter, C. N., these proceedings.

Toomre, A., and Toomre, J. 1972, *Ap. J.* **178**, 624.

Unger, S. W., Pedlar, A., Axon, D. J., Whittle, M., Meurs, E. J. A., and Ward, M. J. 1987, *M.N.R.A.S.* **228**, 671.

van Breugel, W. J. M., Heckman, T. M., Butcher, H. R., and Miley, G. K., 1984, *Ap. J.* **277**, 82.

van Breugel, W. J. M., Miley, G. K., Heckman, T. M., Butcher, H. R., and Bridle, A. H. 1985a, *Ap. J.* **290**, 496.

van Breugel, W. J. M., Filippenko, A. V., Heckman, T. M., and Miley, G. K. 1985b, *Astrophys. J.* **293**, 83.

van Breugel, W. J. M., Heckman, T. M., Miley, G. K., and Filippenko, A. V. 1986, *Ap. J.* **331**, 58.

van Breugel, W. J. M. 1986, in *Jets from Stars and Galaxies*, eds. R. N. Henriksen and T. W. Jones; *Can. J. Phys.* **64**, 392.

van Breugel, W. J. M., McCarthy, P., and van Gorkom, J. 1987, In *Cooling Flows in Clusters and Galaxies*, ed. A. C. Fabian; Cambridge, England.

van Breugel, W. J. M., McCarthy, P. 1987, In Proc. of Conference on *Active Galactic Nuclei*, Atlanta, Eds. Miller and Wiita (in press).

van Breugel, W. J. M., Filippenko, A. V., McCarthy, P. J., Miller, J. S., and Miley, G. K., 1988, in preparation.

Wehinger, P., Wyckoff, S., and Spinrad, H. 1984, in *B.A.A.S.* **16**, vol. 2, 250.

Whittle, M. 1985a, *M.N.R.A.S.* **213**, 1.

Whittle, M. 1985b, *M.N.R.A.S.* **213**, 33.

Whittle, M., Haniff, C. A., Ward, M. J., Meurs, E. J. A., Pedlar, A., Unger, S. W., Axon, and D. J., Harrison, B. A. 1986, *M.N.R.A.S.* **222**, 189.

Wilson, A. S., and Ulvestad, J. S. 1987, Ap. J. **319**, 105.

Wlérick, G. *et al.* (1985), In *IAU Symposium 119 on Quasars*, edited by G. Swarup and V. K. Kapahi, p. 129.

Wyckoff, S., Johnston, K., Ghigo, F., Rudnick, L., Wehinger, P., and Boksenberg, A. 1983, *Ap. J.* **265**, 43.

Highly ionized gas in PKS 2152–69

Clive N. Tadhunter[1]

Space Telescope-European Coordinating Facility, European Southern Observatory,

Karl-Schwarzschild-Straße-2, D-8046 Garching bei München, FRG

Summary. A gas cloud 10 *arcsec* to the NE of the nucleus of PKS 2152-69 exhibits many features that are more commonly associated with active nuclei. These include an extremely high ionization state with relatively strong emission lines from species like Ne^{+4}, He^+, Fe^{+6} and Fe^{+9}, blue asymmetric wings to the forbidden lines and a continuum that rises steeply to the blue. The proximity of the cloud to the radio axis and the tangential ionization gradients indicate that the cloud is energized by a narrow beam of radiation or plasma from the nucleus. Future studies of the cloud will have relevance to a number of long-standing problems in active galaxy research.

1 Introduction

Large scale regions of radio-emitting plasma are only one of the manifestations of activity in early-type galaxies. When looked at using sensitive detectors in the light of prominent emission lines like [OIII] $\lambda5007$ and Hα, extended emission line nebulosities on scales of $5 - 100$ *kpc* are often revealed (*e.g.*, Fosbury 1986), especially in galaxies where there are already signs of strong activity in the nuclei. These nebulosities are sometimes, but by no means always, associated with extended radio structures (see van Breugel, these proceedings). They embrace a wide range of energetic phenomena, as I mean to demonstrate in this short contribution with the example of a single galaxy: PKS 2152–69.

2 The high ionization cloud in PKS 2152-69

PKS 2152–69 is the brightest radio source in the southern constellation of Indus. Of moderate power ($P_{5GHz}^{tot} = 5 \times 10^{25}$ *W* Hz^{-1}, $H_o = 50$ *km* $s^{-1}Mpc^{-1}$), observations with the Molonglo Synthesis Telescope at 843 *MHz* resolve it into a fairly compact double in PA 23±3° with approximate separation 70±30 *kpc*. The associated elliptical galaxy has a typical optical luminosity for such a radio source ($M_b \sim -21.4$). Its relative proximity ($z = 0.0282$) gives us the opportunity to study the extended ionized gas in some detail and with reasonable spatial resolution. We find three main components of ionized gas. First, a nuclear component of intermediate ionization and weak broad components. Second, an extended low ionization component that covers a large projected area of the galaxy at low surface brightness out to a maximum radius of 20 *arcsec* (16*kpc*). Third, a high ionization cloud situated ~ 10 *arcsec* to the NE of the nucleus and with $\sim 1/4$ the Hβ luminosity of the nucleus. In detail, this cloud has the following properties:

---

[1]Affiliated to the Astrophysics Division, Space Science Department, European Space Agency

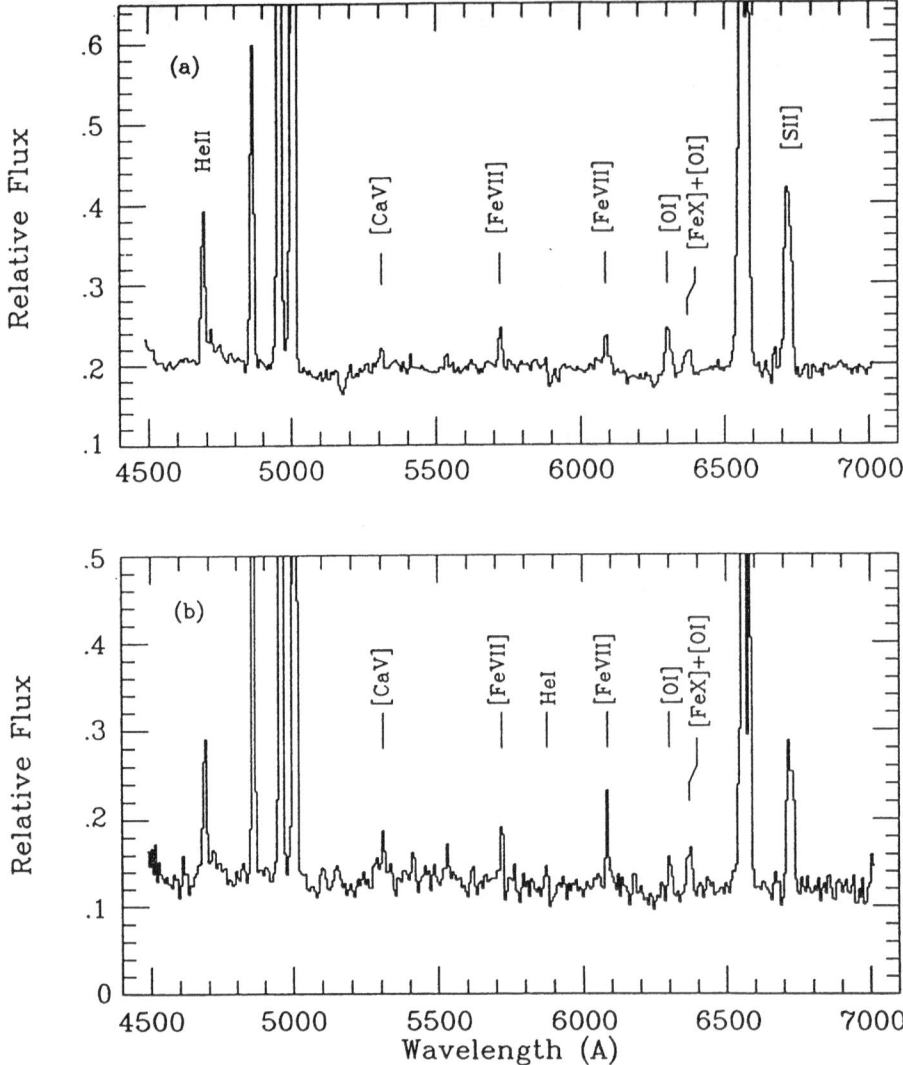

Figure 1: Low resolution spectra of the high ionization cloud in PKS 2152-69: (a) spatially integrated spectrum, (b) spectrum of the highest ionization part of the cloud. These result from a totals of 12 800s and 3 600s integration respectively with the ESO 3.6m telescope, using an RCA CCD as detector.

- An extremely high ionization state with relatively strong emission lines of [NeV], HeII, [CaV], [FeVII] and [FeX]. Figure 1a shows a spatially integrated low resolution spectrum of the cloud.

- A bluer continuum than any of the other components in the same galaxy (see di Serego Alighieri *et al.* , these proceedings, for details).

- Strong ionization gradients. The highest ionization states are associated with both the peak of the local blue continuum and the side of the cloud facing the nucleus. Figure 1b shows that, in the highest ionization part of the cloud, [FeX] λ6374 is of comparable strength to [OI] λ6300.

- Faint wings extending up to 3000 $km\,s^{-1}$ to the blue of the [OIII] line cores.

Detailed results on PKS 2152–69 and its associated emission line regions are presented in Tadhunter *et al.* (1987, 1988) and di Serego Alighieri *et al.* (1988).

3   The energization of the cloud

Both the continuum results (di Serego Alighieri *et al.* , these proceedings) and the fact that the highest ionization states are associated with the side of the cloud facing the nucleus, show that the nucleus is the source of energy for the cloud. However, the presence of low ionization gas at other locations in the galaxy and strong tangential ionization gradients within the cloud itself, show that the energy cannot be emitted isotropically by the nucleus; it must be directed towards the cloud. There are two main candidates for the energy transport mechanism:

1. **The plasma jet.** The offset between the cloud/nucleus axis and the radio axis is only 15 — 25°. Therefore the cloud might represent a spectacular example of the type of jet/ISM cloud seen in other radio galaxies (van Breugel, these proceedings). In this case, the ionizing continuum would be generated *in situ* at the jet/cloud interface, thus providing a natural explanation for the association between the ionization gradients and local optical continuum within the cloud. The line profiles can also be explained by such an interaction if the radio jet can entrain significant amounts of ISM.

2. **A beam of radiation from the nucleus.** This is also supported by the association with the radio axis, because in relativistic beaming models the continuum is expected to be beamed along the inner VLBI jet. In this case, the observed optical continuum would represent light scattered from the beam by material in the cloud.

It is not yet possible to distinguish between these alternatives, although the continuum results (di Serego Alighieri *et al.* , these proceedings) seem most consistent with the latter. A key future observation will be high resolution radio synthesis mapping using the Australia Telescope. If a radio knot or hot spot is observed at the site of the cloud, then it is likely that at least some of the energy is supplied by the plasma jet. On the other hand, the absence of such a hot spot, and any signs of a jet leading to the cloud, would lend strong support to the second hypothesis.

Whatever the exact means of energy transport, the cloud is a potentially excellent laboratory for testing our ideas about physical mechanisms in active galaxies in general. Elements of what we observe in the cloud are also seen in the extended regions around other radio galaxies and, of course, active nuclei. On a detailed level, the cloud is relevant to some of the long-standing problems in AGN research, like the formation of the coronal lines and the formation of broad wings to the forbidden lines. More fundamentally, if the cloud is interacting with an intense radiation beam from the nucleus, future multi-wavelength observations of PKS 2152–69 will place important constraints on models for relativistic beaming in extragalactic radio sources.

Acknowledgements This work was carried out in collaboration with R. Fosbury and S. di Serego Alighieri. I acknowledge the support of an ESA external fellowship.

References

Fosbury, R.A.E., 1986. In: *Structure and Evolution of Active Galactic Nuclei*, Giuricin, G. *et al.* (eds), Reidel, p297.

di Serego Alighieri, S., Binette, L., Courvoisier, T., Fosbury, R.A.E. & Tadhunter, C.N., 1988. Submitted to *Nature*.

Tadhunter, C.N., Fosbury, R.A.E., Binette, L., Danziger, I.J. & Robinson, A., 1987. *Nature*, **316**, 733.

Tadhunter, C.N., Fosbury, R.A.E., di Serego Alighieri, S., Bland, J., Danziger, I.J., Goss, W.M., McAdam, W.B. & Snijders, M.A.J., 1988. *Mon. Not. R. astr. Soc.* in press.

Clive Tadhunter

Ringberg castle - access to the lecture hall

Peter Scheuer

Models of Hot Spots

Peter Scheuer

Mullard Radio Astronomy Observatory
Cavendish Laboratory
Madingley Rd., Cambridge CB3 OHE, U.K.

1. Why bother with models?

This session on "models" is followed by one on "simulations". The fluid flows in and around hot spots are not soluble analytically, so surely we ought to simulate them numerically; making crude approximations in order to simplify the problem to back-of-envelope status is bad physics if it is not forced upon us. So why waste time on "models" ? In many fields of physics, in which one knows what one is talking about, I should agree with those sentiments. But hot spots in radio sources are much too complicated for numerical simulations to have predictive value. They are even more dependent on the assumptions that one builds in, often tacitly, than on the sophistication of the fluid dynamic codes. I believe that simulations are good for testing whether we have understood a particular piece of physics adequately, and sometimes for rubbing our noses in some phenomena that we had forgotten or dismissed (wrongly) as unimportant. I do not believe that we could make a proper comparison between "theory" and observation if only we used a sufficiently wonderful code; the codes determine the quality of the hi-fi, but the music is still in the models we put in.

2. The standard model

In the beginning, Blandford & Rees said, let there be fire and let it squirt out through de Laval nozzles, and it will make jets. And there were jets, even in the early 1970s there were a few. And they said, at the ends of the jets, let there be working surfaces, beyond which there will be lots and lots of radio emission, and they estimated that, to within astrophysical accuracy, it was good.

Now I seem to remember that Blandford & Rees (1974) were not too specific about the occupation that these working surfaces worked at, but once you have a catchy phrase it catches on, especially if it is coined by famous men. And when, some 4 years later, at least four people or groups of people simultaneously invented shock acceleration, it was immediately obvious that the most important occupation of the working surfaces was diffusive shock acceleration (though they were probably not very efficient and made a lot of heat as well). So we got the Standard Model.

The Standard Model brought with it the Standard Uncertainties about the speed, density and composition of the jets. Of protagonists from the limiting schools of thought, we have with us Wolfgang Kundt who believes that $\rho \simeq 0$, $v \simeq \infty$; I had hoped that we should also have with us Jacques Roland who believes that $\rho \simeq \infty$, $v \simeq 0$. There are further uncertainties regarding the formation of the hot spots which are our particular concern here. Most model predictions refer to shocks normal to the jet. Real sources are not at all axisymmetric; the jets weave about, and strike the IGM at an oblique angle, forming a "compact hot spot", after which they often seem to bounce off to make a bigger hot spot further on and/or to one side. Fluid simulations show that even in axisymmetric models most of the shock surfaces are quasi-conical rather than normal. And the few hot spots that have been mapped in real detail are more complex than either of these pictures. Also, we don't know where the magnetic field comes from, nor much about its structure. Even the best polarization maps can only give a few of the answers we need to know; in particular, they can, for the foreseeable future, tell us rather little about the small-scale structure of the field.

3. A few non-standard ingredients

This talk being the aperitif to the session, I see it as my job to throw as many doubts as possible on the Standard Model, in order to stimulate the jaws and the digestive juices. Some of these doubts have been provoked by the simulations, which Alan Matthews started about 3 years ago, of what might happen if you arbitrarily threw an isotropically tangled magnetic field into the standard model at some point, and simply watched it distort.

(i)   One of the things we could not help noticing was that, even if you threw in the magnetic field and all the fast electrons at the base of the jet, the hot spot was still extremely conspicuous. The mere compression in the shocks just before the hot spot was enough to increase the radio brightness so much. Simple-minded estimates, such as those shown in Table 1, confirm that the surface brightness could easily go up by a factor of 100 - *without any diffuse shock accelera-tion at all.*

Table 1

Spectral index α	0.5	0.6	0.7	0.8	0.9	1.0
Increase in surface brightness	38	47	59	73	90	111

Table I shows the factor by which the surface brightness is expected to increase for adiabatic compression by a factor 4, for various spectral indices α ($S \sim \nu^{-\alpha}$). The calculation assumes that the field increases by a factor $[(1^2+4^2+4^2)/3]^{1/2}$ and that the relativistic electron population remains isotropic. The compression factor at a strong shock could be a little higher if the "thermal" electrons, but not the protons were relativistically hot after the shock. If the jet moves at high γ, and the shock is fully relativistic, the compression factor is less than 4, but then the dimming of the jet by relativistic aberration, discussed under (iii), necessarily comes into play. In real (as opposed to axisymmetric) sources, the jet may suffer two (or more) strong oblique shocks in succession; in that case total compression factors much larger than 4 may be reached.

Now when we look at high dynamic range VLA maps we find that, in a large proportion of quasars one can see a jet - often a feeble-looking jet, but one whose surface brightness is nevertheless a few tenths of a percent of the maximum surface brightness of the hot spot.

TABLE 2

Surface brightness of hot spot
─────────────────────────────
Surface brightness of jet

3C9	strong jet	3C263	1000
3C68.1	50	3C334	fairly strong jet
3C208	25	3C336	20-100
3C215	strong jet	3C432	≥1000

This Table is constructed largely from VLA maps kindly sent to me by Alan Bridle. They were made with the highest possible dynamic range, in a search for counterjets. The estimates are not only rough but resolution-dependent, so only orders of magnitude are given in the table.

So, if we supposed that the vast majority of the electron acceleration took place at the "working surface" we might actually get *too much* radio emission in the hot spot. That is not to say that diffusive acceleration does not occur at the shock; it probably does. In fact, for electrons that are already highly relativistic, with γ = thousands, whose Larmor radius exceeds the shock thickness, adiabatic compression cannot be the right physical picture (though in some circumstances the results are similar - see e.g. Hudson, 1965). The point I want to stress is that it is probably wrong to think of the relativistic electron population as being made from essentially thermal plasma at the "working surface". Particle acceleration that occurs before (perhaps long before) the jet reaches the "working surface", is important; it may, for example, play a decisive part in determining the energy spectrum in the hot spot.

(ii) For the moment, let us assume that jets must be in pressure balance with the ex-hot spot material that surrounds them (variously called backflow, cocoon, ...). Then why can we see jets at all? If the fraction of pressure due to goodies (by which I mean magnetic field energy and relativistic electrons) is the same (or greater) in hot spots as that in jets, the luminosity per m^3 ought to be the same. But since the lobe diameter is ≥10 times the jet diameter, the jet must in fact have several times the lobe's volume emissivity to show up at all.

It could be argued that the comparison is not quite fair. Consider an adiabatic expansion from the hot spot pressure to lobe pressure; there is then a downward shift in the whole particle energy distribution, and we therefore get less emissivity at a given frequency than we should by comparing independent minimum energy calculations for each (with the same low-energy or low-frequency cut-off). If the bulk of the gas is at non-relativistic temperatures, this is more than made up for by the fact that, in an adiabatic expansion, the (ultrarelativistic, r=4/3) goodies cool less than the non-relativistic (r=5/3) gas. If (as I think more likely) the gas is relativistically hot, one loses a significant factor of

$$\text{(lobe pressure/hot spot pressure)}^x.$$

where x = 0.1±0.1, depending somewhat on the spectral index and the detailed assumptions fed into the calculation.

There is of course an entirely separate way to escape from the difficulty: the jet may be encircled by magnetic hoops, and not at all in pressure balance with its surroundings. I do not want to repeat the arguments for and against that idea here, as I am sure you are familiar with them.

I do want to talk about relativistic beaming, which is at first sight the most appealing argument against *both* heresies (i) and (ii). The argument is that relativistic beaming of the one jet we see gives it a much greater surface brightness than the intrinsic, rest-frame value. That argument can work for any one source, but not for a large fraction of sources. The ratio of surface brightnesses, (observed: intrinsic), is $(\gamma(1-\beta \cos\theta))^{-(2+\alpha)}$, and (unlike the comparison between jet and counterjet) we must not forget the factor γ here. Consequently in most sources the apparent brightness of even the *approaching* jet is "Doppler-enfeebled", not "Doppler-boosted".

Table 3 shows the percentage of (randomly oriented) sources in which the brighter jet is boosted by a factor of at least f by relativistic beaming, for various Lorentz factors γ. From the Table we see that, *regardless of the γ we assume*, less than 9% of jets are boosted by a factor 10 or more, and less than 20% are boosted even by a factor 4 or more. With the Lorentz factors >10 that are probably typical if we are to account for superluminal velocities, >90% of jets are made *fainter* by relativistic beaming. Thus relativistic beaming can account for the existence of some rather bright jets, but in the majority of cases it actually makes my argument *stronger*.

TABLE 3

f \ γ	1.1	1.3	1.5	2.0	3	5	10
0.1					80	47	24 %
0.5					40	25	13 %
1	78	64	55	42	29	18	10 %
2	27	36	34	29	21	14	7 %
3	3	22	25	22	17	11	6 %
4		14	18	18	15	10	5 %
6		4	11	14	12	8	5 %
8			6	11	10	7	4 %
10			3	8	8	6	4 %
15				5	6	5	3 %
20				3	5	4	3 %
40					3	3	2 %

(iii)   Finally, there is one crushing argument for the hot spots as the dominant site of particle acceleration: - the beautiful observations of synchrotron light by our hosts and organisers (See papers by Meisenheimer and Röser in this volume.) These short-lived particles must surely tell us where the particle accelerators really are, and they appear to be all over the hot spots, not even just in the working surface. That may be true; quite probably it is true, but let me conclude my list of doubts by showing you that it ain't necessarily true.

One of the obvious things that we were forced to notice by Alan Matthew's simulations is the tremendous variation in the amount of magnetic field stretching: in some places the input field is stretched to the limit where it becomes dynamically important, while in others it is not amplified much. A little thought shows this to be insensitive to the details of the fluid dynamics; it is a result of topology and common sense: wherever the fluid flow divides, as at a stagnation point in a steady flow, there is enormous stretching. And if you prefer to ignore theory altogether, a look at the best-resolved hot spots will show abundant filamentary structure, reaching right out into the lobes in Rick Perley's famous map of Cyg A. Now everyone knows that a region of low magnetic field can act as a "freezer" in which we can keep very energetic particles fresh for quite a long time. What may not be so widely appreciated is just how effective a freezer it can be. Suppose the particles inside one of the regions of stronger magnetic field, B, are showing signs of synchrotron loss at frequency ν_{max} after time t; thus

$$1/\gamma_{max} = C_1 B^2 t \text{ and therefore } \nu_{max} = C_2 B \gamma_{max}^2 = C_3 B^{-3} t^{-2}$$

(where C_1, C_2, and C_3 are universal constants). Particles on a weaker flux tube, at say 0.1 B, will at that time have maximum energy given by $1/\gamma_{max}^{*} = C_1 (0.1B)^2 t$ and therefore, when some of these particles diffuse into field B, they radiate up to frequencies

$$\nu_{max}^{*} = C_2 \ B \ (\gamma_{max}^{*})^2 = C_3 B \ (0.1B)^{-4} t^{-2} = 10^4 \nu_{max} \ .$$

For example, when most of the hot spot radiation is already showing a steepening spectrum at 10 GHz, particles which diffuse in from regions with only 0.1 of the typical magnetic field will radiate quite happily at 10^{14} Hz. As the field structure is almost certainly very complex, and there is probably a great deal of reconnection of field lines going on, there are plenty of possibilities for such diffusion.

4. Summary

1)   Though hot spots look a lot brighter than the jets feeding them, the ratio is not very much greater than passive adiabatic compression in a strong shock would have produced. The ratio predicted for adiabatic compression becomes greater (i) for a sequence of strong oblique shocks (ii) in most cases, if the effects of relativistic beaming are included.

2)   The fact that jets are visible against the background radiation from much deeper lobes at the same (?) pressure indicates that the "working surface" puts a smaller fraction of the energy dissipated into cosmic ray electrons than dissipative processes occurring further back (in the jet?).

3)   Inhomogeneity and diffusion could account for the synchrotron light observed in hot spots; the light does not necessarily occur precisely where the accelators are. (This may have particular rele-vance to the extended light-emitting region behind the preceding hot spot in Pictor A, which Dr. Röser showed at this meeting.)

References

Blandford, R.D. & Rees, M.J., 1974, Mon.Not.R.astr.Soc. **169**, 395.
Hudson, P.D., 1965, Mon.Not.R.astr.Soc. **131**, 23.
Meisenheimer, K., 1988, (This volume).
Röser, H.-J., 1988, (This volume).

Stefan Appl   Wolfgang Kundt

THE TERMINAL SHOCK IN JETS

S. Appl, M. Camenzind

Landessternwarte Königstuhl
D-6900 Heidelberg 1

1. Introduction

There is growing observational evidence that jets in FR II sources move with relativistic speeds /1/ and also strong support for the idea that magnetic fields are dynamically important, at least for relativistic jets /2/. In particular, the magnetic fields will change the nature of the shockfronts which are believed to exist in hot spots. These are commonly considered to be the downstream flow of such a shock resulting from the interaction of the jet with the IGM. When magnetic fields are involved in these shocks, collisionless effects are much more important than collisional ones.

In this context we are interested in the physical conditions in hot spots and study therefore the jump conditions of relativistic MHD shocks connecting the states upstream and downstream of the discontinuity. The details of this work are found in /3/. Assuming first order Fermi acceleration mechanisms, the observed synchrotron spectra are closely related to the jump in the physical quantities across the shockfront /4/.

2. Characteristic Speeds

The problem will be treated in the context of ideal relativistic MHD in a one-fluid approximation. Important for the occurence of shocks are the characteristic speeds in MHD, namely the Alfven speed (A) and the slow (SM) and fast (FM) magnetosonic speeds. As these are not simple extensions of the corresponding Newtonian expressions /5/, we give here the 4-velocities evaluated in the shock frame. The Alfven speed is a solution of

$$(1 + \frac{B'^2}{4\pi n \mu}) u_A^2 - \frac{B_n'^2}{4\pi n \mu} = 0 \quad , \tag{1}$$

and the magnetosonic speeds follow from the equation

$$(1 - c_s^2) u_{FM,SM}^4 - (c_s^2 + \frac{B'^2}{4\pi n \mu}) u_{FM,SM}^2 + c_s^2 \frac{B_n'^2}{4\pi n \mu} = 0 \quad , \tag{2}$$

where n is the particle density, μ the specific enthalpy, c_s the sound speed (in units of c) and B' the magnetic field $(B'_\alpha = u^\beta * F_{\beta\alpha})$, all measured in the plasma frame. $B'^2 = -B'^\alpha B'_\alpha$ and $B'_n = B'^\alpha l_\alpha$ is the component along the propagation direction l.

3. The Jump Conditions

The jump conditions across a shock front are derived from the conservation laws for particle number, energy-momentum and magnetic flux. We evaluate the shock frame compression ratio $r = n_2 \gamma_2 / n_1 \gamma_1$

$(\gamma = (1 - \beta^2)^{-1/2})$ as a function of the jet velocity β_1, σ (= Poynting flux/total mass-energy flux), a quasi-Newtonian Alfven Mach number $M = 4\pi n_1 \mu_1 u_1^2 / B_\parallel^2$ and the polytropic index of the downstream plasma. The unprimed $B_\parallel (B_\perp)$ denotes the magnetic field in the shock frame parallel (orthogonal) to the jet velocity, which is taken to be perpendicular to the shock front. The thermodynamic pressure in the jet is neglected against the ram pressure and the magnetic pressure. All downstream quantities can be derived from the compression ratio r, for which we get a polynomial of seventh degree in the most general case

$$\sum_{k=0}^{7} c_k(\gamma_1, \sigma, M, \Gamma) r^k = 0 \quad . \tag{3}$$

The coefficients c_k are themselves polynomials in the upstream variables γ_1, σ, M and the postshock polytropic index Γ. For a Synge gas Γ is given by /6/

$$\Gamma(\Theta) = 1 + \left(\frac{1}{\Theta} \left(\frac{K_1(1/\Theta)}{K_2(1/\Theta)} - 1 \right) + 3 \right)^{-1} \tag{4}$$

$(\Theta = kT/mc^2)$ and is therefore dependent on r. The nonlinear system consisting of equs. (3) and (4) is then solved numerically.

4. Fast Magnetosonic Shocks

Equs. (3) and (4) have in general several solutions corresponding to the different transitions possible in MHD. Fig. 1 shows the different transitions for various choices of parameters.

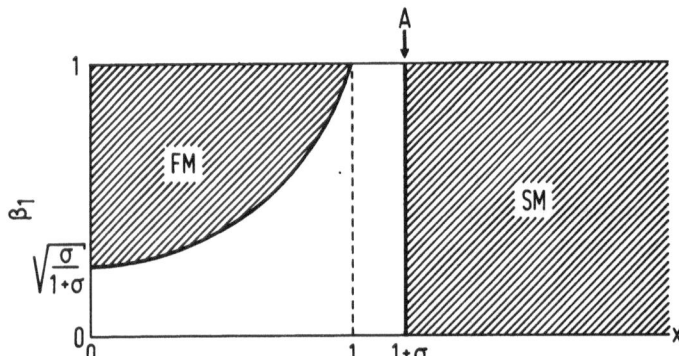

Fig. 1: Possible transitions in parameter space. FM, A and SM are the regions, where fast sonic shocks, Alfven shocks and slow sonic shocks are possible. σ is held fixed and $x = 1/M^2$.

The Alfven Mach number and the fast magnetosonic Mach number are related to our parameters via the following relations.

$$\frac{M_A^4}{M^4} - \frac{M_A^2}{M^2} \left(1 + \sigma + (M^{-2} - 1)\beta_1^2 \right) + \frac{\sigma \beta_1^2}{M^2} = 0 \tag{5}$$

$$M_{FM}^4 \left(\frac{1}{M^2} + \frac{\sigma}{\beta_1^2} \right) - M_{FM}^2 \left(1 + 2\sigma + (M^{-2} - 1)\beta_1^2 \right) + \sigma \beta_1^2 = 0 \tag{6}$$

M and M_A have the same Newtonian limes.

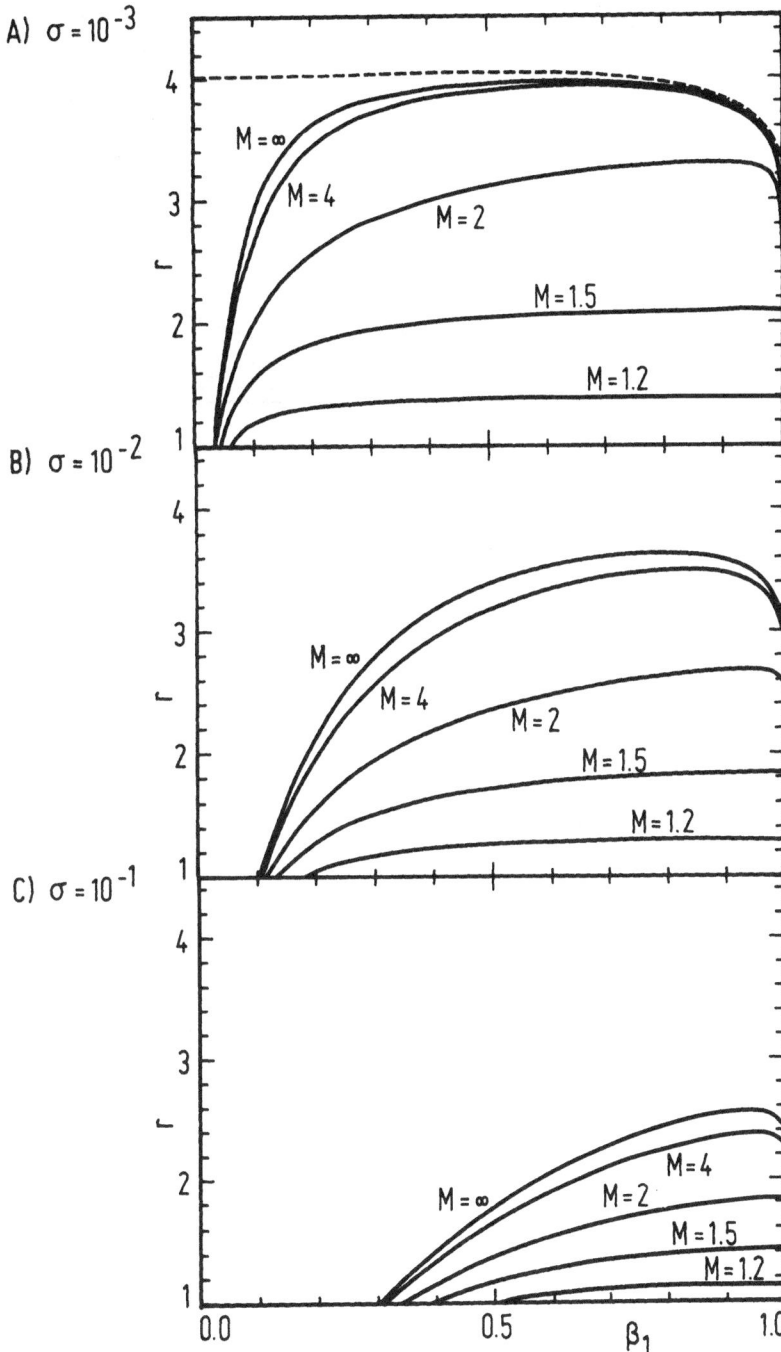

Fig. 2: A) - C) Shock frame compression ratio as a function of the jet velocity, for various upstream Mach numbers and Poynting fluxes. The dashed line in A) corresponds to the purely hydrodynamic case.

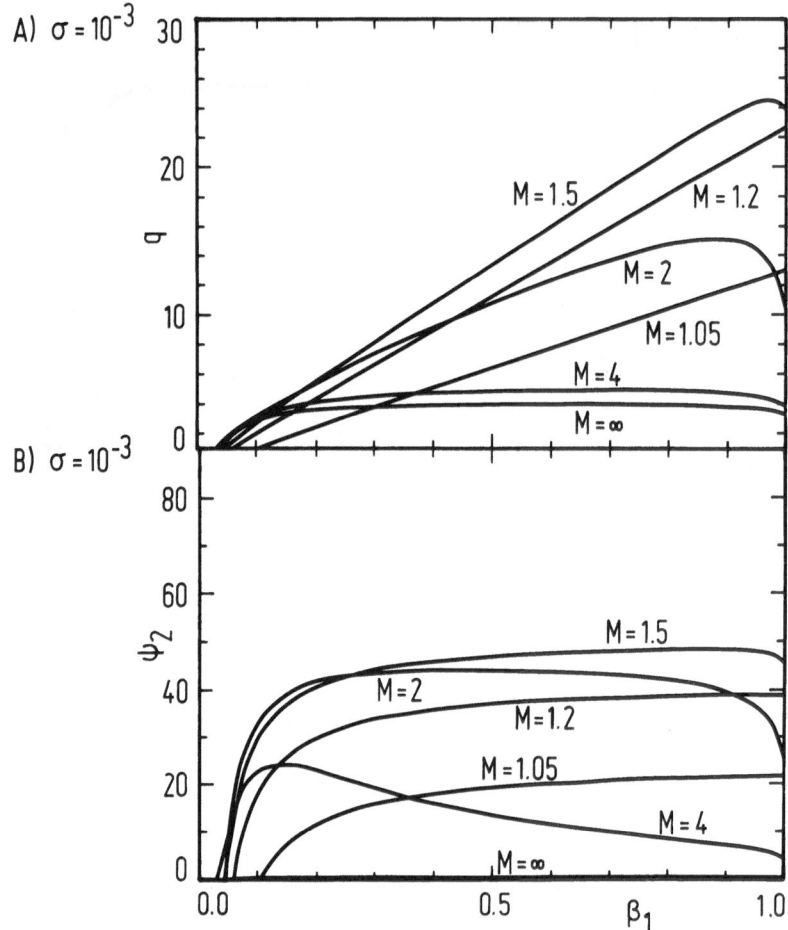

Fig. 3: A) Amplification of the transverse magnetic field in the shock
frame, B) Refraction angle between the upstream and downstream plasma
flow.

Furthermore we give an estimate of the magnitude of σ and M for the magnetic fields and densities
found in strong hot spots (e.g. Cyg A)

$$\sigma = \frac{B_{1\perp}^2}{4\pi n_1 \mu_1 \gamma_1^2} = 0.05 \left(\frac{B_{1\perp}}{0.1mG}\right)^2 \left(\frac{n_1^S}{10^{-5}cm^{-3}}\right)^{-1} \frac{1}{\gamma_1} \tag{7}$$

$$M = \frac{(4\pi n_1 \mu_1)^{1/2} u_1}{B_{1\parallel}} = 4.3 \left(\frac{B_{1\parallel}}{0.1mG}\right)^{-1} \left(\frac{n_1^S}{10^{-5}cm^{-3}}\right)^{1/2} \beta_1 \sqrt{\gamma_1} \quad , \tag{8}$$

where n_1^S means the jet density in the shock frame.

In Fig. 2 we show the compression ratio for different values of σ and M as a function of the
jet velocity. This figure demonstrates two important properties of fast magnetosonic shocks. First, a
threshold in β_1 occurs; for β_1 smaller than a critical value depending on σ and M, a fast magnetosonic
shock cannot occur. This threshold always exists, even for extremely small values of σ. Secondly, a

high Poynting flux reduces the shock strength considerably. This effect has already been pointed out in /7/ and /8/ in the context of pulsar physics. The compression of the plasma in the shock transition will also amplify the component the magnetic field parallel to the shock front. This amplification of the magnetic field, which in the case of a purely transverse field equals the compression ratio, is plotted in Fig. 3a. For small σ and M, the amplification of $B_\perp$ can exceed the compression ratio enormously. Finally, one sees from Fig. 3b that for these values of σ and M the plasma flow is refracted considerably through the shock. The refraction angle can be as much as $40° - 50°$. In an axisymmetric configuration, this would correspond to a relativistically rotating plasma cone.

5. Conclusions

We find that for reasonable choices of our parameters the compression ratio does not depart very much from the hydrodynamical value and stays between 3 and 4. If synchrotron electrons are produced via first order Fermi processes, then observations of spectral indices suggest compression ratios of about 4. If further the estimates (7) and (8) for σ and M are realistic, then according to Fig. 2 the jet velocities in FR II sources are relativistic ($> 0.4c$). Assuming $\sigma = 0.01$, $M = 5$ and $\beta = 0.5$, the pitch angle of the magnetic field in the jet is $45°$, but downstream through the amplification of the transverse field, it will be as much as $75°$. This is in accordance with optical observations, where a quasi-perpendicular magnetic field is seen in hot spots /9/. This could cause problems for first order Fermi acceleration mechanisms. The absence of hot spots in FR I sources could be due to either strong magnetic fields or velocities below the fast magnetosonic speed, in accordance with the opinion that these jets are subrelativistic /1/. In addition, if one assumes high Poynting fluxes in VLBI jets /2/, the observed knots are unlikely to be shocks. In our calculations we have neglected the contribution of the relativistic electrons which could increase the compression ratio somewhat.

This work was partially supported by the SFB 328. We would also like to thank H.J. Röser and K. Meisenheimer for valuable discussions.

6. Literature

1. P.A.G. Scheuer: In **Astrophysical Jets and their Engines**, ed. by W. Kundt,
   NATO ASI Series, Math. and Phys. Sciences, Vol. 208 (Reidel, Dordrecht 1987) p. 129
2. M. Camenzind: Astron. Astrophys. **162**, 32 (1987)
3. S. Appl, M. Camenzind: submitted to Astron. Astrophys. (1988)
4. J.G. Kirk, P. Schneider: Astrophys. J. **315**, 425 (1987)
5. A. Lichnerowicz: **Rel. Hydrod. and MHD**, Benjamin (N.Y. 1967)
6. J.L. Synge: **The Relativistic Gas**, North Holland (Amsterdam 1957)
7. W. Kundt, E. Krotschek: Astron. Astrophys. **83**, 1 (1982)
8. C.F. Kennel, F.V. Coroniti: Astrophys. J. **283**, 694 (1984)
9. K. Meisenheimer et al.: in preparation (1988)

Eduardo Trussoni   Paul Wiita

GIANT RADIO GALAXIES VIA INVERSE COMPTON WEAKENED JETS

Paul J. Wiita and Alexander Rosen
Department of Physics and Astronomy, Georgia State University
Atlanta, Georgia 30303
and
Gopal-Krishna and L. Saripalli
Radio Astronomy Group, Tata Institute of Fundamental Research
P.O. Box 1234, Indian Institute of Science Campus, Bangalore 560 012, India

ABSTRACT

Both analytical and numerical models for the propagation of relativistic jets through a hot interstellar medium (ISM) and into an even hotter intergalactic medium (IGM) have been considered. Earlier work allowed us to explain: the current mean linear-size of double radio sources (~350 kpc); the decrease in linear-size with cosmological redshift, z; and, the increase in linear-size with radio power (at fixed z). We now have extended our models to allow for intrinsically extremely powerful jets, which may start off advancing relativistically through the interstellar medium. Eventually the energy density in the lobes becomes comparable to that of the microwave background, and inverse Compton losses of the synchrotron emitting electrons against the background photons become important. We argue that only powerful radio engines are responsible for giant radio galaxies (GRGs, those whose linear size exceeds 1.5 Mpc) and we can also explain most of the observed peculiarities of the GRGs, such as their rarity, moderate radio flux and relatively strong radio cores.

I. INTRODUCTION

In recent papers (Gopal-Krishna and Wiita 1987 (GW), Wiita and Gopal-Krishna 1987, 1988) we combined observational constraints on the properties of galactic halo and intergalactic media with very simple analytical models for radio jets. Relativistic plasma with a relativistic bulk velocity comprises our model jets, but the speed of advance of the heads (or hot spots) of the jets, v_h, is assumed to be non-relativistic. X-ray observations of relatively isolated elliptical galaxies, which usually produce powerful (FR II) radio galaxies, indicate the presence of an extended, quasi-isothermal halo of hot gas ($kT_{ISM} \approx 1$ keV, $n(r) \approx 10^{-2} cm^{-3}[1+(r/2kpc)^2]^{-3/4}$; $e.g.$ Forman, Jones and Tucker 1985). The bulk of the x-ray background emission is probably best explained in terms of a uniform IGM (with $n \approx 7 \times 10^{-7} cm^{-3}$, $kT_{IGM} \approx 18$ keV; $e.g.$ Guilbert and Fabian 1986).

When such a jet reaches the pressure-matched interface between the halo and the IGM, at R_h, its propagation should be dramatically affected. We considered two limiting analytical models: Model A assumed that the initially constant opening angle, θ, of the jet is unchanged on crossing the interface so that the decrease in density there produces a big drop in the ram pressure slowing the head of the jet, which therefore accelerates; however, Model B had the velocity remaining continuous across the interface so that the jet would flare and actually slow down. We argued that the properties of the halo have not changed very much, at least for redshifts, $z \lesssim 1$ (GW). But because the IGM was both hotter and denser in the past, $R_h \approx R_h(z=0)(1+z)^{-10/3}$. In this earlier work we used average, observationally estimated, values for various parameters, *e.g.*, $\theta=0.04$ rad, $v_h \approx 10^4$km s^{-1}, as well as those given above, yielding $R_h(z=0) \approx 171$ kpc.

A fundamental assumption is that the hot spots and lobes emit much less radio energy once the nuclear engine turns off (at $t_N \approx 10^8$ yr) or the advance of the beam head becomes subsonic with respect to the external medium, whichever occurs first. Using results close to the geometric mean of Models A and B we find the average total linear size for nearby (3CR sample, with $\langle z \rangle \approx 0.15$) radio sources is $\langle 2D \rangle \approx 350$ kpc and the dependence on redshift is predicted to be $\langle 2D \rangle \propto (1+z)^{-3}$. Both this value of current linear size and its dependence on cosmological redshift are in very good agreement with recent analyses of observational datasets (*e.g.* Eales 1985, Gopal-Krishna *et al.* 1986, Kapahi 1985, Oort *et al.* 1987).

II. NUMERICAL MODELING

We have performed both extensive simple numerical simulations which follow the boundary between the jet plus swept up cocoon and the external media as well as preliminary runs using a sophisticated full two-dimensional hydrodynamics code. The boundary following code has been described in a series of papers (Wiita and Siah 1986, Siah and Wiita 1987, Mitteldorf 1987, Mitteldorf and Wiita 1988) and has been shown to quite accurately estimate the size and shape of relativistic jets plowing through external media of different types. Applying this code to over 200 situations involving jets propagating across an interface (Rosen and Wiita 1988a,b) we find excellent support for our analytical model. In particular, the linear-size *versus* redshift relation is strongly confirmed. This numerical work also gives a good match to the observed relation between linear size and radio luminosity, $\langle D \rangle \propto P^{0.3}$ (Kapahi 1986, Oort *et al.* 1987).

Still, the boundary following code cannot provide details of the flow patterns or radiation emission regions or any hope of understanding the hot spot structure. Therefore, it is extremely important to examine this picture in the greater detail made possible by using a proper hydro-code. We have used the *Zeus* code, developed by Norman and collaborators (*e.g.* Norman and Hardee, this volume) to examine the flow of a light (though non-relativistic) jet emerging through a plane-parallel power-law ISM into a constant density IGM. Although our results to date consist of only five abbreviated runs, we do apparently reproduce both the acceleration on crossing the interface predicted by Model A and the flaring predicted by Model B. The ultimate slowing down of the head found in both the analytical and boundary following approaches has not yet been reproduced in the hydrodynamical runs, but a test of this result requires running the code for much longer periods.

174

III. RELATIVISTIC BULK MOTION AND INVERSE COMPTON LOSSES

It must be stressed that our earlier work concentrated on the average properties of FR II radio galaxies and was based on the average observed properties of galactic halos and the IGM. To test our hypothesis that giant radio galaxies are formed by extremely powerful radio jets additional effects must be included. As the beam power, L_b, rises, v_h eventually becomes a significant fraction of c, at least for short periods, and we must generalize our earlier work. Using ram pressure balance for a relativistic beam we can show that (Gopal-Krishna, Wiita and Saripalli 1988 [GWS], Wiita and Gopal-Krishna 1988) within the halo,

$$v_h(D) = cX[1+(D/a)^2]^{3/8}/\{D+X[1+(D/a)^2]^{3/8}\}, \tag{1}$$

with D the distance of the hot spot from the center of the galaxy and
$X = (2/\theta)(L_b/\pi c^3 n_0 m_H \mu)^{1/2}$. After crossing the interface detailed results depend upon the choice of Model A (constant θ) or Model B (continuous v_h):

$$v_h(D) = cX_{A,B}/(D+X_{A,B}), \text{ with } X_A = X\{n_0/[n_{IGM}(0)(1+z)^3], X_B = X(R_h(z)/a)^{3/4}, \tag{2}$$

$$D_A(t) = \{X_A^2 + 2X_Act + R_h^2(z)[1-8X_A^2/(5XX_B)]\}^{1/2} - X_A, \tag{3}$$

$$D_B(t) = [X_B^2 + 2X_Bct - 3R_h^2(z)/5]^{1/2} - X_B. \tag{4}$$

At sufficiently long times we find $D \propto t^{1/2}$, $D \propto L_b^{1/4}$ and $v_h \propto 1/D$, although these limits are often not reached because the advance of the head becomes subsonic or $t_{lim} > t_N$, the period that the nucleus (and beam) remains active.

In estimating the radio emission we will retain the standard equipartition assumption, but modify an assumption implicitly made in most discussions of extended radio sources, which is that the radio emission, P, is a fixed fraction, ϵ of the total beam power ($2L_b$, in our notation). For, in reality, the fraction of power emitted in the radio band will decrease due to inverse Compton scattering of the relativistic electrons off the cosmic background photons. Both very weak and very extended sources can have the magnetic energy density in their lobes, u_B, approach the energy density of the microwave background, $u_{ph}(z) = 4.0 \times 10^{-13}(1+z)^4$ erg cm^{-3}. When this occurs, the efficiency at which a radio source emits radio flux decreases, while its x-ray flux rises. Although this basic point has been made earlier (*e.g.* Rees and Setti 1968, Eales 1985), we note that it is particularly applicable to very extended double radio sources since a large fraction of their emission comes from their lobes and not their hot spots. In the lobes, u_B is lower than in the hot spots, so this inverse Compton "reduced radio efficiency" (RRE) is more important (GWS).

The effect of RRE can be approximated by using observations that indicate the average magnetic energy density of a source, $\langle u_B \rangle \approx u_{B,h}/10$ (Saripalli 1987), with $u_{B,h}$ the magnetic energy density in the hot spot. Then, using equipartition and ram pressure confinement of the hot spot we find (GWS) $\langle u_B \rangle \approx (9/70) \rho_{ext} v_h^2$. However, we must also note that $\langle u_B \rangle$ cannot fall below the limit set by the static confinement of the lobe due to thermal pressure of the external medium, so $\langle u_B \rangle_{min} = (9/14)p_{ext}$. Then we can write

$$L_R(t) = \epsilon'(t)L_R(t=0) = \epsilon'(t) \epsilon 2L_b, \text{ with} \tag{5}$$

$$\epsilon'(t) = L_R(t)/[L_R(t)+L_X(t)] = [1 + (u_{ph}(z)/\langle u_B \rangle)]^{-1}, \tag{6}$$

with the larger of the two values of $\langle u_B \rangle$ inserted in (6), for radio power at any stage for our beam model. The observable radio flux density, S_y (in Jy) is then found if the distance of the source, a cosmological model (values of H_0 and q_0) and a spectral index are specified.

IV. GIANT RADIO GALAXIES

Under a score of known extragalactic radio sources have total linear extents, 2D, exceeding 1.5 Mpc (with H_0 taken as 50 km s^{-1}Mpc^{-1}), thereby qualifying them as giant radio galaxies (GRGs). These sources have other properties in common besides their rarity and immense sizes (Saripalli *et al.* 1986). Their radio luminosities cluster in a narrow range ($P_{408} = 10^{25}$–10^{27} W Hz^{-1} at 408 MHz), just above the FR luminosity transition. GRGs also have a higher ratio of core radio luminosity to total radio luminosity than do typical double radio galaxies, although the ratio for quasars is even larger.

Using the beam model, the explanation for the modest (observed) powers, despite the huge sizes of the GRGs, could be understood in two ways: either GRGs are associated with moderately powered, but very long-lived, nuclear engines, or GRGs are produced by powerful engines over relatively short times, but with lowered radio emission from the lobes. Our analytical and numerical models strongly support the second hypothesis.

We have computed v_h, D, L_R, and S_y as functions of time for models with total beam powers ($2L_b$) in the range 10^{36} W through 10^{41} W and redshifts between 0 and 1. Weak sources never become GRGs because they either become subsonic in the IGM shortly after leaving the halo or because their initial advance is so slow that they take more than t_N (chosen as 1–3×10^8 yr) to reach linear distances of D $\geqslant$ 750 kpc. While beams with somewhat higher intrinsic powers can make it out that far without fading completely, we still won't observe their sources as GRGs. The effect of inverse Compton losses on such spread-out sources can cut the actual L_R to the extent that the observed radio flux, S_y, falls below a detection threshold. For example, the largest sensitive survey used for discovering GRGs employed the 6C synthesis array at 151 MHz which appears to be confusion limited below a flux density of ~5 Jy (Saunders, Baldwin and Warner 1987). In our papers we refer all flux estimates to 1 GHz, using a spectral index of $\alpha = -0.8$ ($S_y \propto v^\alpha$), so this limit corresponds to a minimum detectable flux density of $S_{lim} \approx 1$ Jy at 1 GHz.

Obviously, the fraction of currently detectable giant sources would be smaller at earlier cosmological epochs. A greater distance yields a smaller S_y for a given L_R, and because the microwave background energy density was higher in the past, we expect the RRE to have set in sooner, cutting down L_R. Besides, our basic model for propagation across the halo/IGM implies R_h was smaller at earlier epochs and c_s was larger, further reducing the number of possible GRGs. We find the minimum beam power needed to produce an observable GRG is about $10^{37.5}$ W for a redshift of only 0.05; but, for z = 0.45, we require about $10^{39.2}$ W.

The final goals of these computations are the total numbers and powers of observable GRGs in each redshift interval. To obtain them, we convolve our estimates of the minimum powers required for the formation of an observable GRG at a given redshift as well as their sizes and radio flux densities with the appropriate radio luminosity function. We use the recent determination of the co-moving density of steep spectrum radio sources given by Peacock (1985) for $q_0 = 0$ (the final results are nearly the same for $q_0 = 1/2$) to obtain the radio luminosity function (RLF), $\rho(P,z)$. Unfortunately, this is not immediately useful, for this RLF is based upon *observed* radio fluxes, while our models predict $P(t)$ for initial P (the flux corresponding to $L_R(0) = \epsilon 2 L_b$). Fortunately, we can effectively redefine the RLF as a function of L_b instead of P because log P remains nearly constant and $\approx$ log $(2\epsilon L_b)$ until a time $t_{1/2}$, after which RRE causes a rapid drop of over an order of magnitude. So we approximate $P = P_0$ for $t \leqslant t_{1/2}$ and $P = 0$ for $t > t_{1/2}$. These arguments lead to (GWS)

$$N(z)\Delta z = \Delta V(z) \int_{P_*}^{\infty} [t_*(P,z) - t_G(P,z)]\ \rho(P,z)/t_{1/2}(P,z)\ dP\ \Delta z, \tag{7}$$

with P_* the minimum initially radiated power required to produce an observable GRG in a given Δz bin and $\Delta V(z)$ the volume contained in that bin (*cf.* Windhorst 1984).

We can also predict the mean radio luminosity, $\langle P \rangle$, of GRGs as a function of redshift by considering the emitted radio power, P_{1Mpc}, at the value D = 1 Mpc, which is the typical half-size of known GRGs (Saripalli *et al.* 1986). We find $\langle P \rangle$ by inserting P_{1Mpc} into the integrand of the RHS of (7) and dividing through by $N(z)\Delta z$. The results for the geometrical mean between Models A and B (see §I and Rosen and Wiita 1988a,b) are compared with observations below.

z-range	$\Delta V(Gpc^3)$	N_{pred}	N_{obs}	$\log\langle P(W/Hz)\rangle_{pred}$	$\log\langle P(W/Hz)\rangle_{obs}$
0.0–0.1	0.79	13	12	26.0	25.6
0.1–0.2	4.74	14	4	26.6	26.4
0.2–0.3	11.1	12	2	26.9	26.7
0.3–0.4	18.7	10	0	27.3	----
0.4–0.5	27.0	7	0	27.5	----

The agreements at low redshifts are certainly very good, and we predict the existence of discoverable GRGs at moderate redshifts.

V. CONCLUSIONS

A simple model, essentially analytic, but buttressed by numerical simulations, seems to provide an excellent way to understand many facts about extragalactic radio sources. It is based on relativistic jets emerging through power-law atmospheres and entering a hotter IGM. Most simply (GW), this picture yields good values for the typical current linear size of double radio galaxies as well as for the linear-size *vs.* redshift relation, using only the single, best supported by observation, value of each parameter that enters into the model. The approximate numerical models verified the basic picture and also provide a good fit to the linear-size *vs.* power relation (Rosen and Wiita 1988a,b). One prediction of this scenario is that AXAF should find smaller galactic halos at larger redshifts because the IGM exerted much higher pressures in the past.

Including the inverse Compton losses of the synchrotron emitting relativistic electrons against the microwave background and modeling the initially relativistic motion of the hot spots we can argue that GRGs are indeed produced by intrinsically powerful AGNs whose radio luminosities suffer declines of about an order-of-magnitude between the time of their turning-on and their qualifying as giants. This single idea thus explains the key facts about GRGs: their moderate apparent powers; their rarity; and their relatively high core/lobe flux ratio. Further, when we consider what should happen to weak beams that actually become subsonic within the extended galactic halos we also appear to be able to explain why there is a break in the local radio luminosity function (Gopal-Krishna and Wiita 1988, Wiita and Gopal-Krishna 1988). We thus feel that these results are very strong evidence in favor of the fundamental propositions that most strong jets are predominantly relativistic even on large scales, that extended hot galactic halos are ubiquitous, and that a cosmologically significant ($\Omega \approx 0.25$) IGM exists.

REFERENCES

Eales, S.A. 1985, *Mon. Not. Roy. astr. Soc.*, (*M.N.R.A.S.*) **217**, 179.

Forman, W., Jones, C. and Tucker, W. 1985, *Astrophys. J.*, **293**, 102.

Gopal-Krishna, Saripalli, L., Saikia, D.J. and Sramek, R.A. 1986, in **Quasars, IAU Symposium No. 119**, eds. G. Swarup and V.K. Kapahi (Reidel: Dordrecht), p. 193.

Gopal-Krishna and Wiita, P.J. 1987, *M.N.R.A.S.*, **226**, 531 (GW).

Gopal-Krishna and Wiita, P.J. 1988, *Nature*, in press.

Gopal-Krishna, Wiita, P.J. and Saripalli, L. 1988, *M.N.R.A.S.* in press (GWS).

Guilbert, P.W. and Fabian, A.C. 1986, *M.N.R.A.S.*, **220**, 439.

Kapahi, V.K. 1985, *M.N.R.A.S.*, **214**, 19P.

Kapahi, V.K. 1986, *Highlights of Astronomy*, **7**, 371.

Mitteldorf, J. 1987, *Ph. D. thesis, University of Pennsylvania.*

Mitteldorf, J. and Wiita, P.J. 1988, in **Active Galactic Nuclei: Georgia State U. Conf.**, eds. H.R. Miller and P.J. Wiita (Springer Verlag: Berlin), p.378.

Oort, M.J.A., Katgert, P., Steeman, F.W.H., and Windhorst, R.A. 1987, *Astron. Astrophys.* (*A&A*), **179**, 41.

Peacock, J.A. 1985, *M.N.R.A.S.*, **217**, 601.

Rees, M.J. and Setti, G. 1968, *Nature*, **219**, 127.

Rosen, A. and Wiita, P.J. 1988a, *Astrophys. J.*, **330**, in press.

Rosen, A. and Wiita, P.J. 1988b, in **Active Galactic Nuclei**, *op cit* p.383.

Saripalli, L. 1987, *Ph. D. thesis, Indian Institute of Science* (in preparation).

Saripalli, L., Gopal-Krishna, Reich, W. and Kühr, H. 1986, *A&A*, **170**, 20.

Saunders, R., Baldwin, J.E. and Warner, P.J. 1987, *M.N.R.A.S.*, **225**, 713.

Siah, M.J. and Wiita, P.J. 1987, *Astrophys. J.*, **313**, 623.

Wiita, P.J. 1978, *Astrophys. J.*, **221**, 436.

Wiita, P.J. and Gopal-Krishna 1987, in **13th Texas Symposium on Relativistic Astrophysics**, ed. M.P. Ulmer (World Scientific: Singapore), p. 355.

Wiita, P.J. and Gopal-Krishna 1988, in **Active Galactic Nuclei**, *op cit*, p.388.

Wiita, P.J. and Siah, M.J. 1986, *Astrophys. J.*, **300**, 605.

Windhorst, R.A. 1984, *Ph. D. thesis, University of Leiden.*

JET SPEED, BEAMING & SIDEDNESS, AND ALL THAT

Wolfgang Kundt
Institut für Astrophysik der Universität Bonn
Auf dem Hügel 71, D-5300 Bonn 1

Abstract: It is argued that the astrophysical jets consist of quasi-monoenergetic relativistic pair plasma, of Lorentz factor $\gamma = 10^{5\pm3}$, performing a quasi-lossfree $\vec{E} \times \vec{B}$-drift in equi-pressure convected fields. The Lorentz factor derived from the naive beaming formula (in unified schemes) has no physical meaning. Hydrodynamical and hydromagnetic jet simulations agree with reality in as much as that they satisfy the conservation laws. Multiple hotspots may result from beam swinging.

1. Estimates of jet speeds

In a few (extragalactic) cases, velocities of jet material have been measured via optical emission lines; the speeds range from hundreds of Km s^{-1} to $\lesssim 10^3$ Km s^{-1}. Such speeds may belong to condensed halo matter, the 'channel walls'; they may not tell us anything about the propagation speed of the (true) jet substance.

Another approach to jet speeds is by pulling the square root from the ratio of some equipartition pressure and a mass density derived from depolarization data. Such speeds can be meaningless if the mass density of the jet substance is some 10^8 times lower than that of its (depolarizing) walls. A similar remark applies to jet speeds derived under the assumption that gravity can influence the jet shapes.

That the jet substance moves relativistically is already suggested by the following resonings [Kundt & Gopal-Krishna, 1981]:

1) The largest jets have projected lengths up to 10^7 lightyears; their central engines are still active. If the supply were nonrelativistic, these sources would be much older than 10^8 yr, in conflict with fuelling demands.

2) For given power, the rigidity of a supersonic beam scales as β_{beam}^{-1}. Eilek et al (1984) have shown that (in particular) the jets of the radio galaxy 3C 465 would not be bent if β_{beam} were $\ll 1$.

3) Seemingly one-sided sources, like 3C 273, ask for head speeds $\beta_{head} \gtrsim 0.7$ [Kundt & Gopal-Krishna, 1986]. For 1510-089, O'Dea [1988] has even found $\beta_{head} \gtrsim 0.9$. Clearly, such enormous head speeds demand relativistic ramming.

4) The sidedness of jets and hotspots, with the sharp structure on the side with the lower depolarization and rotation measure, is a signature of relativistic beaming.

<u>5</u>) There are indications that 'superluminal' motions are not restricted to the scale of $\lesssim 10$ pc but can reach and exceed 10^2 pc (Eckart et al [1985] , Waak et al [1985]). In view of their difficult detectability, they may be the rule rather than the exception. Superluminal appearance with non-relativistic supply would ask for Maxwell's demon.

It is more difficult to arrive at reliable estimates for the bulk Lorentz factor of the flow. In the literature, it is often concluded that relativistic electrons can be accelerated efficiently wherever observed, to Lorentz factors γ exceeding 10^6. such 'conclusions' are based on lifetime arguments of the radiating charges. Instead, the jet of the quasar 3C 273 manifests in-situ deceleration: the radio-to-X-ray spectrum softens almost monotonically from the inner to the outer end of the jet, [Kundt & Gopal-Krishna, 1986]; for slight deviations see Hayes & Sadun [1987]. When I compare the efficiency of sub-relativistic plasma shocks with that of the central supermassive magnetised rotator (near its speed-of-light cylinder), I cannot be convinced of the relevance of in-situ acceleration [Kundt, 1987]. Presently, worst cases are the extended optical hotspot of Pictor A [Röser et al, these proceedings] which asks for continual replenishment (by a swinging beam) and the optical jet of PKS 2152-69 discussed by the ESO group (in these proceedings).

If in-situ acceleration is ignorable, the bulk Lorentz factor in the beams must (essentially) equal the average Lorentz factor of the radiating electrons (and positrons). This leads to the estimate $\gamma = 10^{5\pm3}$ (for different sources). This conclusion is consistent with the following further considerations:

<u>1</u>) Individual superluminal sources imply minimum bulk Lorentz factors $\gamma \lesssim 20$ (50 Km s^{-1} Mpc^{-1}/H$_o$), but their (cumulative) distribution suggests $\gamma \gtrsim 50$ H$_{-17.8}^{-1}$ (based on Porcas [1987]). Their brightness distribution suggests $\gamma \gg 20$ (Barthel, these proceedings).

<u>2</u>) Both Sgr A* and a $\lesssim 10^2$ pc-region around our Galactic center emit synchrotron radiation corresponding to electrons (and positrons) of Lorentz factor $\gamma \gtrsim 10^4$ [Reich et al, 1988]. The radio spectral index $\alpha := \partial \log L_{\nu}/\partial \ln \nu = 0.3$ is equally found in the cores of M 81, M 124, and 3C 123.

<u>3</u>) Lorentz factors $\gamma > 10^3$ are inferred from the radio luminosities of several hotspots, including those of Cyg A.

<u>4</u>) AGN spectra $\nu L_{\nu}(\nu)$ often peak near 1 MeV, reminiscent of relativistic pair creation. Since these pairs find themselves inside the BLR whose pressure exceeds the average galactic pressure by a factor of $\gtrsim 10^{11}$, they are likely to be extremely relativistic (still) on escape.

<u>5</u>) The bolometric lobe/core power ratio is $10^{-2\pm2}$, corresponding to an average 1% efficiency of jet formation (which is reasonable). If protons were the compensating charges of electrons, this efficiency would have to be multiplied by $m_p/m_e = 10^{3.3}$. I conclude that the jet substance is only mildly polluted by protons.

2. Properties of pair-plasma jets

Once we take pair-plasma jets seriously, their (low) particle number densities n can be calculated:

$$n = L/Ac\gamma m_e c^2 = 10^{-9.5} \ cm^{-3} \ (L/A)_o \ \gamma_5^{-1}, \qquad (1)$$

where L = jet power, A = jet cross section. Their mean free path λ_{free} under Coulomb collisions is huge:

$$\lambda_{free} = 1/n\sigma \ \approx \ 10^{45} \gamma_5^2 \ cm \ , \qquad (2)$$

i.e. collisions (and annihilations) do not occur outside the sources.

Instead, the charges will communicate via their convected fields. Such fields are likely to be left-over fields from the central booster. They could, however, also be regenerated by charge-asymmetric interactions with the (quasi-cylindrical) channel walls. Once there are toroidal magnetic fields, the required radial electric fields are a consequence of the Hall effect. In this way, equi-pressure coexistence of $\vec{E} \times \vec{B}$-drifting pairs and their guiding fields may well be a stable configuration. Additional beam-parallel $\vec{B}$-fields are not excluded. Phase coherence of fields and charges fixes the bulk speed at $\beta = E/B_\perp$; consequently, along every streamline β must be uniform. Charges with different (bulk) Lorentz factors γ - if they persist - may segregate at different radii (from the cylinder axis). Clearly, jets of this constitution are not well modeled by hydrodynamic or hydromagnetic simulations.

In real life, (unperturbed) loss-free motion is only an approximation. Even if magnetic fields can isolate the jet substance from its partially ionised surroundings, photons and neutral particles will enter the beam. The 3K background radiation can seriously reduce the power of a (long) jet if its bulk Lorentz factor is too high, and neutral hydrogen atoms will reduce the pair power of jets in (gas-rich) spiral galaxies.

With this proviso, beams won't radiate unless they hit obstacles. This expectation agrees well with the presence of emission gaps in jets and with the bizarre morphology of the jet in M 87 presented by Owen at this workshop - which shows edge brightening, central darkening, bright and dark spirals, and beam flattening.

Are 'infinite' Lorentz factors compatible with wide-angle beaming? An answer to this question has been given by Kundt & Gopal-Krishna [1981], see also Kundt [1982,

1987], and independently by Lind & Blandford [1985]: We see a distribution of particle velocities which is far from isotropic in any Lorentz frame. Thermal obstacles (of small volume-filling factor) divert a small minority of pairs through large angles - like the kick-magnets in terrestrial particle accelerators - whose excessive radiation makes the beam visible, cf. figure 1. The angular spread of the emitted radiation is unrelated to any particular Lorentz factor; it reflects the distibution of perturbed particle orbits.

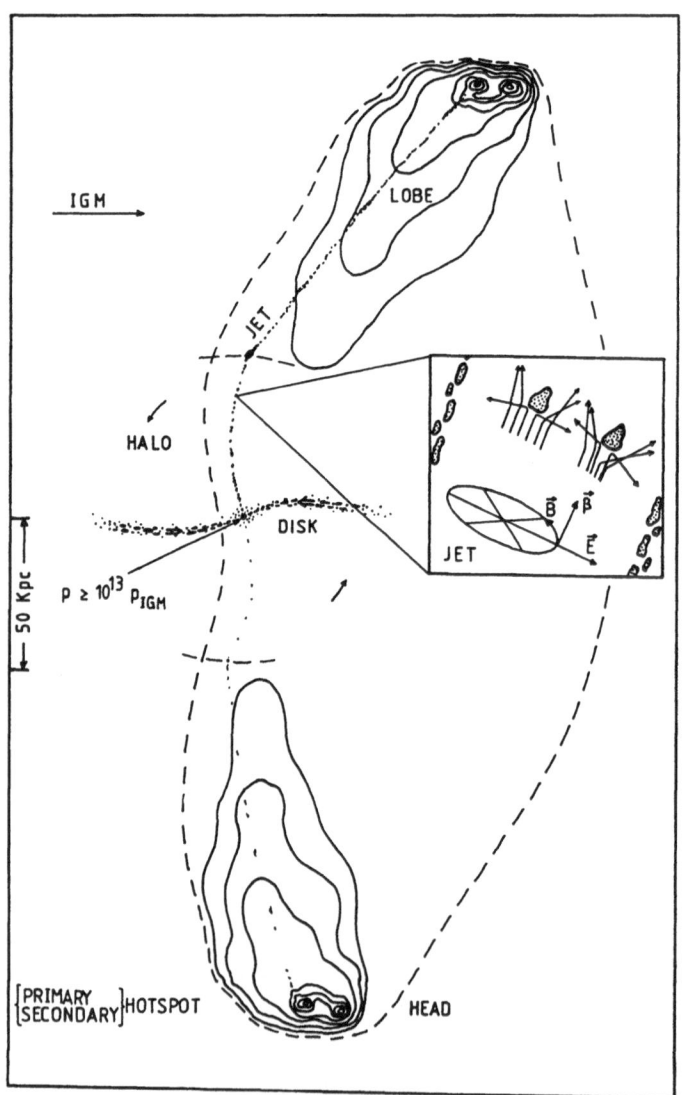

Figure 1: Scheme of a radio galaxy (of class FR II) with swinging beams. $e^{\pm}$-pair plasma with Lorentz factor $\gamma = 10^{5\pm3}$ is squeezed out of the central overpressure region and $\vec{E} \times \vec{B}$-drifts down the jet, all the way to the outer primary hotspot. A minority of pairs still arrive at the earlier dumping place (= the secondary hotspot) and the bridge region. The channel walls consist of whatever matter occupied that region before the ramming jet enhanced the pressure. Synchrotron radiation is emitted by charges which are deflected off heavy obstacles, preferentially into the forward hemisphere.

3. Further jet properties

Are there any intrinsicly one-sided sources, i.e. sources with only one jet? The best example would be 3C 273. But I see no way to block the extreme pressure near the central engine towards one of the two half spaces defined by the feeding disk: How should the shutter be realised? Moreover, too many sources show comparable radio powers from two lobes, and sidedness of compact features has never been seen to flip.

A major concern of this workshop has been the existence of primary and secondary hotspots (as defined by Laing): The primary ones are the endpoints of the jet (if such is seen). They are more compact ($\lesssim 0.5$ Kpc to 2 Kpc), located on the edge of the lobe, have equipartition pressures $p \lesssim 10^{-6}$ dyn cm^{-2} $H_{-17.5}^{4/7}$, less flux and total energy ($\leq 10^{-1}$) than the secondary hotspots, and a harder spectrum ($\Delta\alpha = 0.3$). Why are there secondary hotspots at all? Laing sketched four typical geometries of primary and secondary hotspots most of which showed a faint, concave emission bridge connecting the two. But Cyg A looks different again: its 'bridges' between the hotspots look convex (in projection). If explained by beam-bending, the beam would have to bounce (which is difficult).

If Cyg A can lead the way, I prefer an explanation by beam swinging for which there is ample evidence in a large number of sources, among them 3C 273 [Kundt & Gopal-Krishna, 1986] and Cyg A [Kundt & Saripalli, 1987]. In this interpretation, evolving jets tend to straighten, whereby the beam snaps sideways through its cocoon. The secondary hotspots are the older beam-dumping sites; they have served for longer epochs, hence have larger energies, softer spectra, and magnetic fields along their edges. The bridges mark intermediate termination sites of the swinging beam. In Cyg A, the former jet channel, leading to hotspot A, can still be recognised.

A further point of concern is the width of the jets. Bridle [1984, 1986] has found the empirical law $\Delta\Phi \sim P^{-1}$ (after slight 'refinement' by Gopal-Krishna) for the beam opening angle Φ , or more quantitatively:

$$\Phi = 0.02 - 0.03 \log (P_{\nu}/10^{28} \text{ Watt s }) \tag{3}$$

where P_{ν} is the total spectral power at 1.4 GHz. Apparently, the stronger the power the narrower the beam (and hence the faster the head). Can this law account for O'Dea's fast jet 1510-089?

This communication has been devoted to the extragalactic jet sources. Sould the reader have had the stellar jet sources in the back of her mind, she would not be largely mistaken: Blome & Kundt [1988] apply essentially the same model to the bipolar flows. The most important difference of the stellar sources is a much denser ambient medium.

Acknowledgement: My thanks go to Gopal Krishna for stimulating conversations.

References

Blome, H.J., Kundt, W., 1988: Astrophys. Sp.Sci., submitted
Bridle, A.H., 1984: Astron. J. 89, 979
Bridle, A.H., 1986: Canad. J. Phys. 64, 353
Eckart, A., Witzel, A., Biermann, P., Pearson, T.J., Readhead, A.C.S., Johnston,
    K.J., 1985: Astrophys. J. 296, L 23
Eilek, J.A., Burns, J.O., O'Dea, C.P., Owen, F.N.: 1984, Astrophys. J. 278, 37
Hayes, J.J.E., Sadun, A.C., 1987: Astron. J. 94, 871
Kundt, W., 1982, in: Extragalactic Radio Sources, IAU 97, eds Heeschen & Wade,
    Reidel, p. 265
Kundt, W., 1984: J. Astrophys. Astron. 5, 277
Kundt, W., 1987: Astrophysical Jets and their Engines, NATO ASI C 208, ed. Kundt,
    Reidel, p. 1
Kundt, W., Gopal-Krishna, 1981: Astrophys. Sp. Sci. 75, 257
Kundt, W., Gopal-Krishna, 1986: J. Astrophys. Astron. 7, 225
Kundt, W., Saripalli, L., 1987: J. Astrophys. Astron. 8, 211
Lind, K.R., Blandford, R.D., 1985: Astrophys. J. 295. 358
Morrison, Ph., 1981: Evening Lecture, Socorro, August
O'Dea, C.P., 1988: these proceedings
Porcas, R. 1987: private communication
Reich, W., Sofue, Y., Wielebinski, R., Seiradakis, J.H., 1988: Astron. Astrophys.
    191, 303
Waak, J.A., Spencer, J.H., Johnston, K.J., Simon, R.S., 1985: Astron. J. 90, 1989

HOW IMPORTANT ARE CURRENTS AND FIELDS IN RADIO SOURCES?

Jean A. Eilek
New Mexico Tech, Socorro, NM, USA

ABSTRACT

While fluid dynamical models have had quite a bit of success in explaining large-scale properties of radio sources, several small-scale and microphysical phenomena seem to suggest or even to require electrodynamic processes. I discuss four such mechanisms – particle injection, magnetic confinement, tearing instabilities as the origin of luminous filaments, and twisting of current-carrying jets. From this evidence, I suggest that currents and fields may well be as important as hydrodynamics in the large-sale dynamics and energetics of the sources.

Several authors over the years have suggested the possible importance of currents and fields in radio sources. For instance, dynamo models have been proposed for the core source (*e.g.*, Lovelace *et al.*, 1987); Benford (1978) suggested that the jets are current-carrying beams. Magnetic confinement has been suggested (*e.g.*, Benford, 1978), and force-free, relaxed magnetic field configurations have been suggested by Konigl and Choudhuri (1985). Borovsky (1986) has considered the energetics and dynamics of current-carrying jets. If the currents and fields are important dynamically or energetically, electrodynamics ("ED") could have major effects on the structure and evolution of radio sources (as is the case in, for instance, solar flares and the earth's magnetotail).

Now, no complete ED models of radio sources exist yet, to my knowledge; nor do complete hydrodynamic ("HD") models yet exist, although much more progress has been made in this area. Although it is easy to build qualitative HD or ED scenarios, we do not seem to have a quantitative test of the relative importance of HD or ED in any object. Nonetheless, my feeling is that several small-scale and microphysical phenomena strongly suggest that ED effects are important, or even critical, in the sources. In this contribution I discuss four such effects: the initial injection of relativistic electrons, magnetic self-confinement, the origin of the luminous filaments, and twisted, current-carrying beams. None of these four subjects address explicitly the dynamics of hot spots, the topic of this meeting, although electron injection can certainly be taking place in the hot spots. However, all four topics are consistent with a picture in which a net current propagates down the beam, and returns "to mother" (Benford, 1984) *via* the working surface/hot spot, and through the lobe or on the surface of the lobe.

A. INJECTION OF RELATIVISTIC ELECTRONS

The first topic is the initial acceleration of the relativistic electrons which are seen *via* their radio (and sometimes optical) synchrotron radiation. Since we see synchrotron radiation from the cores and the jets of the sources, we know that some acceleration must be taking place in the core. But also, unless the magnetic field has a very non-umiform structure within the jet (*e.g.*, Owen, Scheuer, these proceedings), the electrons are also likely to need *in situ* acceleration or reacceleration at the hot spots. This is especially true in those objects with optical synchrotron radiation (Röser, these proceedings).

Now, shock and turbulent acceleration are the most commonly invoked mechanisms for accelerating the relativistic electrons in extended radio sources. The electrons might be accelerated from a cool, thermal background such as the ISM or ICM which feeds the source. However, neither shock-Fermi nor turbulent acceleration will pick up subrelativistic electrons and accelerate them to high energies; these processes only act on electrons which are already mildly relativistic.

The reasons for this are as follows (*c.f.* Eilek and Hughes, 1988, for a fuller discussion). Both shock and turbulent acceleration rely on pitch-angle scattering of the particles by resonant Alfven waves; these waves may be self-generated for shock acceleration, but must be externally generated for turbulent acceleration. The resonance condition necessary for an electron of momentum $p = \gamma \beta m c$ to "see" a wave of wavenumber k and frequency $\omega = k_{\parallel} v_A$ (where v_A is the Alfven velocity) is

$$\omega - k_{\parallel} v_{\parallel} + \frac{\Omega_e}{\gamma} = 0$$

(where $k_{\parallel}$ and $v_{\parallel}$ are the components of k and $v = \beta c$ along the magnetic field; $\Omega_e = eB/m_e c$ is the electron cyclotron frequency. But Alfven waves only exist for frequencies $\omega \leq \Omega_i$, the ion cyclotron frequency. This limit translates to a minimum electron energy, γ, that can resonate with Alfven waves. For a low-density plasma, with $v_A/c \gg m_e/m_i$, this limit is

$$\gamma \geq \gamma_{min} = \frac{v_A}{c} \frac{m_i}{m_e}$$

This is the "injection problem": electrons with $\gamma < \gamma_{min}$ will not interact with Alfven waves, so cannot be accelerated by shock-Fermi or turbulent acceleration. (The equivalent minimum proton energy is lower, predicting that thermal protons can be accelerated by these mechanisms). If the relativistic electrons have been accelerated from a thermal background, then some "first stage" mechanism must act to accelerate them up to $\gamma \geq \gamma_{min}$.

This first stage mechanism almost certainly involves a DC electric field. Such fields can occur several places in radio sources. They may well exist in the core, as a fundamental part of the nuclear dynamo. If a net current flows in the source (as would be a logical outcome of a nuclear dynamo), Ohm's law tells us that electric fields must exist throughout the current path. Benford (1984) has suggested that the hot spot/working surface at the end of a current-carrying jet will be the site of an inductive electric field which plays a rule in establishing the return current. Other ED acceleration mechanisms which might exist in hot spots are double layers (which can occur as part of the circuit; *e.g.*, Borovsky, 1986) or shocks propagating across the local magnetic field (which support inductive electric fields in the shock face; *e.g.*, Eilek and Hughes, 1988).

Another possible origin of DC electric fields is local reconnection of magnetic field lines. An electric field is maintained in the reconnecting region, and both electrons and ions can be energized in such a situation (*e.g.*, Spicer, 1982). Transient, explosive reconnection events could provide strong local injection (as with solar flares). Alternatively, several models of radio sources suggest that MHD turbulence is maintained throughout some regions of the sources. This would create local, stochastic reconnection of magnetic field lines, and this lead to a stochastic particle energization by the electric fields (*c.f.* Christiansen, these proceedings; also Konigl and Choudhuri, 1985, and Turner, 1986).

Finally, a caveat is necessary. Some nuclear models predict electron-positron pair production. This is often predicted as a consequence of dynamo models (*e.g.*, Burns and Lovelace, 1982, or Kundt and Gopal-Krishna, 1981), but can also arise from thermal accretion flows (*e.g.*, Eilek, 1980, or Fabian *et al.*, 1986). If this pair production is strong enough to supply the radio source, one may not need to accelerate the electrons from a thermal background, and the injection problem may be irrelevant.

This first point tells us that electric fields are very likely to be energetically important in particle acceleration, somewhere in the source. It does not address large scale currents or ED effects, however. The next three topics will address small-scale dynamics, and may be seen as suggesting that a net current flows in the source.

B. OVERPRESSURE STRUCTURES

Magnetic confinement of high-pressure structures – jets and filaments – has been suggested several times in the literature (e.g.Benford, 1978; Eilek, 1985; Hardee, 1985; Owen, 1986). The observational need for confinement by something other than external thermal pressure is becoming well-established as high-quality, high-resolution interfo-metric images become available. The filaments in the M87 jet (Owen et al., 1988), the jet and filaments in Cyg A (Perley et al., 1984), and several quasar jets (Owen, 1986) are examples of structures which do not appear to be transient and which have minumum pressures (that is, the lowest pressure in magnetic fields and relativistic elec-trons necessary to account for the radio luminosity) well in excess of the background pressure.

I have not included hot spots in this list. They also have a minimum pressure well above the inferred lobe or background pressure. In this case, however, one is not necessarily led to consider magnetic confinement. The usual HD models of Fanaroff-Riley type II sources involve highly supersonic jets (e.g., Norman, these proceedings), so that the hot spots would be confined by the ram pressure of the end of the jet impinging on the external medium. On the other hand, referring to the discussion above on the possibility of net currents in the jets and hot spots, magnetic confinement is certainly a possibility for the hot spots as well as for the jets.

Since magnetic confinement is a familiar topic, I simply recall the basics of con-finement of a cylindrical system (envisioning a jet or a linear filament). An azimuthal magnetic field, B_ϕ, provides a local inwards Lorentz force ($\mathbf{j} \times \mathbf{B}$) on the plasma. A purely azimuthal field is highly unstable, and the addition of an axial field, B_z, will help to stabilize the system (e.g., Benford, 1978). Thus, the net field is probably helical. (Mass loading the system, such as by immersing the beam or filament in a plasma, will also help the stability). This field configuration implies a net current, given by $I = \frac{1}{2}B_\phi rc$, flows down the beam. The return current must necessarily flow outside of the jet or filament. As well as being dynamically important, this current may be energetically important in heating the plasma or in accelerating particles, as noted above.

Finally, it should be noted that this picture of magnetic confinement is essentially a local one, in that the current and current path must be maintained by a dynamo and by some inertial effects, such as the propagation of the beam through the external plasma. Alternatively, one can say that the field lines must be "tied down" to inertial matter somewhere – in the core and at the working surface, in the case of a jet.

C. FILAMENTS IN RADIO SOURCES

It is becoming apparent that radio sources are not uniformly filled with radio-luminous plasma, but rather that the radio emission comes from apparently linear structures which occupy a small part of the total volume of the source. I call these structures filaments. They occasionally appear disordered and turbulent, such as in the lobes of M87 (Hines et al., 1988), or Pictor A (Perley, in preparation). More often, they seem to know about the large-scale source axis, as in Cyg A (Perley et al., 1984), the jet of M87 (Owen etal, 1988) or the Wide-Angle Tailed sources 0110+152 and 1159+583 (O'Donoghue, 1988). The structures in the lobes of M87 are consistent with pressure confinement, in that $p_{min} \sim p_{ext}$; but the filaments in the jet of M87, as well as those in

Cyg A, are clearly overpressure: $p_{min} >> p_{ext}$. Polarization and Faraday information exist for these latter two sources, and in both cases, the projected magnetic field tends to lie along the luminous filaments.

Of course, many mechanisms can be invented to explain these structures. I will argue here that two attractive explanations – cooling instabilites and purely hydrodynamic mixing at the end of a jet – are unlikely to account for most of the observations. We seem to be led, again, to an explanation based on currents; although I do not yet have quantitative support for this idea.

The first thing one might think of to make field-aligned filaments is a cooling instability. Now, cooling instabilities can be driven by synchrotron radiation in a plasma dominated by the internal energy of relativistic electrons or pairs (Eilek and Caroff, 1979). However Hines et al.(1988) show that this instability is unlikely to give luminous filaments like those observed, but is more likely to lead to cool, dense, underluminous structures. If the relativistic electrons and field are mixed with thermal plasma, such as the dense ICM in a cluster of galaxies, bremsstrahlung can drive a cooling instability. Hines et al., and also Owen (these proceedings) found that this model is consistent with the filaments in M87. However, the unusually high ICM densities of the innermost Virgo core seem necessary: the bremsstrahlung cooling time must be less that the synchrotron lifetime in order to produce bright filaments. Therefore, I suspect that this model may not be able to account for the filaments in other radio sources which probably live in lower-density environments.

Another tempting model comes from the appearance of filaments in the "passive magnetic field" (i.e., $B^2/8\pi << p$) two-dimensional simulations of Norman et al.(1988; and these proceedings). These filaments reflect the field line stretching and mass entrainment at the head of a "slab jet" propagating into a background. However, these simulations are far from equipartition (which would have $B^2/8\pi \sim p$), and thus the true pressure in the filaments must be much greater than that which would be inferred from minimum-pressure arguments applied to the radio luminosity. But the true pressure in the filaments must be comparable to that in the interfilament ("background") material in these HD models, whereas the *minimum* pressure of the observed filaments is comparable to or greater than the background pressure. Thus, these simulations do not appear to account for the observed filaments.

This leads one, then, to the possibility that the filaments form from resistive tearing instabilities in a current sheet. This has been suggested now and then in the context of radio sources (e.g., Ferrari, 1985; Eilek and Lovelace, 1987; Hardee et al., 1988) but has not to my knowledge been investigated quantitatively. I will therefore review the basic and likely behavior of this instability, and will speculate that this has something to do with the filaments in radio sources (after all, this is a workshop!).

Tearing instabilities reflect the tendency of a current sheet to filament, due to the self-attraction of parallel currents. A finite resistivity allows field lines to move relative to the plasma (Furth et al., 1963). The process produces high-field, current-carrying regions surrounded by low-field regions. The scale of these magnetic islands seems to range upwards from the thickness of the current sheet to a good fraction of the size of the system (e.g., Mattheus and Montgomery, 1981; Biskamp, 1986). The growth rates for tearing instabilities depend on the local value of the restivity (Furth et al., 1963; also Chiueh and Zweibel, 1988). The Coulomb resistivity is of course very small for the nearly collisionless plasmas of radio jets or the ICM, and we might expect the growth time for the tearing mode to be longer than the life of a radio source. However, if the current is restricted to a thin sheet – either at the interface between two regions of differently directed magnetic field (the jet and the lobe?), or if, say, the return current flows along the interface between the lobe and the external material – the resistivity is almost certainly anomalous (e.g., Spicer, 1982). In this case the tearing mode growth rates can be increased by several orders of magnitude, and this instability may well be important during the life of a radio source.

Finally, it seems likely that the end state of this instability might be self-confined (at least locally), over-pressure structures. The flux ropes seen in the atmosphere of Venus (Russell and Elphic, 1979), and also the plasmoids created during substorms in the earth's magnetotail (*e.g.*, Hones, 1984) may be examples of such structures which arise from resistive instabilities. One could speculate that the luminous filaments in radio sources are current bunches/magnetic islands which arise from tearing instabilites in a surface current sheet. This again emphasizes the possibly important role of currents in these sources.

D. TWISTING DOUBLE JETS

Consider a current-carrying beam which has bifurcated – perhaps through a tearing instability – into two adjacent beams. These two beams will attract each other, and one might expect them to merge. But if the beam trajectories are helical – that is, if the mass flowing in them has some slight angular momentum, perhaps imparted from the instability – then a balance of centrifugal and attractive forces is possible. A quasi-steady helical double jet might thus be maintained, as seen in the inner part of the M87 jet (Owen *et al.*, 1988) and the Cyg A jet (Perley *et al.*, 1984).

This has been suggested by Eilek and Lovelace (1987), and independently by Achterberg (1988), who has worked out some dynamical applications of the model in application to 3C75. Achterberg also points out that purely HD interactions of adjacent beams in a turbulent, common wake can also set up wiggles in the beam trajectories. However, this requires fairly high Reynolds number motion of the galaxy through the external medium, and it is not obvious to me that this is case for the galaxies in question (which are cD's and probably have low velocities relative to the local ICM).

Returning to the ED model of twisted double jets, we can make simple quantitative estimates of the effect. If the pitch angle of the helix is ϕ (so that $\tan \phi$ gives the ratio of the beam separation to the helix wavelength), I is the net current and each beam has density ρ, cross section A and velocity βc, then balancing centrifugal and attractive forces gives

$$I^2 = 4\rho A \beta^2 c^4 \tan^2 \phi$$

We can use $I = \frac{1}{2} r c B_\phi$ if B_ϕ and r are the magnetic field and radius of the jet; and we can scale the mass flux by $L_K = \rho A \beta^3 c^3 = 10^{44} L_{44}$ erg/s, the total kinetic energy flux in the beam. The balance conditions become a relation between the azimuthal magnetic field and the energy flux in the beam: $B_\phi \simeq 80 (L_{44}/\beta)^{1/2} \tan \phi / r_{kpc}$ μG. Given the simplicity of this picture, this relation seems pleasantly close to conventional numbers for the jets. Thus, current-carrying beams again seem consistent with the small-scale structure of a couple of the sources.

Finally, a couple of relevant physical points should be pointed out (*c.f.* Achterberg, 1988). First, this picture is consistent only if the return current from the beams does not flow on the surface of each beam, but rather in a diffuse cocoon – which is probably consistent with a simpleminded ED model of Cyg A or M87, but which might be more of a constraint on models of 3C75. Second, this simple configuration is dynamically unstable to radial disruptions of the beam. Achterberg shows that low density beams slow down this instability, so that a self-confined quasi-equilibrium is probably possible; but this question needs further investigation.

Thus, in summary, several small-scale and microphysical phenomena in extended radio galaxies suggest that electric currents and fields are important in the sources. This may be telling us that a net current flows in the sources, and that the dynamics and energetics of the source reflect electrodynamic as well as hydrodynamic phenomena.

Achterberg, A., 1988, *Astr. Ap.*, 191, 167.

Benford, G., 1978, *M.N.R.A.S.*, 183, 29.

Benford, G., 1984, in *Physics of Energy Transport in Extragalactic Radio Sources*, ed. A. H. Bridle and J. A. Eilek (Green Bank: NRAO).

Biskamp, D., 1986, *Phys. Fluids*, 29, 1520.

Borovsky, J. E., 1986, *Ap. J.*, 306, 451.

Burns, M. L. and Lovelace, R. V. E., 1982, *Ap. J.*, 262, 87.

Chiueh, T. and Zweibel, E., 1988, *Ap. J.*, 317, 900.

Eilek, J. A., 1980, *Ap. J.*, 236, 664.

Eilek, J. A., 1985, in IAU Symposium 107, *Unstable Current Systems and Plasma Instabilities in Astrophysics* eds. M. R. Kundu and G. D. Holman, (Boston: D. Reidel), p. 433.

Eilek, J. A. and Caroff, L. J., 1979, *Ap. J.*, 233, 463.

Eilek, J. A. and Hughes, P. A., 1988, in *Astrophysical Jets* ed. P. A. Hughes, (Cambridge: Cambridge University Press).

Eilek, J. A. and Lovelace, R. V. E. L., 1987, *B.A.A.S.*, 19, 731.

Fabian, A. C., Blandford, R. D., Guilbert, P.N., Phinney, E. S. and Guellar, L., 1986, *M.N.R.A.S.*, 221, 931.

Ferrari, A., 1985, in IAU Symposium 107, *Unstable Current Systems and Plasma Instabilities in Astrophysics* eds. M. R. Kundu and G. D. Holman, (Boston: D. Reidel), p. 393.

Furth, H. P., Kileen, J. and Rosenbluth, M. N., 1096, *Phys. Fluids*, 281, 1595.

Hardee, P. E., 1985, in IAU Symposium 107, *Unstable Current Systems and Plasma Instabilities in Astrophysics* eds. M. R. Kundu and G. D. Holman, (Boston: D. Reidel), p. 439.

Hardee, P. E., Owen, F. N. and Cornwell, T., 1988, in *Active Galactic Nuclei: Proceedings of Georgia State University Conference*, ed. H. R. Miller and P. J. Wiita (New York: Springer Verlag), in press.

Hines, D. C., Owen, F. N. and Eilek, J. A., 1988, preprint.

Hones, E. W., 1984, in *Magnetic Reconnection in Space and Laboratory Plasmas* ed. E. W. Hones, (Washington: AGU), p. 264.

Konigl, A. and Choudhuri, A. R., 1985, *Ap. J.*, 289, 173.

Kundt, W. and Gopal-Krishna, 1981, *Ap. Space Sci.*, 75, 257.

Lovelace, R. V. E., Wang, J. C. L. and Sulkanen, M. E., 1987, *Ap. J..*, 315, 504.

Mattheus, W. H. and Montgomery, D., 1981, *J. Plasma Phys.*, 25, 11.

Norman, M. L., Clarke, D. A. and Burns, J. O., 1988, preprint.

O'Donoghue, A., 1988, Ph.D. Thesis, New Mexico Tech.

Owen, F. N., 1986, in IAU Symposium 119, *Quasars*, eds. G. Swarup and V. K. Kapahi, (Boston: D. Reidel), p. 173.

Owen, F. N., Hardee, P. E. and Cornwell, T., 1988, preprint.

Perley, R. A., Dreher, J. W. and Cowan, J. J., 1984, *Ap. J. (Letters)*, 285, L35.

Russell, C. T. and Elphic, R. C., 1979, *Nature*, 279, 616.

Spicer, D. S., 1982, *Space Sci. Rev.*, 31, 351.

Turner, L., 1986, *Ap. J.*, 305, 668.

Jean Eilek

Mike Norman

Numerical Simulations of Hot Spots

Michael L. Norman

National Centre for Supercomputing Applications
and
Department of Astronomy
University of Illinois
Champaign, IL 61820

and

Institute for Astrophysics
University of New Mexico
Albuquerque, N.M. 87131

Abstract

Numerical simulations of hot spots and their associated jets
are reviewed with special emphasis on their dynamical variability.
Two-dimensional simulations incorporate dynamically passive and important magnetic fields in the ideal MHD limit. Distributions of
total and polarized radio brightness have been derived for comparison
with observation. The move toward three-dimensional simulations is
documented, and hydrodynamical models for multiple hot spots are
reviewed. Useful insights can be gleaned from 2-dimensional slab jet
simulations, which relax the axisymmetric constraints while allowing
high numerical resolution. In particular, the "dentist drill" model
(Scheuer 1982) for working surface variability is substantiated and
shown to result from self-excited jet instabilities near the working
surface. Three-dimensional simulations with magnetic fields are now
feasible with the availability of large memory supercomputers, which
should yield the first realistic hot spot geometries. The development
emphasis will then shift to emission physics, which will continue to
be hampered by our lack of understanding of relevant particle
acceleration mechanisms.

1. Introduction

Hot spots locate a region of interaction between a highly collimated stream of plasma emanating from active galactic nuclei and the
intergalactic medium. This interaction characteristically occurs in
the outer extremities of Fanaroff-Riley (1974) class II radio galaxies, and appears as a compact region of high radio surface brightness. Early high resolution observations of hot spots in Cygnus A
(Hargrave and Ryle 1974) and in other powerful radio sources (Dreher
1981; Laing 1982) revealed their basic characteristics. These
include 1) a compact, disk-like geometrical structure, often with a
steep fall-off in radio surface brightness at its outer edge; 2) a
minimum pressure as derived from synchroton theory well in excess of
the surrounding radio lobe by one or two orders of magnitude; and 3)
a flatter spectral index than the surrounding lobe.

The fluid beam hypothesis of Blandford and Rees (1974) and
Scheuer (1974) provided the basic theoretical framework for understanding hot spots that is in use today by many investigators. Their
framework is based on relativistic gas dynamics in the hypersonic
regime, and thus induces us to conceptualize the hot spot interaction
region in terms of the basic nonlinear waves of gas dynamics: shock
waves, rarefaction waves and contact discontinuities. In multi-
dimensions, it is known from decades of laboratory studies of high
speed gas flow that these three fundamental waves combine in bewil-
dering variety, yielding intricate networks of intersecting shock
waves (e.g., Van Dyke 1982) in both steady and unsteady regimes.

Intense radio observational studies by numerous researchers over
the past decade and a half have lent credence to the fluid beam hypothesis through the discovery of radio jets, which are undoubtedly the
energy pipeline connecting the active galactic nucleus and the extended radio lobes in extragalactic radio sources (Begelman, Blandford
and Rees 1984). More than two hundred radio jets are now known; for a
review of their properties and systematics, see Bridle and Perley
(1984). Although the basic features of hot spots outlined above are
unchanged by these observations, additional complexities and peculiarities have been uncovered as reviewed by Perley in this volume,
including protruding hot spots (e.g., Pictor A), "hamburger hot
spots" (Cygnus A), and "splatter spots" (Williams and Gull 1985).
During the same period, numerical hydrodynamical simulations were

performed by a number of groups (e.g., Norman et al. 1982; Wilson and Scheuer 1983; Smith et al. 1985; Williams and Gull 1985) which elaborate and extend the basic fluid model into the nonlinear, multi-dimensional, time-dependent regime. Many familiar, plus some unfamiliar, complex shock configurations have been produced in these simulations.

Since hot spots represent the point of strong physical interaction in a radio source, they serve as a powerful discriminator between different theoretical viewpoints concerning their nature (e.g., hydrodynamical, magnetohydrodynamical, plasma dynamical, etc.). It seems reasonable and timely, therefore, to ask the question: Is fluid dynamics, in all its complexity as revealed by simulations, an adequate theoretical framework for explaining observed hot spot structures?

We shall supply no clear answer to this question here. Rather, we shall attempt to trace the evolution of numerical simulations of hot spots and their associated jets from highly constrained, axisymmetric calculations with highly simplified radio emission models, to unconstrained, nonaxisymmetric calculations with more realistic emission models. The difficulty of answering our central question derives from the variety of jet/ambient medium interaction scenarios that exist in a multidimensional, time-dependent flow when viewed at one instant in time along an arbitrary line of sight. This is illustrated by two examples from our work: 1) the propagation of a two-dimensional, supersonic slab jet through an external medium; and 2) the radio appearance of axisymmetric, supersonic jets with helical magnetic fields. A shortcoming of all the simulations reviewed, which limits observational comparisons, is that no attempt has been made to couple the energy space evolution of the radiating electrons to the flow dynamics. This is because of our current incomplete understanding of the relevant particle acceleration mechanisms. Nevertheless, no large inconsistencies emerge to indicate that fluid dynamics is not an appropriate framework to continue applying to radio source modelling. An inevitable conclusion from all simulations is that hot spots are part of a global flow pattern that is highly variable, and that it is dangerous to read too much into isolated morphological features of hot spots.

2. Basic numerical approach

The basic numerical approach to modelling hot spots in powerful radio galaxies is to solve the interaction of a supersonic jet propagating through a background gas as an initial-boundary value problem in two or three spatial dimensions. The background gas can be uniform or stratified to represent different galactic environments, and fills a finite (usually rectangular) computational domain at t=0. The jet is introduced through an aperture on the boundary of the domain continuously; i.e., jet parameters remain fixed in time, although an arbitrary time profile (periodic, pulsed, ramped) of flow perturbations can also be modelled. The boundary conditions on the remainder of the periphery are generally either reflecting or continuitive (outflow), depending on whether the boundary is an interior symmetry line (axis, equator) or an exterior boundary, respectively.

The nonlinear evolution equations of gas dynamics or, in the case of magnetized flows, magnetohydrodynamics are discretized in space and time through finite differences, and evolved forward in time by a variety of techniques (a compendium of numerical methods for astrophysical fluid dynamics is given by Winkler and Norman (1986), where all the following topics are covered in detail). To date, virtually all simulations of supersonic jets and their hot spots have solved the Eulerian (as opposed to Lagrangian) equations of ideal gas dynamics on a fixed finite-difference mesh using explicit time integration techniques for reasons of computational efficiency and ease of programming. An ideal gas law equation of state is generally used to close the system of dynamical equations. Shock waves which arise in the flow are treated by three basic techniques: 1) artificial viscosity, 2) flux-corrected transport, and 3) explicit nonlinearity (Riemann solvers). Most simulations have been performed in two spatial dimensions, although a few have been attempted in three dimensions.

The evolution problem is parameterized by a few dimensionless numbers expressing the input jet density, velocity, pressure and age in terms of ambient values: thus, if ρ_j, v_j, p_j and r_j are the above parameters and the jet radius, respectively, and ρ_a, c_a, p_a are the ambient density, sound speed and pressure, then $\eta = \rho_j / \rho_a$, $M_s = v_j / c_a$, $K = p_j / p_a$ and $\tau = t c_a / r_j$ are the conventional dimensionless density ratio, sonic Mach number, pressure ratio and age, respectively. If B_z and B_ϕ are the input axial and toroidal components of magnetic field strength, respectively, then we augment the pure hydrodynamical parameters with two additional parameters: a pitch angle $\gamma=\arctan(B_\phi/B_z)$, and a magnetosonic Mach number $M_{ms}=v_j/(c_a^2+v_A^2)^{\frac{1}{2}}$, where $v_A^2=(B_z^2+B_\phi^2)/\rho_j$ - the Alfven velocity squared. With this choice of parameters, the dimensionless kinetic luminosity is $L=M_s^3/\eta^{\frac{1}{2}}$.

A variety of approaches have been taken to derive radio emission distributions from the fluid variables that are actually evolved in numerical simulations. One starts with some approximation to the synchrotron emissivity formula that involves the zonal variables, and performs ray integrations through the computational domain at a specified viewing angle. Simulations without magnetic fields must introduce additional approximations (typically flux freezing or equipartition) that reduce the angle-dependent synchrotron emissivity to an isotropic form, usually involving the gas pressure and/or density. Although this approach benefits from simplicity, no polarization information can be derived from such simulations, which limits comparison with observations. In the following, the emission model employed in each investigation will be mentioned.

3. Calculations assuming axisymmetry

3.1 Hydrodynamical simulations

Norman et al. (1982, 1983) simulated axisymmetric, pressure-matched (i.e., K=1), nonrelativistic supersonic jets in two dimensions for a range of η and M_s, and illucidated the basic flow structures which are now familiar from subsequent similar studies. The flow in the vicinity of the hot spot is shown schematically in Fig. 1, which agrees substantially with Figure 2 of Blandford and Rees (1974), but elaborates the terminal shock structure, and indicates the existence of backflow into the cocoon and distortions of the interface (contact discontinuity) between the jet and ambient gases.

Norman et al. found that the flow and interface geometry at the end of the jet was highly nonsteady. Quasi-periodic vortex-shedding events were observed which had a large influence on the terminal

shock front geometry. Typically, in the course of one of these
events, the terminal shock front would alternate between an oblique
incident shock making a regular reflection on the symmetry axis to a
triple shock configuration making a Mach reflection on the symmetry
axis (cf. Fig. 1). This result has been confirmed by Kössl and Müller
(1988).

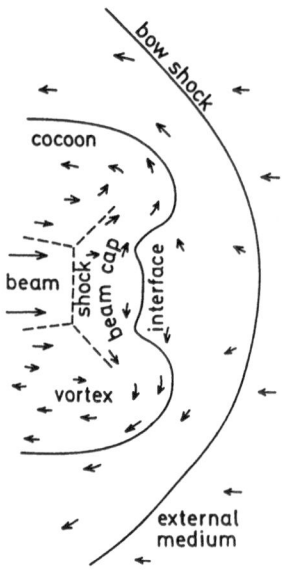

Fig. 1: Working surface schematic and terminology (from Smith et al. 1985).

The hydrodynamical results of Norman et al. were analyzed by
Smith et al. (1985), who derived simulated radio surface brightness
distributions for the jet with initial parameters $\eta=0.1$, $M_s=6$ at
different evolutionary ages τ. They showed that grid resolution and
accurate numerical techniques have an important effect on the structure and dynamics of simulated hot spots. Figure 2 of their paper is
reproduced here to illustrate this point and to serve as a point of
reference for other simulations discussed in this paper. One can see
the basic flow structures come into focus as grid resolution is increased from $\Delta x=r_j/4$ to $r_j/16$, the accuracy of the advection method
is raised from first to second order-accurate, and an interface
tracking technique for keeping the contact discontinuity bounding the
jet sharp is employed.

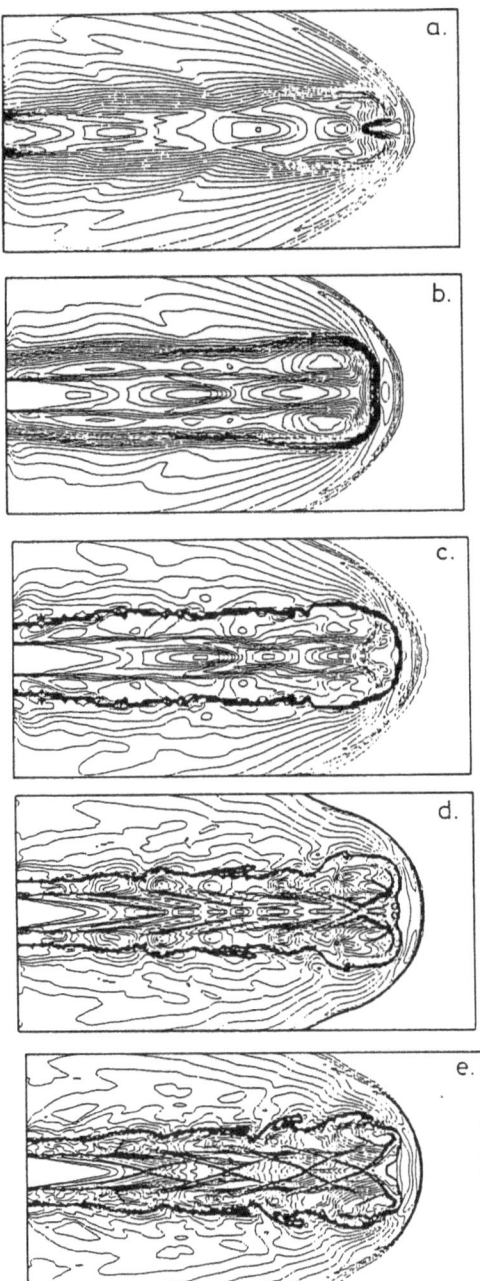

Fig. 2: The effects of numerical techniques and resolution on the structure of a $\eta=0.1$, $M_s=6$ jet and terminal shock structure. a) first order-accurate donor cell advection, low numerical resolution ($\Delta x=r_j/4$); b) second order-accurate monotonic advection, low resolution; c) same as b) but with interface tracking the jet boundary; d) same as c) but at twice the resolution ($\Delta x=r_j/8$); e) same as c) but at four times the resolution ($\Delta x=r_j/16$). Contours of gas density are shown (from Smith et al. 1985).

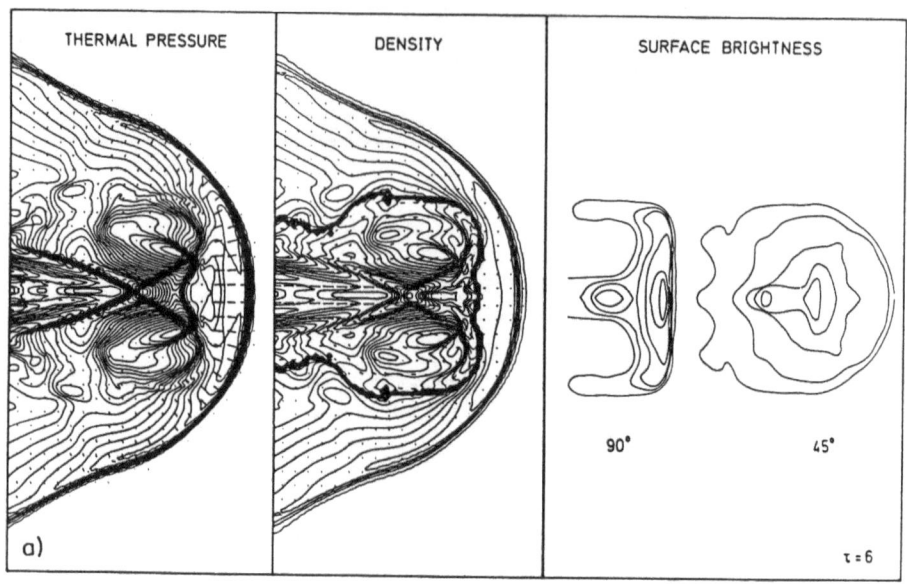

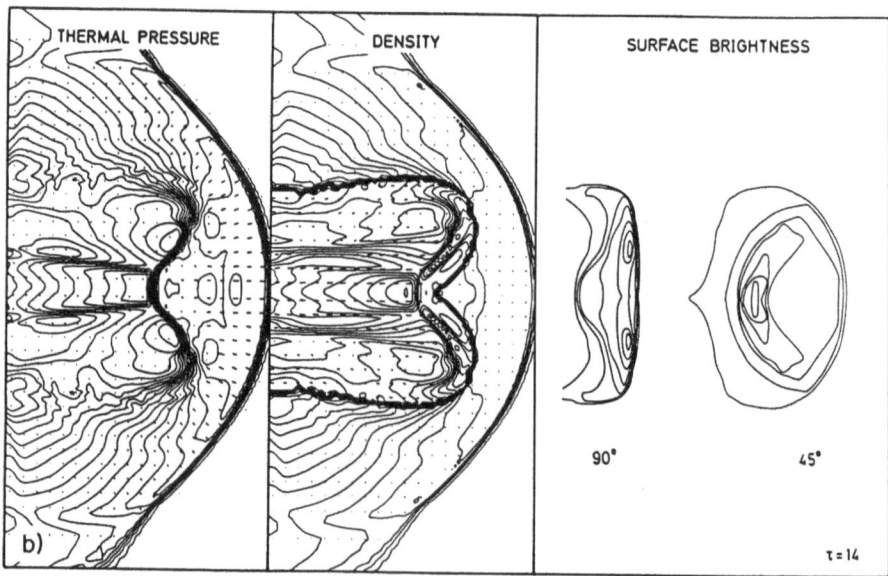

Fig. 3: Simulated hot spots at two epochs in a $\eta=0.1$, $M_s=6$ jet showing the effects of variations in terminal shock structure (pressure panel), contact discontinuity shape (density panel) and viewing angle on radio appearance. a) $\tau=6$; b) $\tau=14$ (from Smith et al. 1985).

Smith and coworkers found that the hot spot's brightness, flatness, and sharpness of the leading edge depends strongly upon viewing angle and time. This is illustrated in Figure 3, which shows predicted hot spots for two terminal shock geometries and two viewing angles. The synchrotron emissivity was assumed to be proportional to the square of the gas pressure inside the jet, and zero outside. Seen edge-on, a hot spot appears brighter and flatter because of the longer line of sight through the beam cap (high pressure region at the end of the jet). Projection effects were shown to make hot spots appear rounder, have smoother edges, and have lower dynamic range. The simulated hot spots reproduced several quite common characteristics of hot spots, such as: 1) a compressed head with sharp outer edge and a length several times longer than the width; 2) twin wings and single tails extending backwards in the general direction of the jet source. The twin wings are identified with the primary backflow channel in the cocoon, whereas the single tail is identified with the incident jet; 3) complex subcomponent geometries, including two bright peaks transverse to the jet axis.

Wilson and Scheuer (1983) investigated the effects of relativistic beaming on the luminosity ratio of approaching and receding hot spots of relativistic twin jets possessing strongly backflowing cocoons, in order to determine if statistically-derived speed limits to 3C radio source expansion based on plasmoid models (Mackay 1973; Macklin 1981) would substantially be altered. They performed non-relativistic jet simulations analogous to Norman et al. (1982), but mapped hot spot radio surface brightness distributions as if they were advancing relativistically, incorporating the effects of Doppler boosting and aberration. Their numerical simulations of a $\eta=0.05$, $M_s=10$ jet reproduced the basic structures found by Norman et al. (1982) including strong backflow. They found that, for a given expansion speed, beam models with backflow have ratios of radio powers only slightly smaller than plasmoid models. This can be understood by noting that the bulk of the luminosity comes from the high pressure beam cap, which is advancing with the same speed as the working surface, and not from backflowing gas, which has lower pressure by Bernoulli's theorem. In addition to these integrated properties, the detailed distributions of radio brightness derived by Wilson and Scheuer are matched quite well by Fig. 3, including two brightness maxima transverse to the source axis at $\theta=90°$ and the V-shaped structure at $\theta=45°$. Both of these features reflect the structure of the Mach disk terminal shock configuration visible in the pressure field of Fig. 3.

3.2 Magnetohydrodynamical Simulations

Incorporating self-consistent magnetic fields into numerical jet simulations accrues two benefits to the study of hot spots: 1) one has a better approximation to the synchrotron emissivity based on the local magnetic field strength and direction, permitting one to compute not only total intensity distributions, but also polarized intensities and electric vector positions; and 2) one can investigate the effects of dynamically important magnetic fields on the structure of the working surface.

Clarke, Norman and Burns (1988) computed the radio appearance of a sequence of jets with passive helical magnetic fields of varying pitch angles γ for various viewing angles. The anisotropic, monochromatic, synchrotron emissivity was evaluated in each zone assuming a strict proportionality between gas and relativistic electron densities and the local magnetic field. The Stokes I, Q and U parameters were then integrated along lines-of-sight, and convolved with a gaussian smoothing beam to produce intensity and polarization distributions comparable to high-resolution VLA maps. The negligible field strength used implies that the jet was effectively hydrodynamical, permitting a direct comparison with earlier works using the isotropic emission law $\epsilon = p^\alpha$.

Figure 4 shows the case $\eta = 0.1$, $M_s = 6$ considered by Smith et al. (1985) at $\tau = 14$, corresponding to Fig. 3b. The panels show the total intensity and polarization E-vector distributions, for the quasi-longitudinal magnetic field case $\gamma = 10°$. As can be seen, the total intensity distributions are quite similar in Figs. 3 and 4, which indicates that working surface geometry (i.e., line-of-sight effects) is the primary determinant of radio appearance. Note, however, that the ring of emission indicated by two bright peaks in Fig. 3b becomes two ridges of emission in Fig. 4b; i.e., the ring is broken at the bottom and top. This is because the angle ψ between the line of sight and the local magnetic field, which is essentially toroidal after the Mach disk, is small at the top and bottom of the hoops. Consequently, so is the synchrotron emissivity, which scales as sin ψ to the 3/2 power in this example. The two ridges of emission separated by an emission trough in Fig. 4b is reminiscent of the "hamburger hot spot" seen in Cygnus A (Laing, Perley and Scheuer 1984).

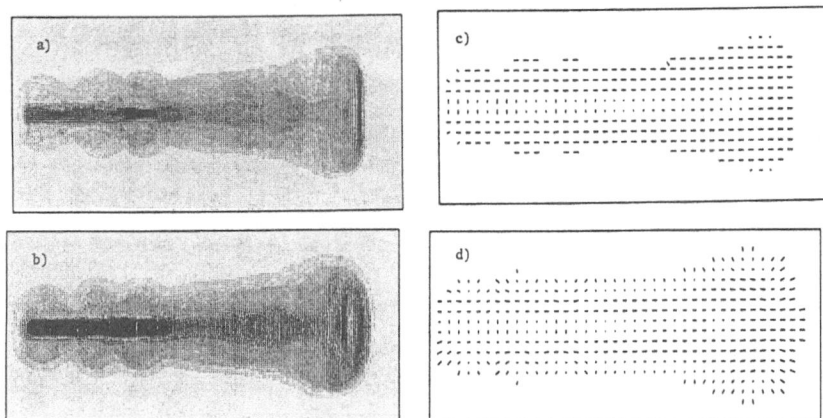

Fig. 4: Total intensity gray scale and polarization E-Vector distributions for a
η=0.1; M_s=6 magnetized jet at r=14. The magnetic field distribution is computed
self-consistently assuming ideal MHD and an input passive helical field of pitch
angle γ=10°. a,c) θ=90°; b,d) θ=75° (from Clarke, Norman & Burns 1988).

The predicted polarization distribution in Figs. 4c,d match
those commonly observed in Class II radio sources: a polarization
flip from longitudinal to transverse apparent field at the hot spot,
and a circumferential apparent field at the hot spot, and a circumferential apparent field at the edge of the cocoon/lobe. Clarke,
Norman and Burns (1988) show that a circumferential apparent field is
a general result for a wide range of input pitch angles (except γ=0°)
and viewing angles, despite the fact that the actual magnetic field
orientation in the cocoon is quasi-toroidal. This is understood on
the basis of how the Stokes Q and U parameters add for magnetic field
loops when seen in projection at an angle.

The results achieved by Clarke and coworkers using a highly
ordered magnetic field bear a striking similarity to the results of
Alan Mathews presented at this conference, who used a very different
magnetic field geometry. Mathews performed purely hydrodynamic jet
simulations, but kept track of how a magnetic unit vector would be
rotated and elongated as it passed along a streamline from the jet,

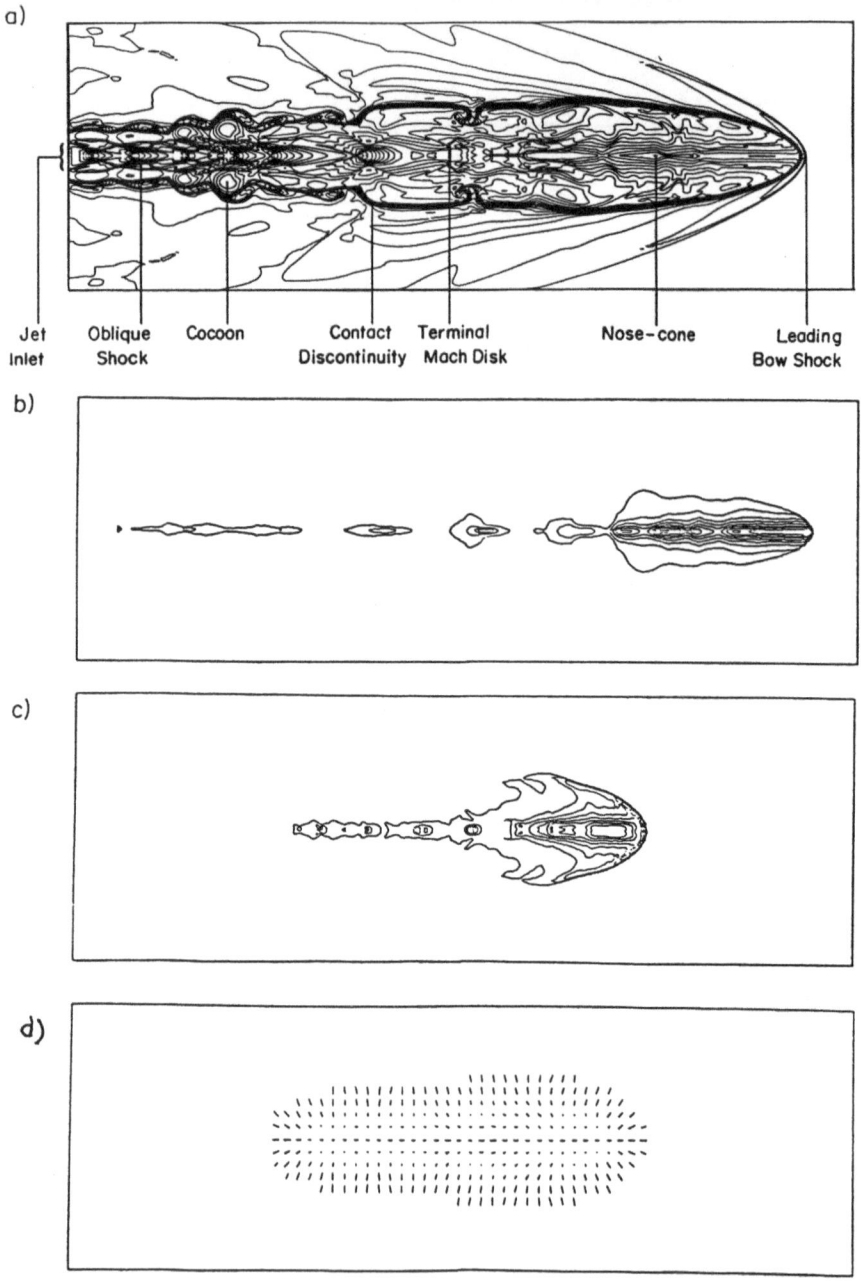

Fig. 5: Structure and radio appearance of the magnetically confined jet of Clarke, Norman & Burns (1986) a) gas density; b,c) total intensity at θ=90° and 30°, respectively; and d) polarization E-vector at θ=30°. Although the magnetic field is purely toroidal in this model, an apparent circumferential field is seen in d) due to projection effects (from Clarke 1988).

through the hot spot into the cocoon. The initial field components were chosen to be randomly orientated, nevertheless the compression at the terminal shock front, and the shear as the flow enters the cocoon, organized the field such that polarization distributions similar to Fig. 4 were produced.

Clarke, Norman and Burns (1986), Lind et al. (1988) and Kössl and Müller (this meeting) have simulated magnetic jets with strong toroidal fields which induce appreciable changes to the flow structure at the working surface. Although somewhat different assumptions were made by the three groups concerning the axial current profile across the jet at the inlet, both groups found that for sufficiently strong toroidal fields, the jets possess a "nose cone". A nose cone is an extension of compressed gas at the head of the jet that is longitudinally confined by ram pressure and radially confined by the self-consistent **JxB** force (Fig. 5a). The terminal shock can be considerably upstream of the contact discontinuity. This means that whereas the hot spot of a hydrodynamic jet is disk shaped, the hot spot of a magnetically confined jet should be cylinder shaped (Fig. 5b,c). Clarke et al. found that the cylinder's aspect ratio depends directly on the input ratio of magnetic to gas pressures. One would expect that long nose cones in three dimensions would be susceptible to nonaxisymmetric instabilities, which has yet to be simulated.

The total intensity and polarization E-vector distributions for a $\eta=0.1$, $M_s=6$, $M_{ms}=2.3$ jet with a purely toroidal field from Clarke (1988) are shown in Fig. 5. The radio mapping procedure is as described above. We see a thin, knotty jet feeds a luminous nose cone. The knots result from internal shock waves in the jet, including the terminal shock. The rapid brightening of the nose cone corresponds to the point where radial force balance is established between gas pressure and magnetic tension. Generally, magnetically confined jets lack the voluminous radio lobes of classical doubles, thus appear as naked jets, quite reminiscent of 3C273 and 0800+608. What evidence do we have that these sources may indeed be magnetically confined?

The total intensity distributions of both 3C 273 (Conway et al. 1981) and 0800+608 (Shone and Browne 1986) at 6 and 20 cm match quite favourably to Figs. 5b,c. Conway (1987) presented observations of 3C273 showing a cross-jet gradient of Faraday rotation, which he interpreted as evidence for a toroidal magnetic field. The observed radial E-vectors at the end of the radio jet strongly indicate a toroidal magnetic field component seen in projection as in Fig. 5d. Finally, Röser and Meisenheimer (1986) observed a color gradient along the optical jet in 3C273 indicating spectral ageing of the synchrotron-emitting electrons as one approaches the end of the jet. This last effect is consistent with the recessed terminal shock front of a magnetically confined jet, provided the terminal shock is the last site of particle reacceleration. Simulations incorporating synchrotron ageing would be desirable to test this model.

4. Non-axisymmetric calculations

Although axisymmetric simulations have led to greater insight into certain flow structures likely to exist in extragalactic radio sources with jets (e.g., internal shocks, vortices, backflow), and have indicated their probable dynamic nature, many radio sources contain highly nonaxisymmetric structures which are not addressed by these simulations. Of interest to this conference are the multiple hot spots revealed by recent observations at sub-kpc resolution, as discussed by Laing (1982). On intuitive grounds, one has the uneasy feeling that perhaps totally unconstrained jets behave very differently from axisymmetric jets, especially at their ends where hot spots are formed. Is this in fact the case? Several recent hydrodynamical studies have broken the bonds of axisymmetry with fascinating results. They fall into two categories: 1) 3-dimensional, cylindrical jets; and 2) 2-dimensional, slab jets.

4.1 3-d cylindrical jet simulations

A useful check on the axisymmetric jet simulations is the work of Arnold and Arnett (1985). They performed 3-d simulations of a $\eta=0.1$, $M_s=6$ cylindrical jet in a cartesian mesh, with assumed quadrantal symmetry; i.e., one quarter of the jet cross section was computed. This symmetry assumption rules out jet bending or deflection, thus their calculation could not address the formation of multiple hot spots by the splatter spot mechanism (see below). Arnold and Arnett's 3-d simulations reproduced and confirmed the essential flow structures found in 2-d, and added a new one: azimuthal vortices in the cocoon. They speculated that the combination of azimuthal and poloidal vortices know from 2-d studies could produce vortex tubes that spiral around the jet axis, and that these could be responsible for filaments in radio lobes by an unspecified mechanism.

Williams and Gull (1985) modelled the formation of multiple hot spots by propagating a jet out to a certain distance and then suddenly changing its input direction. Simulations were performed in the upper half plane of the jet, removing a key symmetry plane found in the work of Arnold and Arnett (1985). Although formally 3-d, the number of computational zones they used normal to the base plane was so low (two zones across the jet radius) that much internal structure was undoubtedly suppressed (cf. Fig. 2). Nevertheless, they were able to reproduce an essential feature of double hot spots in many power-ful extragalactic radio sources (see Laing 1982); namely, that one component is extremely compact (size < 1 kpc), and the other is more diffuse, whose axis often points to the compact component. They pro-posed that the compact component is the site of current jet impact, and that the diffuse component is a "splatter spot" formed by jet material deflected asymmetrically at the compact component. A repro-duction of their results is shown in Fig. 6. Here, the gas pressure is integrated along lines of sight passing through the jet interior to compute the radio surface brightness.

4.2 2-d slab jet simulations

Although 3-d simulations will eventually give us a more reali-stic picture of jet structure and dynamics as they occur in nature, existing 3-d simulations suffer from limited numerical resolution in at least one spatial dimension. This has the unfortunate consequence of suppressing important structures and instabilities, as illustrated in Fig. 2. Simulations in 2-d slab geometry partially avoid this problem by modelling the jet's longitudinal cross-section at high numerical resolution. This permits one to reliably study jet bending, deflection and kink instabilities (Hardee and Norman 1988; Norman and Hardee 1988a), albeit in an idealized geometry. Some qualitative in-sights to *realistic* 3-d jet behavior can nevertheless be achieved, as this example demonstrates.

In simulating slab jets, the initial-boundary value problem is as described in section 2, except that the equations are expressed in 2-dimensional cartesian coordinates. The jet is infinite in extent

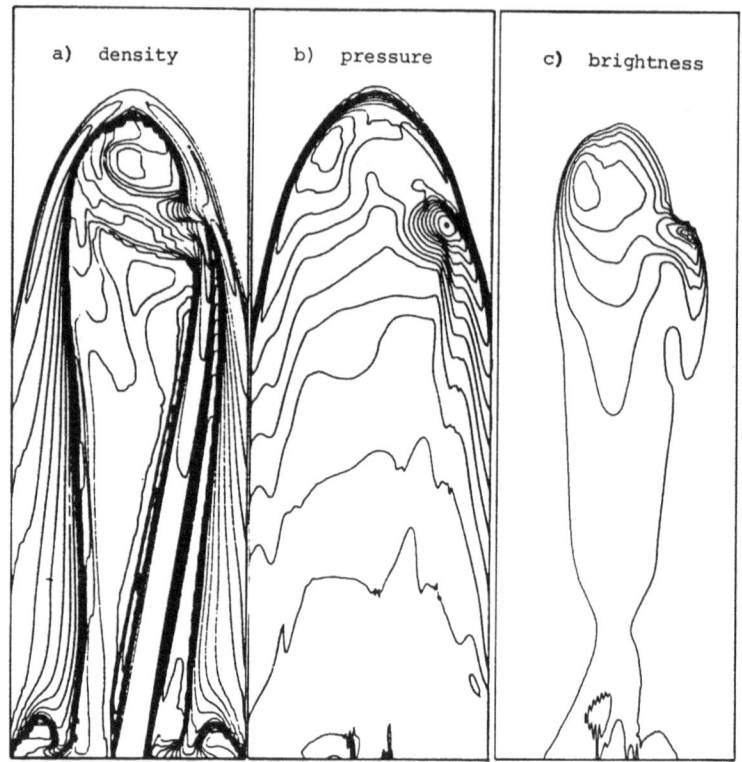

Fig. 6: A nonaxisymmetric double hot spot produced by the "splatter spot" mechanism of Williams & Gull (1985). The jet impinges on the compact subcomponent at right, and ricochets to the other side of the cocoon, where it thermalizes its remaining energy at a more diffuse subcomponent (courtesy S.F. Gull).

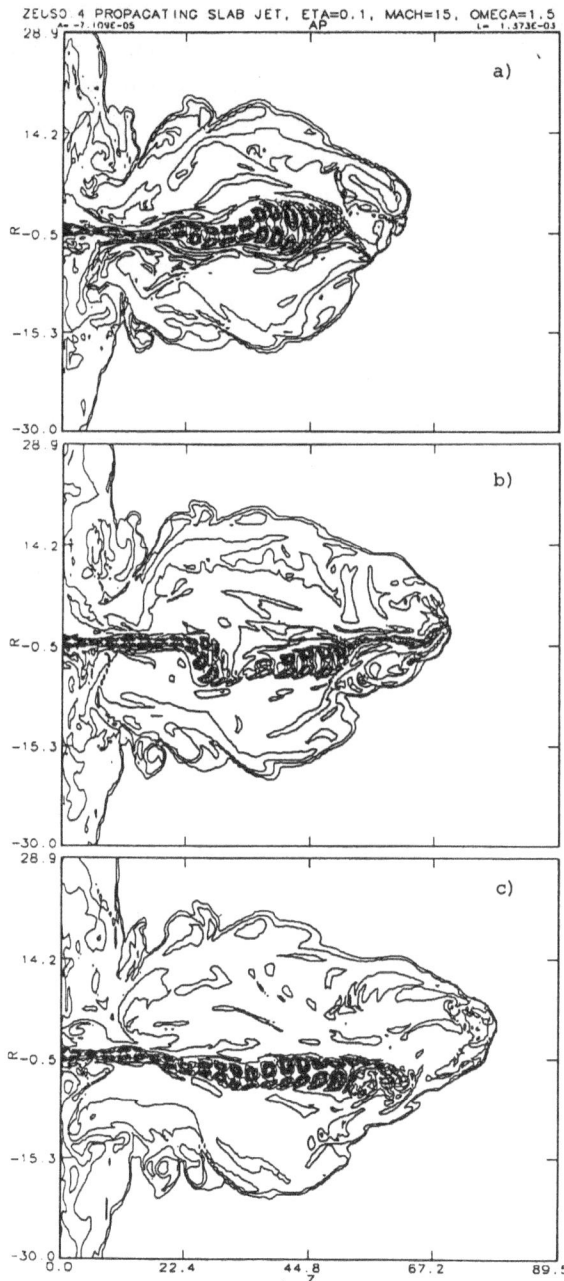

Fig. 7: The propagation of a 2-d slab jet ($\eta=0.1$, $M_s=15$), showing the "dentist drill" (Scheuer 1982) in action. Dynamically passive magnetic flux loops, shown here, trace the jet channel at three closely spaced instants in time: a) $\tau=7.5$; b) $\tau=8.5$; c) $\tau=9.5$. The jet is continually buffeted by strong fluid currents in the cocoon, which redirect the jet on a dynamical timescale. A computational mesh of 320x600 zones was used, with 40 zones spanning the jet (from Norman & Hardee 1988b).

normal to the plane modelled, and is thus geometrically a slab. Fig. 7 shows the results of such a simulation done by Phil Hardee and myself (Norman and Hardee 1988b) to explore the spatial stability of a propagating slab jet. We were interested in seeing if the jet end could be deflected by perturbations arising in the cocoon, which could offer a natural explanation for why many classical radio doubles with observable jets possess jets with strong deflections or sharp bends only as they approach the hot spot (e.g., Cygnus A, 3C175). We hypothesized that an imbalance of ram pressure from strong asymmetric backflow would be sufficient to do this. To break the flow symmetry in our simulation, the jet direction was perturbed sinusoidally at the resonant frequency for the fundamental non-axisymmetric mode with an amplitude of 5%. A passive magnetic field in the form of flux loops was injected along with the fluid in order to trace the jet channel.

Fig. 7 shows the magnetic field distribution at three closely separated times, revealing very different jet structures near the working surface. One must see a complete animation to fully appreciate just how variable the jet flow is. We find that the jet impinges on the ambient medium at one point for awhile (Fig. 7a), and then it is suddenly redirected by a perturbation applied upstream (Fig. 7b), whereupon it begins impinging at a new point (Fig. 7c). In contrast to the splatter spot picture of Williams and Gull (1985), we find a behavior more similar to Peter Scheuer's (1982) "dentist drill" model, wherein the drill tip is lifted between applications. In our simulation, we have discovered a self consistent mechanism for moving the dentist's hand, one unrelated to nuclear activity. The jet is continually buffeted from lateral perturbations as it approaches the head. The section of the jet downstream of the deflection point continues to discharge its momentum to the existing hot spot while a new one is created on a different axis. The latter hot spot eventually propagates ahead of the old hot spot, and even protrudes beyond the broad jet front, reminiscent of the hot spots in Pictor A (Röser, Perley and Meisenheimer 1987).

5. Future prospects

As simulations have advanced from low resolution to high resolution, and from two dimensions to three dimensions, we have seen an increase in the variability of the working surface geometry and the terminal shock front. If the 2-d slab jet results just described are an indication of behavior in three dimensions, even the position of the working surface is highly variable. Large memory supercomputers, such as the CRAY-2, are now available that will make feasible 3-d simulations of resolution approaching that of Fig. 7. Such simulations will give us our first good look at the dynamic nature of the interaction between radio jets and ambient plasma. In three dimensions, arbitrary magnetic field geometries can be incorporated in simulations as passive or dynamic fields, which should yield the first realistic hot spot geometries. In order to tap the information content of spectral index variations, the development emphasis is expected to shift to emission physics in the coming years, even though progress will continue to be hampered by our lack of understanding of relevant particle acceleration mechanisms. Perhaps an inferential approach will be required for progress here, much as the current models are being used to infer radio sources' underlying dynamical laws.

Acknowledgements

I wish to thank my collaborators, both past and present, for their contributions to the work reviewed here: David Clarke, Phil Hardee, Jack Burns and Michael Smith. The author thanks Steve Gull for permission to reproduce Fig. 6. A number of the calculations reported here were performed at the National Center for Supercomputing Applications, University of Illinois at Urbana-Champaign under partial support from NSF grants AST-8611511, AST-8516921 and EPSCoR grant RII-8610669.

References

Arnold, C.N. and Arnett, W.D. 1986, *Ap.J.Lett.* **305**, L57.

Begelman, M.C., Blandford, R.D. and Rees, M.J. 1984,
   Rev.Mod.Phys. **56**, 255

Blandford, R.D. and Rees, M.J. 1974, *Mon.Not.R.Astron.Soc.* **169**, 395.

Bridle, A.H. and Perley, R.A. 1984, *Ann.Rev.Astron.Astrophys.* **22**,319.

Clarke, D.A. 1988, Ph.D. Thesis, Univ. of New Mexico, Albuquerque.

Clarke, D.A., Norman, M.L. and Burns, J.O. 1986, *Ap.J.Lett.* **311**, L63.

Clarke, D.A., Norman, M.L. and Burns, J.O. 1988, *Ap.J.*, submitted.

Conway, R.G. 1987, in *Magnetic Fields and Extragalactic Objects*,
   eds. E. Asseo and D. Gresillon, (Cargese: Editions de Physique
   No.5), 317.

Conway, R.G., Davis, R.J., Foley, A.R., and Ray, T.P. 1981,
   Nature **294**, 540.

Dreher, J.W. 1981, *Astron. J.* **86**, 883.

Fanaroff, B.L. and Riley, J.M. 1974, *Mon.Not.R.Astron.Soc.* **167**, 31P.

Hardee, P.E. and Norman, M.L. 1988, *Ap.J.*, in press.

Hargrave, P.J. and Ryle, M. 1974, *Mon.Not.R.Astron.Soc.* **166**, 305.

Kössl, D. and Müller, E. 1988, Max-Planck-Institut für Astrophysik
   preprint MPA 340.

Laing, R.A. 1981, *Ap.J.* **248**, 87.

Laing, R.A. 1982, Proc. IAU Symp. 97, *Extragalactic Radio Sources*,
   eds. D.S. Heeschen and C.M. Wade, Dordrecht: Reidel.

Laing, R.A., Perley, R.A. and Scheuer, P.A.G. 1984, in *Physics of
   Energy Transport in Extragalactic Radio Sources*, eds. A.H.
   Bridle and J.A. Eilek, (Greenbank: NRAO), cover photo.

Lind, K., Payne, D. and Blandford, R.D. 1988, preprint.

Mackay, C.D. 1973, *Mon. Not. R. Astron. Soc.* **162**, 1.

Macklin, J.T. 1981, *Mon. Not. R. Astron. Soc.* **196**, 967.

Norman, M.L. and Hardee, P.E. 1988a, *Ap. J.*, in press.

_____. 1988b, in preparation.

Norman, M.L., Winkler, K.-H.A., and Smarr, L.L. 1983, in
   Astrophysical Jets, eds. A. Ferrari, and A.G. Pacholczyk,
   (Dordrecht: Reidel), 227.

Norman, M.L., Smarr, L.L., Winkler, K.-H.A. and Smith, M.D, 1982,
   Astron. Astrophys. **113**, 285.

Norman, M.L. and Winkler, K.-H.A. 1986, in *Astrophysical Radiation
   Hydrodynamics*, eds, K.-H.A. Winkler and M.L. Norman,
   (Dordrecht: Reidel), 182.

Röser, H.-J., Meisenheimer, K. 1986, *Astron. Astrophys.* **154**, 15.

Röser, H.-J., Perley, R.A. and Meisenheimer, K. 1987 in *Magnetic Fields and Extragalactic Objects*, eds. E. Asseo and D. Gresillon, (Cargese: Editions de Physique No. 5), 361.

Scheuer, P.A.G. 1974, *Mon. Not. R. Astron. Soc.* **166**, 513.

_____. 1982, in *Extragalactic Radio Sources, IAU Symp. No. 97*, eds. D.S. Heeschen and C.M. Wade, (Dordrecht: Reidel), p. 163.

Shone, D. and Browne, I.W.A. 1986, *Mon.Not.R.Astron.Soc.* **222**, 365.

Smith, M.D., Norman, M.L., Winkler, K.-H.A. and Smarr, L. 1985, *Mon. Not. R. Astron. Soc.* **214**, 67.

Van Dyke, A. 1982, *Album of Fluid Motion*, (Stanford: Parabola Press).

Williams, A.G. and Gull, S.F. 1985, *Nature* **313**, 34.

Wilson, M.J. and Scheuer, P.A.G. 1983, *Mon. Not. R. Astron. Soc.* **205**, 449.

Winkler, K.-H. and Norman, M.L. 1986, *Astrophysical Radiation Hydrodynamics*, (Dordrecht: Reidel).

Mike Wilson   Alan Heavens
Klaus-Dieter Fritz   John Kirk

Wall jets

M. J. Wilson,

School of Applied Mathematical Studies,
The University of Leeds, Leeds LS2 9JT, U.K.

Observations of jets in extragalactic radio sources have suggested a model for their behaviour in which, as a result of their interaction with dense clouds in the surrounding medium, they are 'bent' or deflected through varying angles (Burns 1986). Furthermore, the detection of multiple components in hotspots (typically a compact primary hotspot towards which the jet, if one is visible, usually points, and a larger, more diffuse secondary hotspot) has been explained by supposing that the primary hotspot marks the point of initial impact of the jet with the denser material composing the wall of the cavity around the jet. The jet is then deflected from the cavity wall, it is supposed, and the secondary hotspot marks the impact of this deflected jet with another part of the cavity (Laing 1982). This model for the interaction is supported by the numerical simulations of Williams & Gull (1985) who investigated the behaviour of gas jets that are precessed through a small angle, after having been allowed to 'inflate' for themselves a cavity in a much denser medium.

However, in interpreting observations and numerical work, it is worth remembering the behaviour of supersonic gas jets in the laboratory. In particular, I would like to draw attention to a paper by Lamont & Hunt (1980, 'The impingement of underexpanded, axisymmetric jets on perpendicular and inclined flat plates', henceforth LH) which describes experiments in which supersonic gas jets were allowed to impinge on flat plates at various angles of incidence. I have, with the publisher's (CUP) permission, taken the liberty of reproducing some of their results as they are relevent to the theme of this workshop. Figure 1 (LH Figure 4a) shows what typically happens when a supersonic jet (initial Mach number 2.2) impacts normally upon a flat plate: there is a shock in the jet just above the plate, which extends across the width of the jet. Gas travelling down the jet passes through this shock and into a high pressure region above the plate's surface. The gas is then accelerated through a transonic rarefaction wave and out into a thin, cylindrically symmetric layer of supersonically moving material which flows away from the point of impact. This layer of gas is known as a wall jet. The flow within the wall jet is not, in general, in pressure equilibrium with the surrounding gas, so that a repeating pattern of shocks and rarefaction waves can be set up within it, not unlike those wave patterns found in axisymmetric gas jets exiting from nozzles.

Figure 2 shows what happens when the flat plate is inclined at a more oblique angle (30°, LH Figure 25b) to the jet axis. Again a wall jet is set up as a result of the jet's impact with the plate: post-collision, the gas flows in a thin layer along the surface of the plate away from the point of impact. Incidentally, note that in Figure 2b there is a Mach-disc-like feature present in the flow, even though there is significant deviation from axisymmetry. For further information about these experiments, and a detailed description of the flow fields, the interested reader should consult the paper by Lamont & Hunt and the references therein. The important point to note is that after impact with the wall, the subsequent fate of the gas jet is to run along surface of the target object and not 'rebound' off it. This is true for both normal and oblique impacts, though there is a difference in that for a normal impact the wall jet runs off equally in all directions, while for an oblique impact there is an obvious preferred direction to the path of the wall jet along the plate.

Now the relevance of this work to astrophysical jets is this. If supersonic gas jets will not 'bounce' off solid plates (but form wall jets instead), then they are unlikely to do so when they impact on gaseous targets - we can regard a flat plate as being a gas cloud or target medium in the limit of infinite density contrast. Furthermore, if one considers the Williams & Gull model in the light of these experimental results one must say that it is improbable that the jet will 'bounce' off the walls of the cavity as it is precessed to one side, but will run along the inside of the cavity as a wall jet instead. However, one could envisage the following simple modification to the basic model. As before, the primary hotspot is associated with the point where the jet initially impacts with the cavity wall; thereafter, however, material runs along the inside of the cavity wall in a supersonically moving wall jet, the wall jet being thin in

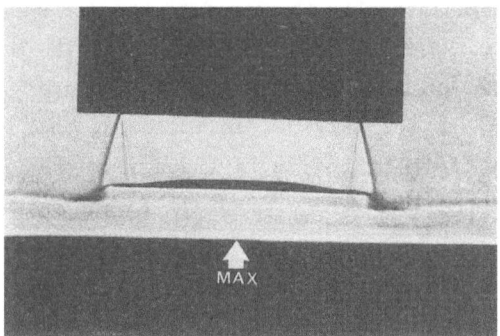

Figure 1

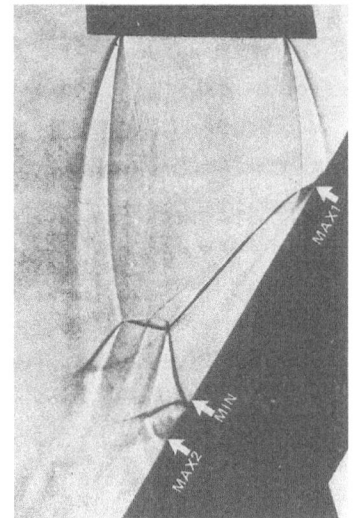

Figure 2

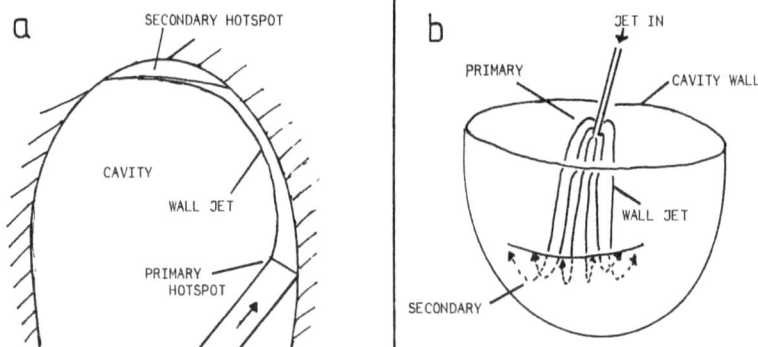

Figure 3. Schematic illustration of wall jet formation in a radio source

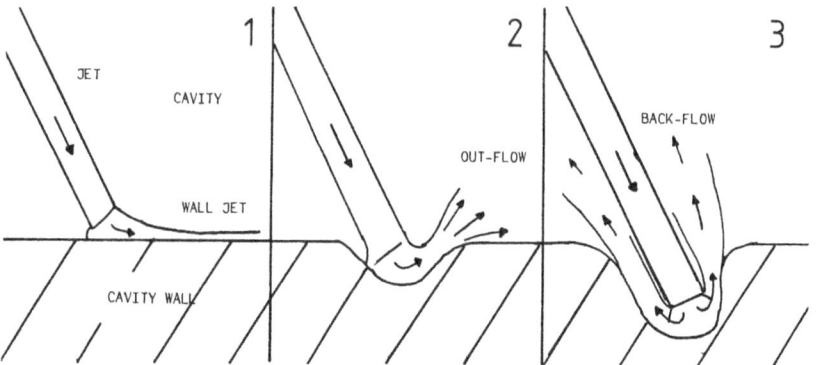

Figure 4. Schematic illustration of a jet impacting with a gaseous 'wall'

a direction normal to the wall but large laterally (compared to the jet's radius prior to impact with the wall) . The secondary hotspot marks the point where the wall jet is strongly shocked as the result of some subsequent interation; this will occur, if the wall jet is following the concave cavity wall, since there is a limit to the angle through whicha supersonic stream can be turned by a compressive bend without the formation of a shock (Figure 3).

Can the jet in any circumstances be said to rebound? Consider Figure 4 which shows the stages during the impact of a jet with a cavity wall composed of material with a much higher density than that of the jet, but not infinitely higher. In stage 1, the jet has not made much impression on the cavity wall, and a wall jet is formed. In stage 2, the wall jet has started to excavate a depression in the cavity wall and the possibility arises that the forming depression may tend to throw the post-shock jet material away from the wall. This cannot be ruled out, but the resulting flow is unlikely to be very well collimated, and having a low Mach number near the point of impact may manage to flow around the 'lip' of the depression and still form a wall jet of sorts. In stage 3 the jet has penetrated a substantial distance into the cavity wall and a well collimated back-flow is the result. In this situation the jet may perhaps be said to be rebounding.

In conclusion then, I would argue that a wall jet is the likely result of a jet's impact with a high density target. Some degree of rebound or 'splash-back' may result if the jet is allowed to penetrate the wall for any length of time, but if the density contrast between the jet and the surrounding medium is high and the jet is continuously precessing then this is unlikely to occur.

I would like to thank Dr S Falle for useful discussions.

References

Burns,J.O. ,1986.Can.J.Phys. ,64,373.

Laing,R.A. ,1982.*IAU* Symp. No.97,167, eds Heeschen,D.S. & Wade,C.M. ,
    Reidel, Dordrecht.

Lamont,P.J. & Hunt,B.L. ,1980.J.Fluid Mech. ,100,471.

Williams,A.G. & Gull,S.F. ,1985.Nature, 313,34.

SIMULATIONS OF SYNCHROTRON LOSS IN HOTSPOTS

Alan P. Matthews

Mullard Radio Astronomy Observatory

Cavendish Laboratory, Madingley Road CB3 OHE

Cambridge, United Kingdom

1. Introduction

According to current theory, a hotspot is associated with a strong shock where a supersonic plasma jet strikes the intergalactic atmosphere. A growing body of numerical simulations have predicted flow patterns and the synchrotron radiation of model radio galaxies which incorporate various elements of the physics which may be relevant. As a contribution to this effort, I present simulations of the radio emission and polarisation of hotspots in which synchrotron losses have been taken into account. They have been modelled on the basis of simulations of an axisymmetric, non-relativistic jet into which a passive, initially randomly oriented magnetic field is introduced; the magnetic field configuration is then distorted by the flow. This work is part of my PhD project under the supervision of Peter Scheuer.

2. Model

Three elements constitute the model: a fluid flow, magnetic field and relativistic charged particles, and synchrotron radiation.

The *fluid flow* is that of an axisymmetric, supersonic jet directed into a uniform ambient medium. It was computed with an axisymmetric version of the code used by Williams and Gull (1985), which is an Eulerian, donor-cell scheme. The flow is non-relativistic, compressible and inviscid, except that a finite cell size in the computing grid introduces an effective viscosity due to numerical diffusion across each cell.

We assume that only material originating in the jet radiates, so *magnetic field and fast particles* are included in the jet where it enters the grid. They are assumed to be dynamically negligible and are convected passively with the flow.

No shock acceleration is explicitly included in the simulations, but we assume that it may have occured lower down the jet to produce a population of *fast particles* with an energy spectrum of the form $dn = K\gamma^{-(2\alpha+1)}d\gamma$, where K is a constant, γ is the Lorentz factor of a particle, α is the radio spectral index (taken here to be 1) and dn is the number of particles per unit volume in the range γ to $\gamma+d\gamma$. This spectrum evolves due to radiative and adiabatic energy losses.

The *magnetic field* is initially random, or tangled on a small scale, and is then "frozen-in" to the fluid and compressed and sheared by its flow. Imagine a cube with sides i, j and k containing a magnetic field $B_0 = (B_{0x}, B_{0y}, B_{0z})$, which is distorted to become a parallelopiped with sides a, b and c. Flux conservation implies that the distorted field is $B = [a.(b \times c)]^{-1}(B_{0x}a + B_{0y}b + B_{0z}c)$. It is then possible to calculate how an initial distribution of B_0 evolves into a distribution of B within the parallelopiped.

To track the fast particles and magnetic field, marker particles were injected into the base of the jet. Each marker particle carries an associated set of three vectors a, b and c which follow the distortion of a fluid element at the particle; a and b are chosen to start poloidal and c toroidal, and in an axisymmetric flow they remain so.

To simulate the *radio maps* (assuming linear polarisation) the Stokes' parameters I, Q and U per unit volume at a frequency ν were obtained by integrating the single particle spectra of Ototal and polarised synchrotron radiation over the evolved particle spectrum and the distribution of projected magnetic field on the sky. The calculation was facilitated by assuming that B_0 was normally distributed. The Stokes' parameters were then integrated numerically along the line of sight by binning the flux from marker particles spread through a cylindrical volume, and then smoothing the resulting map with a Gaussian beam with HWHM = 1 pixel.

3. Results

I simulated a Mach 10 jet with a density 0.1 of the ambient density and pressure equal to ambient pressure. The jet radius was 10 grid cells. Figure 1 shows the source at time t = 612.3 in units of (cell width/jet speed). Figure 1(a) shows pressure contours in the upper half of the diagram and density contours in the lower half, with marker particles superposed. Figure 1(b) shows poloidal vectors in the region of the flow plane used to simulate the hotspots. The toroidal vector is amplified by the factor (radius/initial radius).

Magnetic field is compressed at the jet shock and then into a cylindrical sheath in the "backflow". Poloidal field is predominantly ordered along flowlines, while toroidal field is stretched by flowlines diverging (in cylindrical geometry) after the jetshock. Very large poloidal fields are created in shear layers, both where jet and ambient material meet, and within the "cocoon" where ex-backflow material is circulating between the jet and the backflow.

Simulated radio maps of total intensity with polarisation E-vectors are shown in figure 2, each of which has a different amount of synchrotron loss. Fast particles radiate away energy over a long time, or relatively quickly in strong magnetic fields, so the particle spectrum depends on the history of a particular fluid element. The critical frequency at which the radiation spectrum steepens rapidly, and hence the effect of synchrotron loss, varies within the source but scales inversely with $B_0{}^3t^2$, where B_0 is the input rms field strength and t is the age of the source. Therefore either B_0 or t may be varied to produce the same effect as varying the frequency ν. Figure 2(a) has no synchrotron loss included, but in the other maps B_0 was chosen to vary as follows: (b):1μG, (c):10μG and (d):100μG, with a common source age of 1.5×10^{14}s, viewed at 75° to the jet axis at ν = 1GHz.

4. Discussion

Without synchrotron loss, the brightest feature in fig. 2(a) is a ring seen in projection as an ellipse, which is due to very large poloidal fields in the cocoon interior, between the jet and the backflow. Although fields of this kind may well be generated in shear layers, this bright ellipse is largely a numerical artefact, because the field is artificially localised at a single marker particle (and hence as a ring in cylindrical geometry). In addition, the magnitudes of these fields are not to be trusted, since they are generated in a shear layer in which the numerical algorithm is sensitive to error. But what is more important is that the *physical model* does not take into account processes which could limit the growth of such fields, such as their becoming dynamically significant, reconnecting, or the shear layer becoming unstable, perhaps turbulent.

With synchrotron loss included the ellipse disappears because the fast particles in the ring have lost their energy more quickly than in other parts of the source. What emerges (fig.2(b) and (c)) is a more realistic image of a hotspot arising from compressed fluid just beyond the jet shock. This hotspot is elongated normal to the jet axis, and has a trailing lobe of emission above 1% of the peak intensity in the hotspot. When synchrotron ageing is increased (which is equivalent to mapping at a higher frequency) in fig. 2(d), the lobe of ex-jet material is no longer visible. All that remains is a jet and a hotspot which persists despite no shock acceleration.

These simulations illustrate that synchrotron loss takes its toll not only in old parts of a source, but also in regions of enhanced magnetic fields. Synchrotron loss is only one of several interacting processes which may be essential in a realistic model for the synchrotron radiation we observe. The growth of large fields in the simulation support the case that magnetic forces could be important, while fractional polarisation in the simulated maps which is higher than generally observed may be a clue that order in the field is reduced by reconnection and turbulence.

Williams, A.G. and Gull, S.F. 1985, Nature, 313, 34

Pressure

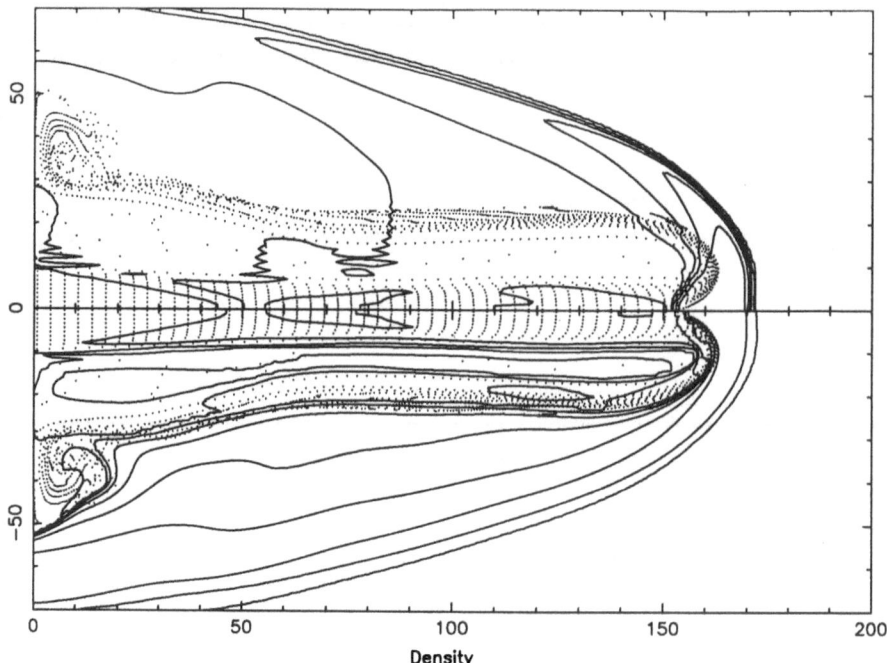

Figure 1(a). *Fluid flow and marker particles in a simulated jet. Contours are logarithmic; the highest level is 0.5 of the peak, and the ratio between successive levels is 0.5.*

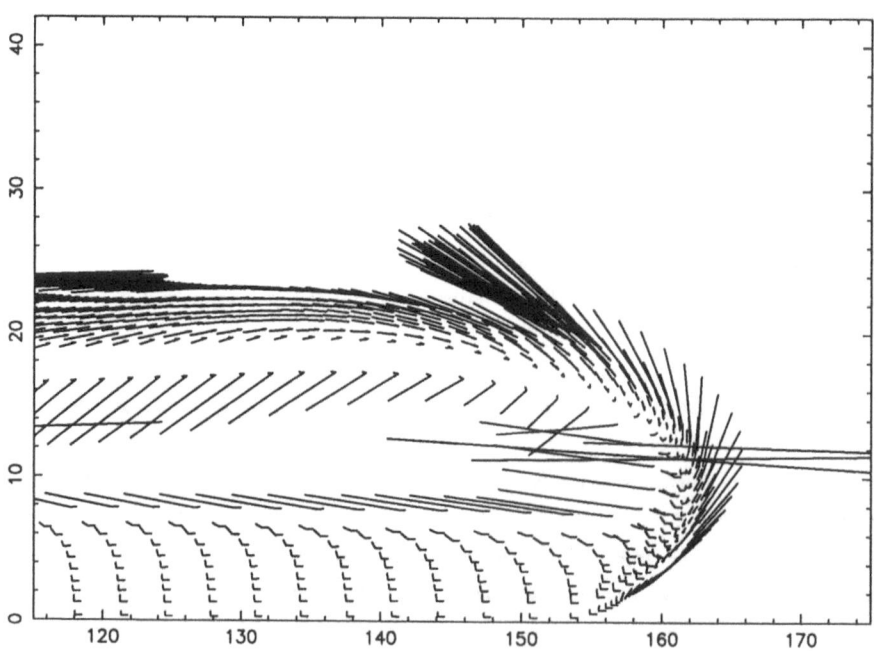

Figure 1(b). *Poloidal vectors tracing distortion of fluid elements in the flow. A unit vector has a length of half a grid cell.*

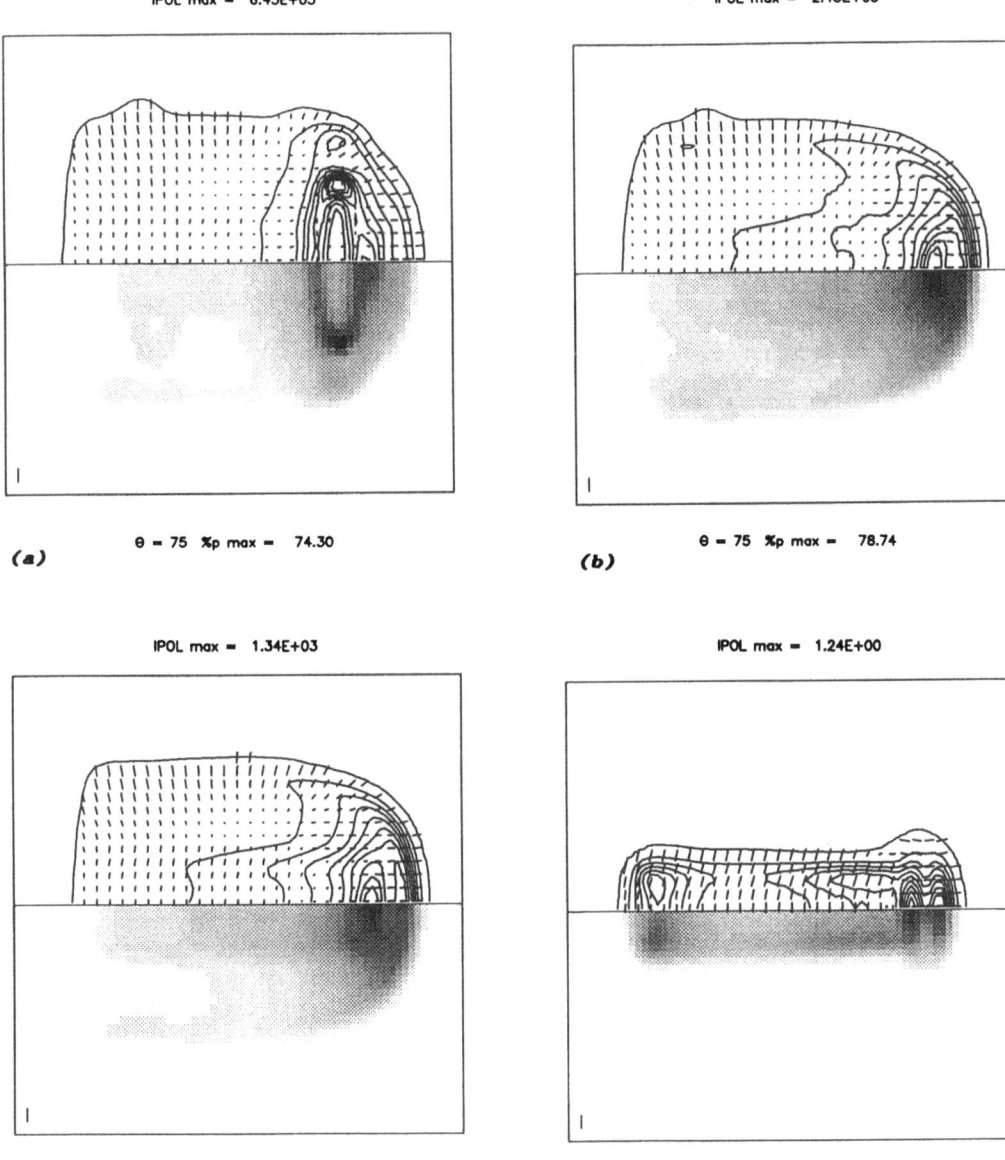

Figure 2. *Simulated radio images of hotspots with different amounts of synchrotron loss. Total intensity is plotted as a greyscale below the jet axis, and as contour levels above the axis. The contour levels are at 0.01, 0.2, 0.3,...,0.9 of the peak. Polarisation E-vectors have length proportional to percentage polarisation; 100% is represented by the bar in the bottom left hand corner. The source age is $1.5 \times 10^{14} s$, and the mapping frequency is 1GHz. Fig. 2(a) has no synchrotron loss, but the other images have input rms field strengths of (b):1µG, (c):10µG and (d):100µG.*

Dieter Kössl

The Influence of Magnetic Fields on the Propagation of Supersonic Jets

Dieter Kössl

Max-Planck-Institut für Physik und Astrophysik

Institut für Astrophysik

D-8046 Garching, Federal Republic of Germany

Abstract

A numerical magneto-hydrodynamical model is described, which is used to study the influence of magnetic fields on the propagation of supersonic jets. Some results and preliminary interpretations of the simulations are discussed.

1. Introduction

Polarisation measurements indicate that most (if not all) extragalactic jets are magnetized [1]. Whether these fields are dynamically important or not, is hard to judge, because of difficulties in estimating field strengths and other physical parameters of jets from the observational data [2]. There are, however, some theoretical arguments in favour of a field strength near the equipartition value [3].

Since the flow pattern in supersonic jets is too complicated to allow analytical solutions, the only way to study the influence of magnetic fields on the propagation of jets is via numerical simulations [4,5,6]. In the following we describe an axisymmetric numerical model, which is used to examine the influence of equipartition-fields of different orientations on the propagation of jets.

2. Basic Equations and Numerical Methods

The equations appropriate for modelling magnetized jets are the ideal magneto-hydrodynamical (MHD) equations [7]. Within the framework of these equations, it is easy to show that the magnetic flux through a closed fluid-line is conserved. In other words, the field lines are "frozen" in the fluid. If the flow includes strong vortices, like e.g. in the cocoon of "light" supersonic jets, the field lines are wound up and regions with a high magnetic energy density are generated. In these regions the field changes direction within extremely small volumes. This leads to reconnection and annihilation of the field within the vortex. The field energy is transformed into thermal energy and the central region of the vortex is heated. This process can be simulated mathematically,

225

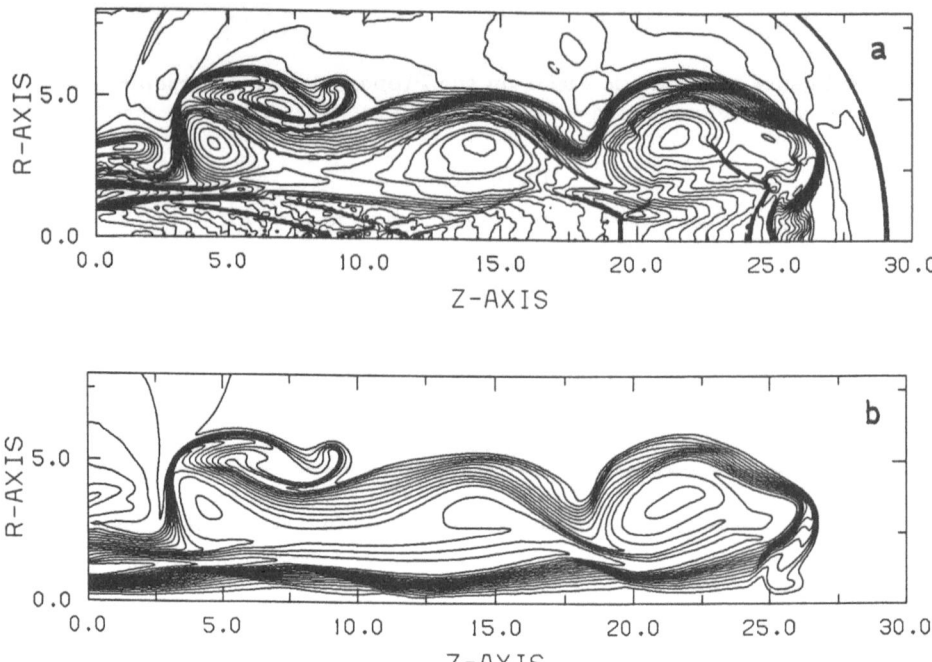

Fig. 1: Density contours (a) and field lines (b) of a jet with a poloidal field
($t = 9.07$). 30 logarithmically spaced contour lines are shown (a).

by averaging the field over small volumes [8]. The averaging process is equivalent to introducing
a turbulent diffusion term to the MHD-equations. In particular did we include a model for the
turbulent diffusion of the poloidal field components.

Another issue of importance in MHD-computations is to guarantee $\mathrm{div}\,\vec{B} = 0$. In axial
geometry the poloidal field components are determined by a single function, namely the toroidal
component of the vector potential A_ϕ [8]. The use of the vector ptotential will guarantee $\mathrm{div}\,\vec{B} = 0$,
but introduces other difficulties, because the magnetic forces have to be computed from the field
components, which are determined by $\vec{B} = \mathrm{rot}\,\vec{A}$. This leads to second-order spatial derivatives,
which can be the source of numerical instabilities. One method to keep the numerical scheme
stable is to introduce an additional diffusion. The results of test calculations have shown that the
turbulent diffusion mentioned above is suffcient for this purpose.

The numerical method used in the jet simulations is a newly developed FCT (flux-corrected-
transport) algorithm. Details of the method are described elsewhere [9,10].

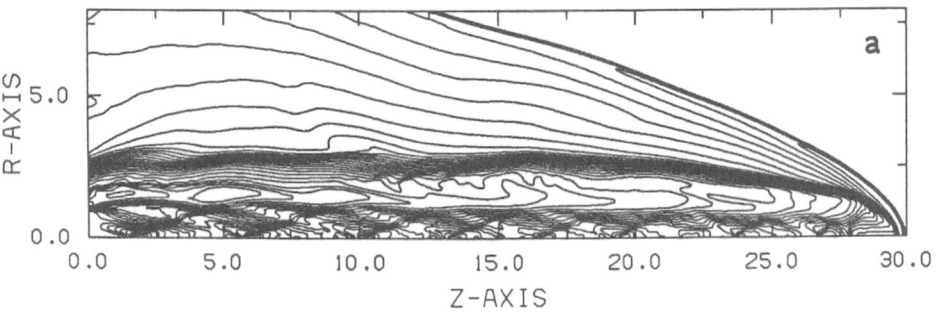

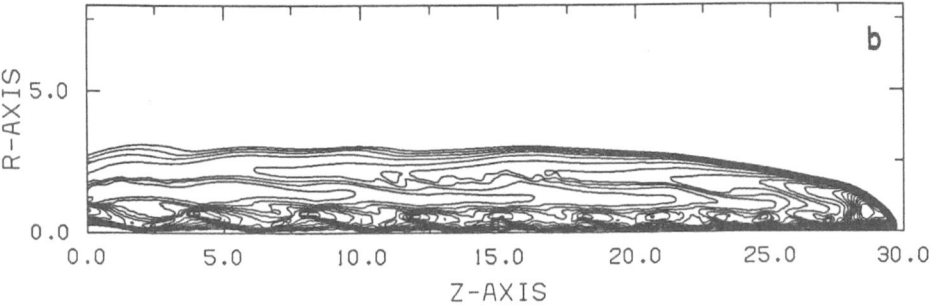

Fig. 2: Density contours (a) and field strength (b) of a jet with a toroidal field
($t = 4.98$). Shown are 30 logarithmically spaced (a) and 20 linearly
spaced (b) contour lines, respectively.

3. Results

(a) Numerical setup: As common in astrophysical jet simulations the MHD equations are written
in normalized form [11]. The units of length, velocity and density are chosen to be the jet (beam)
radius R, the sound velocity c_M and the density ρ_M of the undisturbed medium. The jets studied
in the simulations are characterized by a Mach number of 6 inside the beam and by a pressure
and density contrast of 0.1 and 1.0, respectively, between beam and ambient medium. The field
strength has been chosen such that the average magnetic pressure at the nozzle is equal to the
thermal pressure. In the simulation including a helical field the magnitude of the toroidal and
of the poloidal field components were approximately the same. The initial conditions for the
magnetic field corresponds to a current carrying torus with a large radius of $0.9R$ and a small
radius of $0.4R$. These conditions somewhat resemble the currents flowing in a "thick accretion
torus" [3].

The simulations have been carried out in cylindrical geometry, where the z-axis coincides
with the symmetry axis of the jet. The grid covers a region of $0 \leq z \leq 30R$ and $0 \leq r \leq 8R$, and
the numerical resolution was $\Delta z = \Delta r = \frac{R}{20}$.

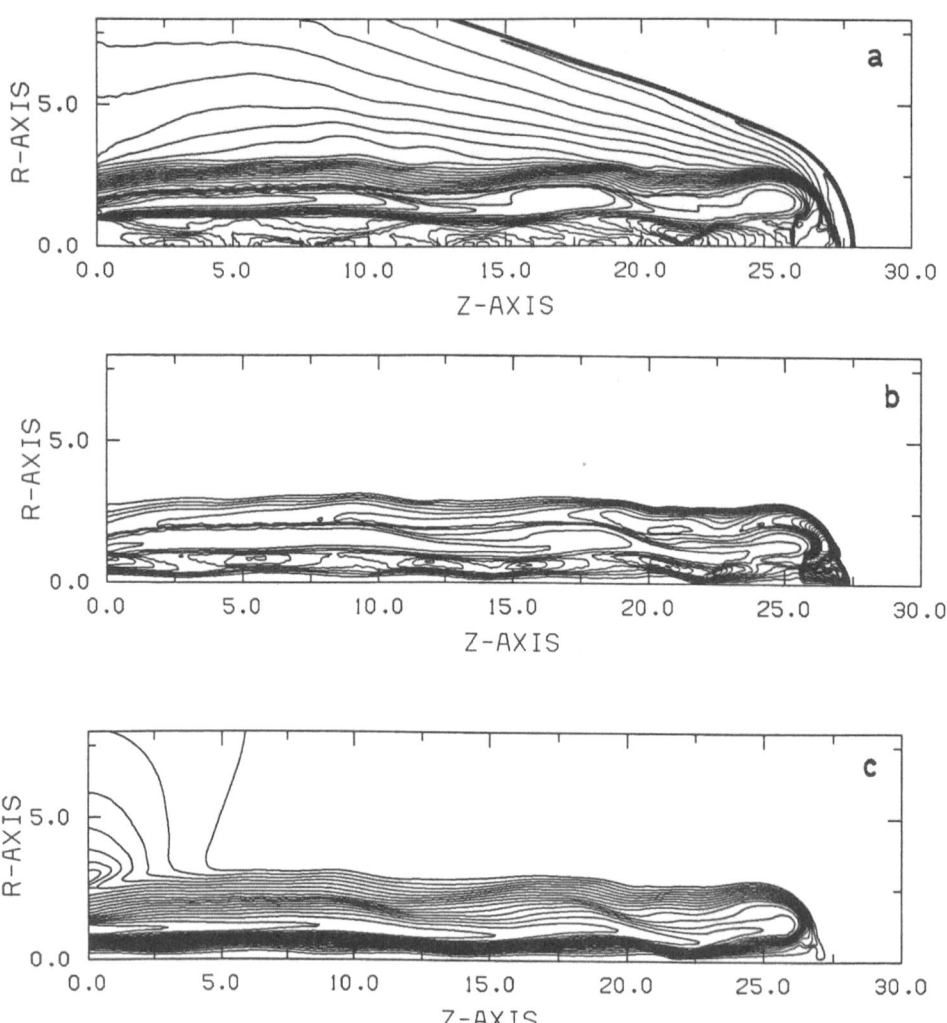

Fig. 3: Density contours (a), field strength of toroidal field (b) and field lines of poloidal field (c) of a jet with a helical field ($t = 5.08$). Shown are 30 logarithmically spaced (a) and 20 linearly spaced (b) contour lines, respectively.

(b) Poloidal field: A purely poloidal field directly influences the propagation of a jet in two ways: (i) the jet is underexpanded, i.e. its pressure is larger than that of the ambient medium, due to the isotropic pressure of the field, and (ii) the propagation velocity is substantially smaller than that of a hydrodynamical jet due to the magnetic tension along the field lines, which are stretched out by the beam (Fig. 1). A more subtle influence arises from the magnetic field annihilation in vortices in the cocoon, where the field energy is transformed to thermal and kinetic energy.

Table 1: Characteristic features of the beam

field	propagation velocity	beam diameter	behaviour of beam diameter
—	4.7	$\simeq 1.4$	oscillating
poloidal	3.2	> 2.0	increasing
helical	5.6	$\simeq 1.2$	oscillating
toroidal	6.0	< 1.0	decreasing

Table 2: Characteristic features of the cocoon

field	backflow velocity	strength of vortices	cocoon diameter	behaviour of cocoon
—	supersonic	medium	1.8	oscillating
poloidal	supersonic	strong	2.6	instabilities
helical	subsonic	weak	1.1	almost smooth
toroidal	subsonic	no vortices	0.8	smooth

This energy release amplifies the strength of the vortices to a great extent eventually leading to a "fat" cocoon with regions of highly supersonic backflow just "above" the vortices. The maxima of the field strength are located at the axis just downstream of the reflection points of the internal shocks in the beam. At these points the magntidude of the field is more than a factor of two higher than at the nozzle (Fig. 1b).

(c) *Toroidal field:* For a purely toroidal field the magnetic forces due to the tension along the field lines are larger than those which stem from the isotropic pressure, thus leading to a net force towards the axis [7]. Therefore the pinch-effect should be able to collimate the beam. The results of the simulations actually show that the beam diameter becomes smaller during the propagation of the jet (Fig. 2). Moreover the field strength of the toroidal field—measured at a constant distance of $r = 0.5R$ from the axis—slowly increases (on the average) along the beam and is finally amplified to three times its initial value at the terminal shock. Oscillations, which stem from the regular shock structure in the beam, are superimposed on this slow increase (Fig. 2b). Contrary to the simulation with a poloidal field the cocoon is now thin and the backflow is almost laminar and subsonic. This result, however, could be a numerical artefact, which is due to the comparatively low grid resolution, corresponding to a large numerical diffusion [9].

(d) *Helical field:* For the initial conditions chosen in this simulation the influence of the toroidal and the poloidal field cancel each other partially. The beam width, the cocoon width and the strength of the vortices are similar to those in the hydrodynamical simulation (Fig. 3). The maximum values of the field components are located at the axis for the poloidal and just ahead of the vortex in the beam cap for the toroidal field (Fig. 3b–c). The absolut maximum of the field strength coincides with the maximum of the poloidal field. Another interesting quantity of this simulation is the ratio of toroidal to poloidal field strength. Within the beam this ratio is slightly less than 1 as expected from the initial conditions. At the terminal shock, however, the toroidal component is amplified while the poloidal compontent is continuous [7]. Downstream the terminal shock the ratio of the field components is 5 on the average, with a huge spike immediately behind the shock $(B_\phi/B_p = 25)$. The ratio is larger throughout the cocoon $(2 \leq B_\phi/B_p \leq 6)$ and has a maximum value of 60 just ahead of the vortex in the beam cap.

4. Conclusions

The tables contain a comparison of the results of the MHD-jet-simulations for the different field orientations. The influence of equipartition magnetic fields on the propagation of a pressure matched jet can be summarized as follows: a poloidal field decollimates and decelerates the beam, whereas a toroidal field collimates and accelerates the beam. The field configuration for a helical magnetic field is predominantly perpendicular to the jet in the cocoon and at the beam cap (i.e. downstream the terminal shock), and it is mainly parallel to the jet within the beam. A more detailed analysis of the results is in preparation [10].

References

1. Miley, G.: 1980, *Ann. Rev. Astron. Astrophys.*, **18**, 165
2. Bridle, A.H., Perley, R.A.: 1984, *Ann. Rev. Astron.*, **22**, 319
3. Begelman, M.C., Blandford, R.D., Rees, M.J.: 1984, *Rev. Mod. Phys.*, **56**, 255
4. Clark, D.A., Norman, M.L., Burns, J.O.: 1986, *Astrophys. J. Letters*, **311**, L63
5. Norman, M.L., Clark, D.A., Burns, J.O.: 1988, preprint
6. Kössl, D.: 1987, in *Interstellar Magnetic Fields*,
    eds. R. Beck & R. Gräve, Springer (Berlin), p. 222
7. Landau, L.D., Lifshitz, E.M.: 1984,
    Electrodynamics of Continous Media, 2nd ed., Pergamon Press (Oxford)
8. Parker, E.N.: 1979, *Cosmical Magnetic Fields*, Clarendon Press (Oxford)
9. Kössl, D., Müller E.: 1988, *Astron. Astrophys.* in press
10. Kössl, D., Müller E., Hillebrandt, W.: 1988, in preparation
11. Norman, M.L., Smarr, L., Winkler, K.H.A., Smith, M.D.: 1982,
    Astron. Astrophys., **113**, 285

Fermi Acceleration

Luke O'C. Drury

Dublin Institute for Advanced Studies,
School of Cosmic Physics,
5 Merrion Square,
Dublin 2, Ireland.

0 Preliminary remarks

As was requested by the organisers this talk is merely intended to establish the framework for contributions from the participants. It should not therefore be read as an objective or detailed review of Fermi acceleration theory (which would in any case be a task for a major monograph). Rather it should be seen as a commentary and supplement to existing reviews. The most recent, to which the reader in search of further references is referred, is that by Eichler and Blandford (1988).

1 A short historical introduction.

The concept of Fermi acceleration we owe, in a negative sense, to E. Teller who at a Chicago symposium in 1948 expounded the theory (along with Alfvén and Richtmeyer) that the cosmic rays were an essentially local solar phenomenon. Fermi disagreed (rightly) and was convinced that the cosmic rays were a Galactic phenomenon. To justify this belief he was forced to develop a theory of cosmic ray acceleration based upon interstellar processes and this formed the background to his famous paper of 1949. His problem was to devise a process which could produce a power-law distribution of energetic charged particles (in the case of the cosmic rays mostly protons) and his solution was based on the then novel idea that the Galaxy is filled with interstellar clouds containing *frozen-in* magnetic fields. Fermi pointed out that these *magnetized clouds* would scatter cosmic ray particles thus rather naturally explaining the long confinement time of the cosmic rays in the Galaxy and their isotropic distribution.

Now in the rest frame of one of these clouds the scattering is purely magnetostatic and therefore energy conserving. But if the clouds are moving with random velocities V, a relativistic particle will have its energy (or equivalently its scalar momentum p) changed by an amount $O(V/c)$ at each scattering,

$$\Delta p = O(V/c)\, p.$$

While this change can be positive or negative Fermi argued that head-on collisions (which lead to an energy gain) are more probable than overtaking collisions by a factor again $O(V/c)$ so that on average each collison leads to a gain

$$\langle \Delta p \rangle = O(V/c)^2\, p.$$

Thus the particles are accelerated at the rate

$$\dot{p} = \alpha \frac{V^2}{c^2} \frac{1}{\tau_{\text{coll}}} p = \frac{p}{\tau_{\text{acc}}}$$

where α is a numerical factor of order unity and τ_{coll} is the collison rate of the particles with the clouds. As Fermi graphically expressed it this can be seen as an attempt by the particles to come into thermal equilibrium with the clouds.

Fermi's second major point was that this process operating on its own would lead to an exponential increase with time in particle energy (at least for an energy independent τ_{coll}). If the particles are simultaneously subject to a loss process which removes them from the system in such a way that their number falls exponentially with age, then the steady solution will give a power-law energy spectrum. More formally, in terms of the phase space density of particles, $f(p, x, t)$, one can write the equation for particle conservation in the form

$$\frac{\partial f}{\partial t} + \frac{1}{4\pi p^2}\frac{\partial \Phi}{\partial p} = Q(p) - \frac{f}{\tau_{loss}}$$

where Φ is the flux of particles through the cylinder $|p| = p$, τ_{loss} is the time scale for particles to be removed from the system (Fermi considered nuclear collisions) and $Q(p)$ is a source term representing injection of fresh particles into the process. In his 1949 theory Fermi obtained

$$\Phi = \dot{p}\, 4\pi p^2\, f(p) = \frac{\alpha_1}{\tau_{coll}}\frac{V^2}{c^2} 4\pi p^3\, f(p)$$

where α_1 is a coefficient of order unity. Thus the steady solution to the equation of particle conservation at energies above the injection energy (*i.e.* where $Q(p) = 0$) is

$$f(p) \propto p^{-q}, \qquad q = 3 + \frac{\tau_{acc}}{\tau_{loss}} = 3 + \frac{c^2\tau_{coll}}{\alpha_1\tau_{loss}V^2},$$

a power-law spectrum in momentum with exponent determined by the ratio of the loss time to the acceleration time.

But clearly the type of random walk in momentum space envisaged by Fermi will in general lead to a momentum space diffusion as well as systematic drifts; thus one should generally write

$$\Phi = \frac{V^2}{c^2\tau_{coll}} 4\pi p^2 \left(\alpha_1 p f - \alpha_2 p^2 \frac{\partial f}{\partial p}\right)$$

where $\alpha_{1,2}$ are dimensionless Fokker-Planck coefficients measuring the importance of the systematic and stochastic energy changes. In most recent discussions of Fermi acceleration the scattering is supposed to be due to an ensemble of Alfvén waves and in this case V is the Alfvén speed V_A and $\alpha_1 = 0$. Indeed the phrase *second order Fermi acceleration* is often used as a synonym for momentum space diffusion. Some general points which should be noted are

- With $\alpha_2 > 0$ particles can be *decelerated* as well as accelerated.

- The acceleration time $\tau_{acc} \approx \tau_{coll}c^2/V^2$ is large compared to τ_{coll} unless $V \approx c$; *i.e.* the process is generally *slow*.

- Because it is slow it is also, in general, *inefficient*. For example let us look at Fermi's simple model. The cloud–cloud collision time is of order $\tau_{coll}c/V$ and this is the time scale for dissipation of the kinetic energy of the clouds into heat.

$$\frac{\text{CR energy production}}{\text{Heat production}} \approx \frac{\text{CR energy density}}{\text{Cloud KE density}}\frac{\tau_{diss}}{\tau_{acc}} \approx \frac{V}{c} \approx 10^{-4}$$

232

where we have used the "fact" that in the Galaxy all energy densities in the ISM are about the same. Similar arguments can be made for other models of Fermi acceleration. The kinetic energy available in turbulent motion is predominantly converted to heat and not into the acceleration of a few relativistic particles.

- But the most severe problem is that the original objective, to obtain power-law spectra, requires unnatural fine-tuning of the process. It is necessary to have $\tau_{\text{acc}} \approx \eta_{\text{loss}}$ at all p and it is not obvious why this should be the case. In particular for acceleration by an ensemble of Alfvén waves with loss by diffusion out of the wave region this is virtually impossible. Whereas for the rather simple geometrical scattering considered by Fermi it was reasonable to assume α_1 and τ_{coll} to be independent of energy there is no reason for this to be the case in Alfvén wave scattering where particles of different energies are scattered by totally different parts of the wave power spectrum.

Thus Fermi acceleration in the classical sense, while it is a mechanism which must occur, is by no means without problems.

2 Shock waves and Fermi acceleration: ten years of optimism

The picture was radically changed about ten years ago by the publication of four seminal papers (Krymsky, 1977; Axford *et al*, 1977; Blandford and Ostriker, 1978; Bell, 1978). These all describe a mechanism variously called diffusive shock acceleration, first order acceleration at shocks and the regular mechanism for acceleration at shocks. The fundamental picture is very similar to Fermi's except that instead of looking at a spatially uniform (on average) system we consider an energetic charged particle being scattered upstream and downstream of a shock.

While being scattered upstream the particle's energy changes only due to the random relative velocity of the scattering centres (which for the moment we ignore) and is otherwise conserved in the frame moving at the mean velocity of the scattering centres which we identify with the bulk plasma velocity flowing into the shock, U_1. Similarly downstream the particle energy is conserved in the frame moving with the velocity U_2 with which the shocked plasma exits the shock front. It is thus convenient to measure p with respect to a frame moving with the local plasma velocity, assumed to be the velocity at which the scattering centres are advected. With this convention a particle crossing the shock front with velocity vector $\mathbf{v}$ experiences a change in p given by

$$\Delta p = p \frac{\mathbf{v} \cdot (\mathbf{U}_1 - \mathbf{U}_2)}{v^2} + O\left(\frac{U}{v}\right)^2 p$$

and so the flux of particles through $|\mathbf{p}| = p$ at the shock is, integrating over all directions at which the particle can cross the shock front,

$$\Phi(p) = \int \Delta p\, p^2 f_s(p)\, \mathbf{v} \cdot \mathbf{n}\, d\Omega$$

where $\mathbf{n}$ is the shock normal. For $U \ll v$ the distribution at the shock, f_s, is close to isotropy and to leading order in the small parameter U/v

$$\Phi(p) = \frac{4\pi}{3} \mathbf{n} \cdot (\mathbf{U}_1 - \mathbf{U}_2)\, p^3\, f_s(p).$$

With this result we can once again write down a conservation equation for particle number, but whereas before we were considering a homogeneous system with distributed acceleration and losses here the acceleration is localised in the shock front and we must consider advective gains and losses as particles are advected into and out of the neighbourhood of the shock. Thus the steady state conservation equation now has the form

$$\frac{1}{4\pi p^2}\frac{\partial \Phi}{\partial p} = Q(p) + f_1 \mathbf{n}\cdot\mathbf{U}_1 - f_2\mathbf{n}\cdot\mathbf{U}_2$$

where f_1 denotes the far upstream distribution (a source through advection into the shock) and f_2 denotes the far downstream distribution (a loss through advection out of the shock). The final point to note is that the only steady spatial structure allowed by downstream diffusion has $f(x,p) = f_2(p)$ throughout the downstream region. Thus $f_s(p) = f_2(p)$ and the equation becomes

$$\frac{1}{3}\mathbf{n}\cdot(\mathbf{U}_1 - \mathbf{U}_2)\,p\frac{\partial f_2}{\partial p} = (f_1 - f_2)\,\mathbf{n}\cdot\mathbf{U}_1 + Q,$$

the standard equation of diffusive shock acceleration in the test-particle limit. Thus in terms of the shock compression ratio, $r = \mathbf{n}\cdot\mathbf{U}_1/\mathbf{n}\cdot\mathbf{U}_2$, the steady state spectrum above the injection energy is a power-law

$$f(p) \propto p^{-q}, \qquad q = \frac{3r}{r-1}$$

with exponent determined only by the compression ratio.

It is worth noting that this derivation says nothing about the obliquity of the magnetic field to the shock normal; the shock can be quasi-parallel or quasi-perpendicular, the magnetic field direction can even vary in space and time; as long as the distributions remain close to isotropy and the scattering centres are advected with the plasma the shock will accelerate the standard spectrum.

To summarize, Fermi acceleration results from the fact that magnetic fields couple microscopic degrees of freedom of individual particles to the macroscopic dynamics of plasma systems. The resulting momentum space fluxes can be expressed for homogeneous turbulence in the form of two distributed terms;

$$\Phi_1 = \alpha_1\frac{V^2}{c^2\tau_{\mathrm{coll}}}4\pi p^3 f,$$

a systematic term (often zero) and

$$\Phi_2 = -\alpha_2\frac{V^2}{c^2\tau_{\mathrm{coll}}}4\pi p^4\frac{\partial f}{\partial p},$$

a stochastic second order term. In addition at a shock front there is a flux,

$$\Phi_{\mathrm{S}} = \mathbf{n}\cdot(\mathbf{U}_1 - \mathbf{U}_2)\frac{4\pi}{3}p^3 f\delta(x - x_{\mathrm{S}})$$

localised to the shock (at x_{S}). Only this flux is naturally linked to a loss process in such a way that power-law spectra are inevitably produced.

The associated time scales are, in the case of homogeneous turbulence, easily seen to be

$$\tau_{\mathrm{acc}} = O\left(\tau_{\mathrm{coll}}c^2/V^2\right) \approx 3\kappa/V^2$$

where κ is the spatial diffusion coefficient. The time scale for shock acceleration is not so obvious, but can be shown to be (in test particle theory)

$$\tau_{\mathrm{acc}} = \frac{3}{\mathbf{n}\cdot(\mathbf{U}_1 - \mathbf{U}_2)}\int_{-\infty}^{+\infty} dx\,\exp\left[-\left|\int_0^x dx'\,\frac{\mathbf{n}\cdot\mathbf{U}(x')}{\mathbf{n}\cdot\kappa(p,x')\cdot\mathbf{n}}\right|\right]$$

which simplifies for a parallel shock and

$$\kappa = \begin{cases}\kappa_1 & x < 0 \\ \kappa_2 & x > 0\end{cases}$$

to the well known result

$$\tau_{acc}(p) = \frac{3}{U_1 - U_2} \left(\frac{\kappa_1}{U_1} + \frac{\kappa_2}{U_2} \right).$$

Shock acceleration is usually much faster than classical Fermi acceleration because $U \gg V$ and because κ near a shock is much smaller than κ in the general medium.

3 Open problems

Fermi acceleration is a process for accelerating particles which already have substantially more than thermal energies, in other words it answers the question *where does the energy of the particles come from* but not *where do the particles come from*. This injection problem is an old one common to all discussions of Fermi processes; in his original paper Fermi proposed an ingenious bootstrapping process whereby the secondary particles produced in high energy collisions could provide the injection source. While this clearly fails for the Galactic cosmic rays it may be an important process in some extra-galactic sources (especially as a source of positrons and electrons). Here again shock acceleration enjoys an advantage over classical Fermi acceleration. While the structure of a high Mach number collisionless shock is very uncertain it is quite plausible that the non-thermal tail of the distribution function of the shocked particles may provide a source of direct injection in the shock front itself. And indeed observations of the Earth's bow shock do suggest that of order 10^{-3} of the incident protons are reflected with enough energy to enter the shock acceleration process. The question of electron injection is even more obscure, but should if anything be easier at relativistic shocks where the difference in rest mass between protons and electrons is less important. And of course a shock in an electron–positron plasma can only inject electrons (and positrons!).

Another well known problem is the question of *reaction effects*; how do the accelerated particles modify the structure of the shock. This has to be considered in any discussion of the efficiency of shock acceleration. There are two main effects, first a smoothing of the shock profile leading to a structure consisting of a subshock and precursor (and the subshock may disappear leaving a totally smooth structure) and secondly a change in the compression ratio of the shock . It should be noted that once reaction effects become important the spectrum will start to deviate from an exact power-law form. There seems to be no reason why relativistic particles, essentially because of their very large mean free path, should not provide the dominant source of kinetic energy dissipation in strong shocks; in principle 98% of the energy could be used for particle acceleration.

A final problem which plagues the theory is the question of what to use for the diffusion coefficient, *i.e.* what determines the level of scattering. It is convenient to scale the diffusion coefficient in terms of the Bohm diffusion coefficient,

$$\kappa_{Bohm} = \frac{1}{3} r_g v$$

where r_g is the particle gyroradius and v its velocity. The idea here is that the shortest mean free path that a magnetic field can produce is of the order of the gyroradius and indeed quasilinear theory gives

$$\kappa_{\parallel} \approx \kappa_{Bohm} \left(\frac{B}{\delta B} \right)^2_{resonant} \qquad \kappa_{\perp} \approx \kappa_{Bohm} \left(\frac{\delta B}{B} \right)^2_{resonant}$$

where $\delta B^2_{resonant}$ denotes the power in waves resonant with the particle. Unfortunately, apart from questions about the applicability of quasilinear theory, this merely changes the question of estimating the diffusion coefficient into one of estimating the level of resonant wave activity. Upstream of the shock the streaming instability (Lerche, 1967; Kulsrud and Pearce, 1969; McKenzie and Völk 1982; Bell 1978) is very strong so that the assumption $\delta B/B \approx 1$ and $\kappa \approx \kappa_{\text{Bohm}}$ is probably good. Indeed various nonresonant instabilities may amplify the effective value of B and give even smaller diffusion coefficients; there is plenty of energy available in a strong shock! Downstream of the shock wave one expects an initially very high level of wave activity as the upstream waves are advected through the shock and amplified; however in the absence of significant resonant excitation damping may present a problem as discussed by Achterberg and Blandford(198*).

4 Recent work

An interesting development in test particle theory is its application to relativistic shocks. This is mainly due to Kirk, Schneider and Heavens (see papers at this workshop), but can be traced back to the pioneering work of Peacock (1981). From basic theory the only way to get a spectral slope other than $3r/(r-1)$ is to induce anisotropies in the distribution at the shock; this can be done by weak scattering and oblique magnetic fields (a case which has not yet been investigated) or by pitch angle dependent losses (obviously relevant to synchrotron losses) or by considering relativistic shocks. Fortunately the resulting changes in the spectral index seem to be quite small.

In a preprint (to be published in Mon. Not. Roy. astr. Soc.) Achterberg considers acceleration in *disordered fields*; by this he means that instead of the usual assumption of a homogeneous large-scale background field (so that all field lines are open) he considers a model in which space is divided into a hierarchy of nested closed magnetic 'boxes' or 'bottles'. If he assumes a self-similar hierarchy of magnetic structures, then on letting a shock propagate through the system he obtains a final average accelerated spectrum which is a power-law, but differs somewhat from that conventionally obtained.

The form of the spectrum when synchrotron losses produce a cut-off at high energies has been considered by among others Biermann, Meisenheimer and Heavens. This effect is important because it provides one of the few observational test of the theory. On the whole these very simple models seem to fit the data remarkbly well (see contributed papers at this workshop).

Schlickeiser and his coworkers have considered the effects of concurrent second order Fermi acceleration and shock acceleration. It is clear that if second order Fermi is allowed to operate throughout the semi-infinite downstream region behind a shock, then it will ultimately be the dominant effect. However if we consider only the neighbourhood of the shock, defined roughly as one diffusion length upstream and downstream, which is where the shock acceleration takes place, then the flux due to second order Fermi can be estimated as of order $\Phi_2\left(\kappa_1/U_1 + \kappa_2/U_2\right)$. Thus the ratio of the second order flux to the shock accelerated flux is

$$\frac{-\alpha_2 V^2 p \frac{\partial f}{\partial p}\left(\frac{1}{U_1} + \frac{1}{U_2}\right)}{f\left(U_1 - U_2\right)}$$

which is of order M_A^{-2} where M_A is the Alfvén Mach number of the shock. Thus for astrophysical shocks where typically $M_A \gg 10$ the effect is small.

Analytic approximations to the spectral slope of particles accelerated in steady modified shocks have been presented in preprints by Pelletier and Roland (in press, Astr. & Ap.) and Schneider and Kirk but these results are only valid for momentum independent diffusion coefficients and thus of limited utility.

Various non-resonant instabilities of strong modified shock structures have been discussed by Drury and Falle (1987), Zank and McKenzie (1985,1987) and by Berezhko *et al* (1987). These are important because they indicate that steady modified shock structures probably only exist as mathematical fictions. In addition they allow amplification of magnetic fields in the precursor and by producing many small subshocks may facilitate the injection and acceleration of low-energy particles.

But the most important recent development in shock acceleration must surely be the publication last year of the first time-dependent numerical solutions (Falle and Giddings, 1987; Bell, 1987). This is the only way to study reaction effects properly and the success of these programs, running on rather small computers, indicates that it should soon be possible to run calculations of jets including particle acceleration. This will allow a much better comparison between theory and observation.

5 Concluding remarks

Fermi acceleration, in its shock acceleration form, is the only particle acceleration process which seems capable of producing anything like the observed spectra with the observationally required efficiency in extragalactic sources. In addition there is good evidence that the process is observed, though in a much weaker form, by spacecraft in the heliosphere (*cf* Kennel *et al*, 1986). But perhaps its greatest attraction is that the theory has a sufficiently precise mathematical formulation to allow (at least in theory) detailed calculations. Within the next decade we may hope to see numerical models incorporating the effects of particle acceleration, fluid dynamics and magnetic fields.

Finally I would like to thank the organisers of this workshop for a most stimulating week and P. Duffy for a critical review of this short paper.

References

Achterberg, A. 1981 *Astron. Astrophys.* **98** 195.

Achterberg, A. and Blandford, R. D. 1986 *Mon. Not. R. astr. Soc.* **218** 551.

Axford, W. I., Leer, E. and Skadron, G. 1977 *Proc. 15th ICRC*(Plovdiv) **11** 132.

Bell, A. R. 1978 *Mon. Not. R. astr. Soc.* **182** 147.

Bell, A. R. 1987 *Mon. Not. R. astr. Soc.* **225** 615.

Berezhko, E. G., Yelshin, V. K., Krymsky, G. F. and Turpanov, A. A. 1987 *Proc. 20th ICRC* (Moscow), **2** 175.

Blandford, R. D., and Ostriker, J. P. 1978 *Astrophys. J.* **221** L29.

Blandford, R. D. and Eichler, D. 1987 *Phys. Rep.* **154** 1.

Drury, L. O'C. and Falle, S. A. E. G. 1987 *Mon. Not. R. astr. Soc.* **223** 353.

Falle, S. A. E. G., and Giddings, J. R. 1987 *Mon. Not. R. astr. Soc.* **225** 399.

Fermi, E. 1949 *Phys. Rev.* **75** 1169.

Kennel, C. F., Coroniti, F. V., Scarf, F. L., Livesey, W. A., Russel, C. T., Smith, E. J., Wenzel, K. P. and Scholer, M., 1986 *J. Geophys. Res.* **91** 11917.

Krymsky, G. F. 1977 *Dokl. Akad. Nauk SSSR* **234** 1306.

Kulsrud, R. M. and Pearce, W. 1969 *Astrophys. J.* **156** 445.

Lerche, I. 1967 *Astrophys. J.* **147** 689.

McKenzie, J. F. and Völk, H. J. 1982 *Astron. Astrophys.* **116** 191.

Peacock, J. A. 1981 *Mon. Not. R. astr. Soc.* **196** 135.

Zank, G. P. and McKenzie, J. F. 1985 *Proc 19th ICRC* (La Jolla) **3** 111.

Zank, G. P. and McKenzie, J. F. 1987 *J. Plasma Physics* **37** 347.

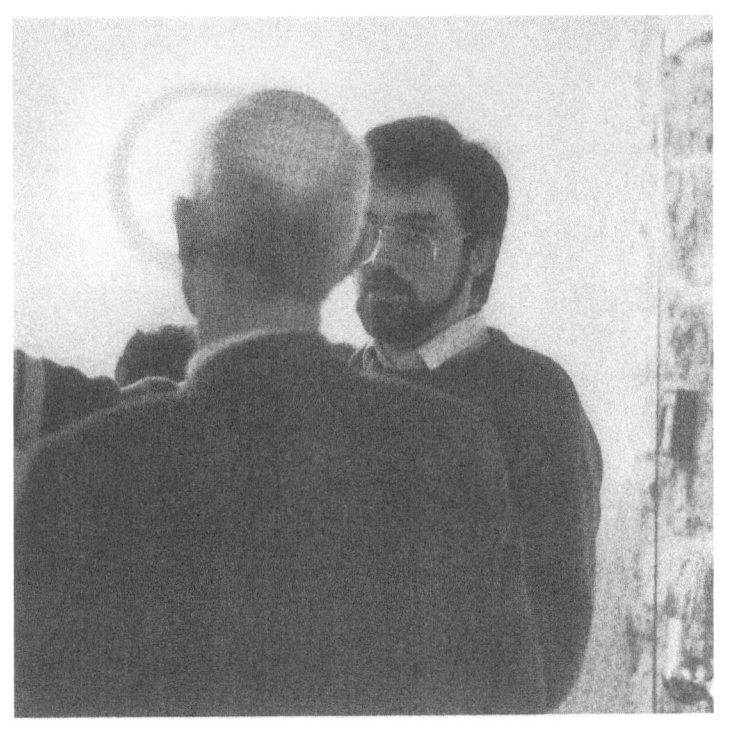

Luke Drury

Dieter Biskamp

Peter Biermann
John Kirk
John Dreher   Klaus Meisenheimer

FIRST-ORDER FERMI ACCELERATION
AT RELATIVISTIC SHOCK FRONTS[*]

J. G. Kirk

Max-Planck-Institut für Physik und Astrophysik
Institut für Astrophysik
Karl-Schwarzschild-Straße 1
D – 8046 Garching
Federal Republic of Germany

1 Introduction

The theory of diffusive acceleration at shock fronts, which applies if the fluid speed is *nonrelativistic*, yields a simple expression for the power-law index of accelerated particles in the approximation that these may be regarded as test particles and that the shock front is a simple discontinuity with its normal parallel to the flow direction and magnetic field. For the index s of the Lorentz invariant phase-space density f ($\propto p^{-s}$, with p the momentum) one has:

$$s = \frac{3r}{(r-1)} \tag{1}$$

where r is the compression ratio of the shock front, defined as the ratio of the velocity of the upstream fluid into the shock (u_-) to the velocity of the downstream fluid out of it (u_+). Although the acceleration process depends on there being effective pitch-angle scattering of the particles in both the upstream and downstream regions, no property associated with this process appears in equation (1). Unfortunately, if the velocity of the fluid through the shock front is *relativistic* this attractive property of first-order Fermi acceleration no longer holds and it becomes necessary to develop specific models describing the scattering process in order to find the index s. The basic reason is that the assumption of near isotropy of the distribution function, implicit in the theory of diffusive acceleration, is no longer valid when the velocity of the particles (in this case approximately that of light) is not large compared to the velocity of the fluid into the shock. An almost isotropic distribution is produced by any kind of pitch-angle scattering, but an anisotropic one reacts differently according to the properties of the scattering.

In this contribution, results will be presented for two types of pitch-angle scattering derived for different values of the spectrum of Alfvén wave turbulence in the quasi-linear theory of plasma physics. The method of calculation is a generalization of that described in [8] — the "Q_J method" and has much in common with the method used in applications involving interplanetary shocks [6]. A full description of the method and presentation of results may be found in [7].

2 Results

A relativistic shock front is completely specified for present purposes by two parameters: the upstream velocity u_- and the downstream velocity u_+. (Units are employed in which the velocity of light is unity.) In the diffusive theory, which applies only to nonrelativistic shocks, one parameter suffices, since equation (1) contains only the combination $r = u_-/u_+$. Several papers on relativistic shocks may be found in the literature [5,2,13,9], but each limits itself to particular cases such as that of ultrarelativistic shocks, or uses approximations concerning the equation of state of the fluid. Here, results are presented for five different physical situations which could occur behind

---

[*] At this conference the author became aware of the independent but essentially identical work of Heavens and Drury on this topic [4]

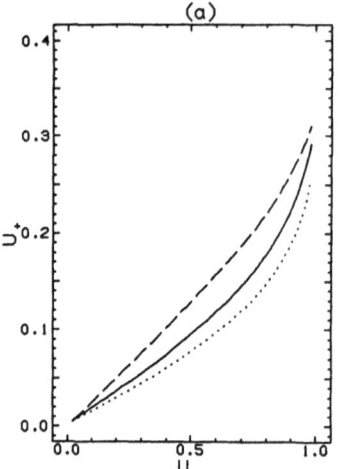 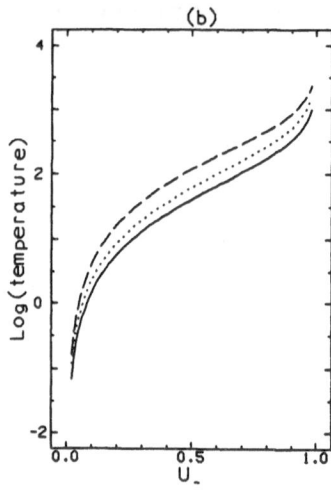

Figure 1: a) The downstream velocity u_+ as a function of the upstream velocity u_- for a strong shock in an ideal gas under the assumption that the pressure is provided either by the electrons alone (dotted line), or by the ions alone (dashed line), or that the gas is in full thermodynamic equilibrium (solid line). b) The logarithm of the temperature (in units of the electron rest mass) behind the shock front. Electrons are relativistic for $\log(T) > 0$, whereas ions require $\log(T) > 3$. When both ions and electrons contribute to the pressure, a lower temperature is achieved behind the shock front.

a relativistic strong shock. In the first, complete thermodynamic equilibrium is assumed to be achieved between all components of the fluid (which consists of fully ionized hydrogen and helium – 25% by mass – together with the associated electrons). Two other possibilities investigated are that all the kinetic energy of the upstream fluid is deposited either into the ions or into the electrons. The relationship between u_- and u_+ which is implied by each of these assumptions is shown in Figure 1, together with the resulting temperature of the downstream plasma. A fourth possibility is to assume that electron–positron pairs are created in the shock, and, together with the electrons, dominate the pressure. This possibility is probably relevant to shocks which occur in the inner regions of AGN's [12], where pairs can be produced by interactions of protons with the local radiation field. Figure 2 shows the relationship between u_- and u_+ in this case.

Finally, there exists a simple formula for the jump conditions across a relativistic shock front:

$$u_- u_+ = 1/3, \tag{2}$$

valid if both the upstream fluid and the downstream fluid obey the equation of state of an ultra-relativistic gas.

The quasi-linear theory of plasma physics predicts that energetic particles resonate with Alfvén waves causing a change of pitch-angle which is diffusive in nature. The pitch-angle diffusion coefficient $D_{\mu\mu}$ (μ is the cosine of the pitch-angle) depends on the spectrum of waves present in the plasma [10]. For a k^{-1} spectrum, where k is the wavenumber of an Alfvén wave, the resulting coefficient is "isotropic" in the sense that the scattering suffered by a particle is independent of its direction: $D_{\mu\mu} = 1 - \mu^2$. For spectra steeper than -1, such as the Kolmogorov spectrum $k^{-5/3}$, a problem arises which has to do with the quasi-linear approximation rather than the underlying physics: the pitch-angle diffusion coefficient develops a "hole" around the point $\mu = 0$ [3]. Nonlinear effects are estimated to be sufficient to fill this hole at least partially and their strength can be parameterized by a quantity ϵ, [14], as shown in Figure 3. The effect of the "hole" is more

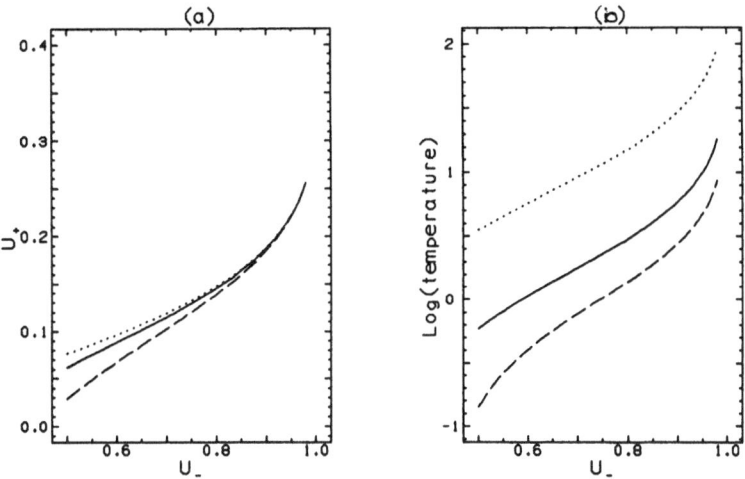

Figure 2: a) The downstream velocity u_+ as a function of the upstream velocity u_- for a strong shock under the assumption that the pressure is provided by the electrons and electron–positron pairs created at the shock front. The number of pairs per proton is approximately 10 (dotted line), 50 (solid line) and 100 (dashed line). b) The logarithm of the temperature (in units of the electron rest mass) behind the shock front.

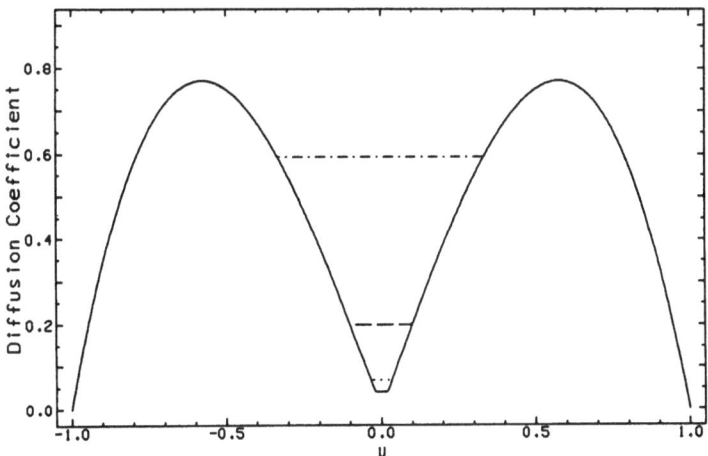

Figure 3: The pitch-angle diffusion coefficient as a function of the cosine μ of the pitch-angle, according to the quasi-linear theory. Alfvén turbulence with a spectrum $q = 2$ is assumed, and the "hole" around $\mu = 0$ is modified according to the prescription of Völk et al. [14], using the parameter ϵ. For $\epsilon = 0$ (solid line) the Alfvén speed is taken to be 1/50 times the particle speed. For $\epsilon = 1/30$ (dotted line), $\epsilon = 1/10$ (dashed line) and $\epsilon = 1/3$ (dot-dashed line), the Alfvén speed is set to zero.

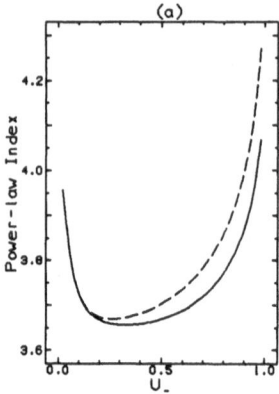
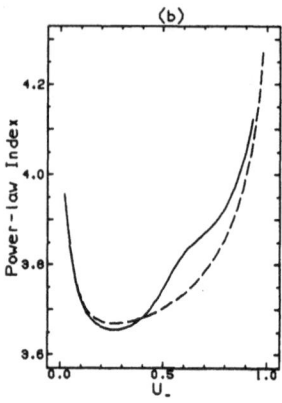

Figure 4: The power-law index of particles accelerated by a relativistic shock front when the particle transport is governed by pitch-angle diffusion. In (a), the pitch-angle diffusion coefficient is isotropic, whereas in (b), a highly anisotropic diffusion coefficient with $q = 2$, $\epsilon = 1/30$ as shown in Figure 3 is assumed. The jump conditions are for a gas which reaches full thermodynamic equilibrium behind the shock front. The solid curve presents the results of the fully relativistic theory; the dashed curve is the result given by the nonrelativistic theory, equation (1).

pronounced for steeper wave spectra. In Figure 3 and in the following a k^{-2} spectrum is employed for illustration.

The power-law index of accelerated particles which results from the combinations of pitch-angle scattering and jump conditions described above is depicted in Figures 4 and 5. For a spectrum of turbulence of the Kolmogorov type ($\propto k^{-5/3}$ one obtains results which lie between the k^{-1} and k^{-2} cases.

3 Discussion

The main conclusion to be drawn from Figures 4 and 5 is that the energetic electrons seen in the hot spots which are the subject of these proceedings could be produced by first order Fermi acceleration at shock fronts. This point deserves mention, since there have been suggestions in the literature that relativistic shocks are intrinsically ineffective accelerators [1]. Almost any value of the power-law index s between 3 and 6 can be obtained by a suitable combination of shock conditions and pitch-angle diffusion coefficients. Furthermore, the results presented apply to strong shocks (except for the case of the ultrarelativistic gas); weaker shocks should produce uniformly softer spectra. Thus, it is not possible to determine that a given population of energetic particles has undergone acceleration at a *relativistic* shock front simply by observing the index of the power law. On the other hand, if this is assumed to be the case, it may be possible to place some constraints on the properties of the fluid in which the shock occurs.

As mentioned in §1, the major difference between acceleration at nonrelativistic shocks and at relativistic ones is that the particle distribution in the latter case is anisotropic in the vicinity of the shock. Thus, if the observed radiation is the result of synchrotron radiation, one may perhaps look for a signature of an anisotropic particle distribution in the emission. The degree of circular polarization seems to be the most promising such signature. In the case of an isotropic distribution, the degree of circular polarization of synchrotron radiation is small ($\sim 1/\gamma$, where γ is the Lorentz factor of the radiating particles). Preliminary estimates suggest that the presence of the hole in the pitch-angle diffusion coefficient shown in Figure 3 enhances the degree of circular polarization by the possibly large factor c/v_A, where c is the velocity of light and v_A is the Alfvén velocity. This

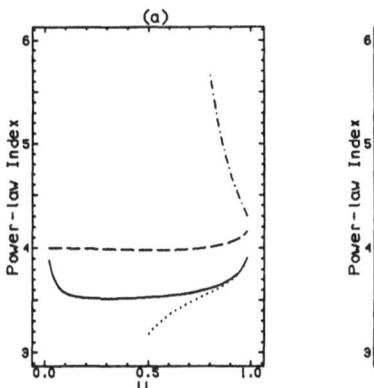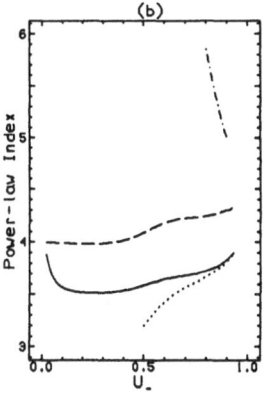

Figure 5: As in Figure 4, but for the jump conditions shown in Figures 1 and 2. The dash-dotted line corresponds to a relativistic gas with jump conditions given by equation (2). The dashed line is for a strong shock with only ion pressure supporting it, the solid line for support by electron pressure only. The dotted line is for a strong shock with pressure provided by electrons and pairs, the number of these being approximately 100 per proton.

would, however, only be detectable at frequencies at which the emission is dominated by particles close to the shock front, which means, in the context of the first order Fermi model, frequencies close to the observed cut-offs [11].

References

[1] Blandford, R. D. and Eichler, D. 1987 *Phys. Rep.* **154**, 1.

[2] Blandford, R. D. and McKee, C. F. 1976 *Phys. Fluids* **19**, 1130.

[3] Davilla, J. M. and Scott, J. S. 1984 *Astrophys. J.*, **285**, 400.

[4] Heavens, A. F. and Drury, L. O'C. these proceedings

[5] Johnson, M. H. and McKee, C. F. 1971 *Phys. Rev.* **D3**, 858.

[6] Kirk, J. G. 1988 *Astrophys. J.*, **324**, 557.

[7] Kirk, J. G. 1988 *Habilitationsschrift*, University of Munich, available as preprint MPA345, Max Planck Institute for Astrophysics, February 1988.

[8] Kirk, J. G. and Schneider, P. 1987 *Astrophys. J.*, **315**, 425.

[9] Königl, A. 1980 *Phys. Fluids* **23**, 1083.

[10] Luhmann, J. G. 1976 *J. Geophys. Res.*, **81**, 2089.

[11] Röser, H.-J. these proceedings.

[12] Sikora M., Kirk J. G., Begelman M. C. and Schneider P. 1987 *Astrophys. J.*, **320**, L81.

[13] Taub, A. H. 1978 *Ann. Rev. Fluid Mech.* **10**, 301.

[14] Völk, H. J., Morfill, G., Alpers, W. and Lee, M. A. 1974 *Astrophys. and Space Sci.*, **26**, 403.

Coffee Break

SPECTRAL INDICES FROM RELATIVISTIC AND NON-RELATIVISTIC SHOCKS

Alan Heavens
Department of Astronomy, University of Edinburgh
Blackford Hill, Edinburgh, EH9 3HJ, U.K.

Summary

This paper consists of two parts; the first part contains the results of work done with Luke Drury on particle acceleration in relativistic shock waves, the details of which will appear elsewhere (Heavens & Drury 1988). The second part summarises the range of spectral indices that have been obtained from this and other calculations of particle acceleration in shock waves.

1 Relativistic shock acceleration

In the non-relativistic theory of diffusive shock acceleration (see Drury 1983 and Blandford & Eichler 1987 and references therein), elastic scattering by magnetic irregularities causes high-energy particles to diffuse in pitch-angle; this continual changing of direction allows some of the particles to cross the shock many times, and gain energy by the first-order Fermi mechanism. In such shocks, the scattering of the accelerated particles can keep the distribution close to isotropy, regardless of whether we look at the distribution in the rest frame of the shock or the up- or downstream fluid. This allows simplification of the transport equation, with diffusion in pitch-angle being replaced by diffusion in space. If the shock is moving relativistically the particle distribution must be anisotropic in at least one of these frames and the problem becomes immediately more complicated. After the pioneering work of Peacock (1981), Kirk & Schneider (1987) recently made an important advance in the treatment of this problem, by expanding the particle distribution in terms of anisotropic eigenfunctions of the scattering equation. We have extended their work in two ways: we have considered shocks propagating into cold material, and we have modified their method to treat more general scattering power spectra. As an example of a power spectrum with more power at longer wavelengths than shorter (not especially motivated by any physical considerations), we have applied it to scattering by magnetic waves with a Kolmogorov power spectrum.

The results for the accelerated particle distribution are shown in Figure 1, given in terms of the synchrotron emission from electrons at the shock: $S_\nu \propto \nu^{-\alpha}$, with α given to an absolute accuracy of 0.01 (for shock speed $u \leq 0.98$) by

$$\text{proton} - \text{electron plasma, D} = \text{constant} : \quad \alpha = 0.50 - 1.58u + 5.43u^2 - 7.50u^3 + 3.73u^4$$

$$\text{(Kolmogorov)} : \quad \alpha = 0.50 - 1.59u + 5.33u^2 - 6.98u^3 + 3.33u^4$$

$$\text{electron} - \text{positron plasma, D} = \text{constant} : \quad \alpha = 0.50 - 0.10u + 0.56u^2 - 1.16u^3 + 0.79u^4$$

$$\text{(Kolmogorov)} : \quad \alpha = 0.50 - 0.18u + 0.79u^2 - 0.98u^3 + 0.49u^4$$

We recommend the use of the Kolmogorov results, as we feel this scattering spectrum is likely to be closer to reality than the isotropic case.

In contrast to the non-relativistic result, the form of the diffusion coefficient does slightly affect the spectral index. However, it is the compression ratio of the shock (measured in the frame in which the shock is stationary) which has the major effect on the spectrum of accelerated particles, as in the non-relativistic case. In fact, applying the non-relativistic result $\alpha = 1.5u_+/(u - u_+)$ gives good results for shock speeds not too close to c (= 1). For the case of a plasma

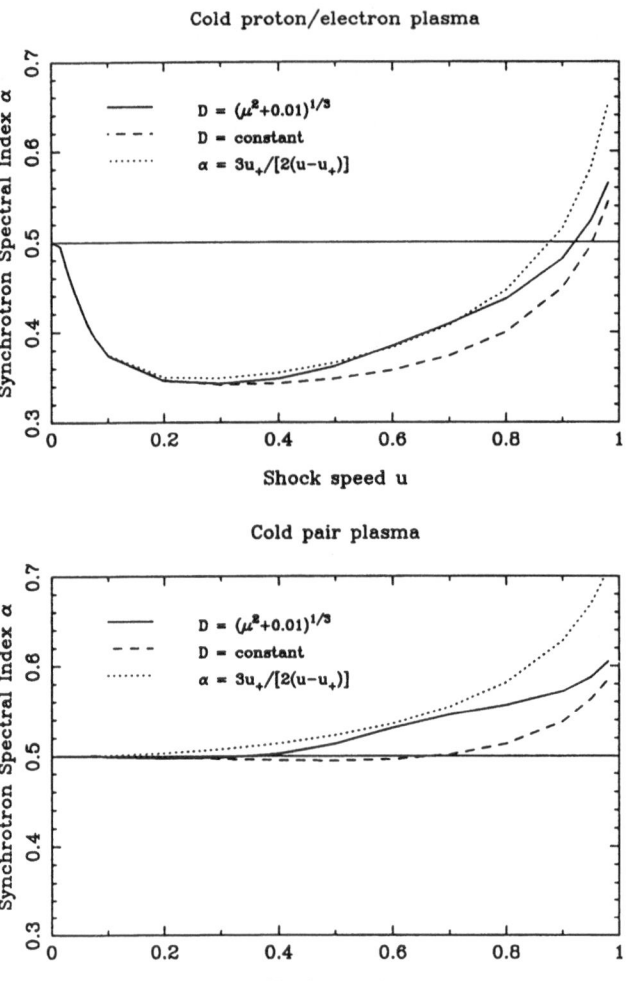

Cold proton/electron plasma

Cold pair plasma

Fig.1(a) The spectral index of the synchrotron emission from Fermi- accelerated electrons at a shock propagating into cold proton-electron plasma. The dashed line corresponds to isotropic scattering, while the solid line corresponds to scattering by magnetic waves with a Kolmogorov power spectrum. The dotted line shows the extrapolation of the non-relativistic result $\alpha = 1.5u_+/(u - u_+)$, which lies close to the Kolmogorov result up to $u \simeq 0.8$.

Fig.1(b) As Fig. 1(a), but for shocks in an electron-positron plasma.

composed of electrons and protons, the adiabatic index changes quite rapidly from $\Gamma \approx \frac{5}{3}$ to $\Gamma \approx \frac{13}{9}$ at $u \approx 0.03$, when the thermal electrons become relativistic (cf Synge 1957). The compression ratio increases from 4 to about 5.5, and the spectral index drops accordingly from 0.5 to about 0.35. As the shock speed approaches 1, relativistic effects reduce the compression ratio, and the spectrum steepens. The anisotropy of the particle distributions then renders the non-relativistic approximation increasingly inaccurate. In the case of electron-positron plasmas, the adiabatic index of the shocked gas changes much more smoothly, and the spectral index varies little, steepening from 0.5 to 0.6 as the shock speed is increased from zero to 0.98. For shock speeds greater than 0.98, the numerical method becomes unreliable, and it is not possible to say with certainty what the behaviour of the spectral index will be for shocks with very high Lorentz factors. However, all our curves lead towards $\alpha \simeq 0.63$ from below, and Kirk & Schneider's curve approaches this value from above, suggesting that this may well be the asymptotic value regardless of the scattering details. It must be stressed that these comments are conjecture – the extrapolation to high γ is made on the basis that nothing unusual seems to happen for γ up to 5. *Caveat emptor.*

Since the spectral index of the accelerated particle distribution is sensitive to the compression ratio of the shock, the results I have presented here will be accurate only if the downstream gas is efficiently thermalised, and magnetic effects do not alter the shock structure significantly. Note also that this is a test-particle calculation; any modifications to the shock structure resulting from the pressure of the accelerated particles will also change the form of the spectrum.

2 Spectral indices from Fermi acceleration at shock fronts

Figure 2 surveys the results of studies of Fermi acceleration in shock waves, with references in the caption. The standard result for strong, non-relativistic shocks in matter with $\Gamma = \frac{5}{3}$ is $\alpha = 0.5$. Thermalised shocks with $u > 10^4 \mathrm{kms}^{-1}$ imply $\Gamma \simeq \frac{13}{9}$ and so $\alpha \simeq 0.35$. Highly relativistic strong shocks give rise to $\alpha \simeq 0.5 \pm 0.1$. Efficient acceleration, in the sense that a large fraction of the bulk momentum flux is converted to accelerated particle pressure, leads to more compressive shocks and a flatter slope. In this case, if the diffusion coefficient increases with momentum, the spectrum should be concave ($d\alpha/d\nu < 0$), but in the case of strong shocks, the spectrum cannot be a true steady-state, as the particle pressure diverges for $\alpha(\nu \to \infty) \leq 0.5$. If synchrotron losses are important within the source, then the spectrum exhibits a high-frequency cutoff, which moves to lower frequencies as the electrons are convected away from the shock. The integrated emission from a finite downstream region then exhibits a steady-state two-power law behaviour, with the familiar break of $\Delta\alpha = 0.5$. Test particles in a strong, non-relativistic shocks under these circumstances have $\alpha = 1$ over part of the frequency range. Included is a simple calculation of a decaying magnetic field configuration simply to illustrate the fact that, with not implausible assumptions, near-power laws with different spectra from standard theory can easily be produced by allowing the magnetic field to vary. The assumptions that have gone into the distributions shown in Fig. 3 are: a population of electrons $f \propto p^{-4}$ up to a sharp cutoff, produced at a shock, and subject to synchrotron losses; a spatially decaying magnetic field $B = B_0(1 + x/x_B)^{-2/3}$, whose form is chosen only for mathematical convenience. The spectrum depends on the relative importance of synchrotron losses and magnetic field decay, *via* the combination $u_+ \tau_{loss}/x_B$, where τ_{loss} is the time for electrons at p_0 to lose half their energy in a field B_0, and u_+ is the speed of the shocked gas relative to the shock.

References

Achterberg, A., Blandford, R.D. & Periwal, V., 1984. *Astron. & Astrophys.*, **132**, 97.
Axford, W.I., Leer, E. & Skadron, A., 1977. *Proc. 15th International Cosmic Ray Conf. (Plovdiv)*, **11**, 132.
Bell, A.R., 1978. *Mon. Not. R. astr. Soc.*, **182**, 147.
Blandford, R.D. & Eichler, D., 1987. *Phys. Reports*, **154(1)**, 1.
Blandford, R.D. & Ostriker, J.P., 1978. *Astrophys. J.*, **227**, L49.
Bregman, J.N., 1985. *Atrophys. J.*, **288**, 32.

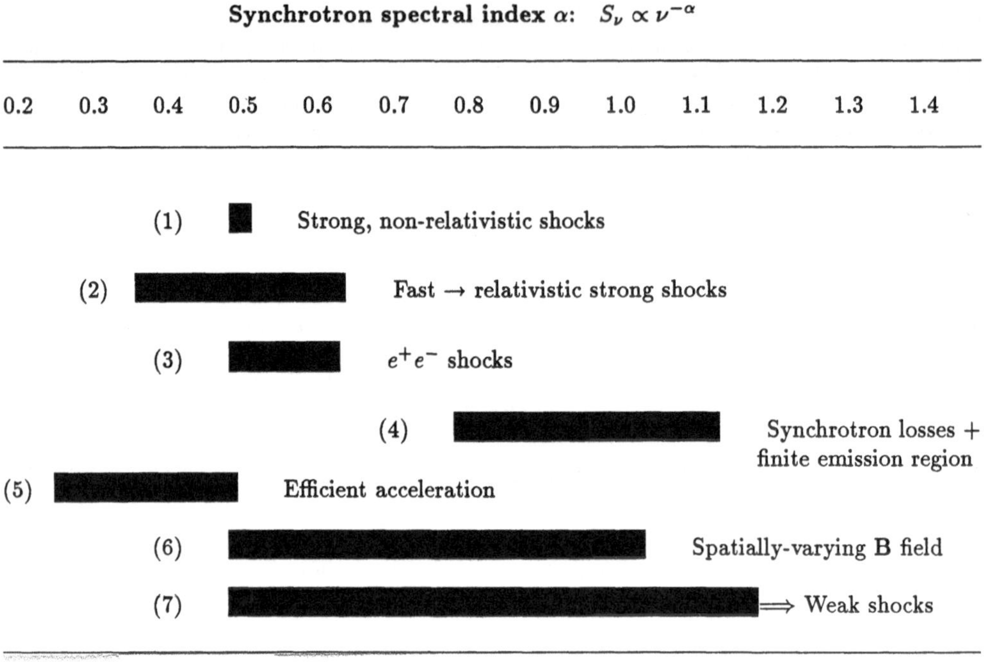

Synchrotron spectral index α: $S_\nu \propto \nu^{-\alpha}$

0.2	0.3	0.4	0.5	0.6	0.7	0.8	0.9	1.0	1.1	1.2	1.3	1.4

(1) ▮ Strong, non-relativistic shocks

(2) ▬▬▬▬▬ Fast → relativistic strong shocks

(3) ▬▬▬ e^+e^- shocks

(4) ▬▬▬▬▬ Synchrotron losses + finite emission region

(5) ▬▬▬ Efficient acceleration

(6) ▬▬▬▬▬▬ Spatially-varying B field

(7) ▬▬▬▬▬▬▬ ⟹ Weak shocks

Fig.2   The ranges of steady-state spectral indices which can be explained by Fermi acceleration at shock waves. The numbers refer to the original papers, as follows:
1) Krimsky (1977), Axford, Leer & Skadron (1977), Bell (1978), Blandford & Ostriker (1978).
2) Kirk & Schneider (1987), this paper.
3) This paper.
4) Bregman (1985), Meisenheimer & Heavens (1986), Heavens & Meisenheimer (1987).
5) Achterberg, Blandford & Periwal (1984), Heavens (1984).
6) This paper.
7) Bell (1978), Kirk & Schneider (1987).

Drury, L.O'C., 1983. *Rept. Prog. Phys.*, **46**, 973.
Heavens, A.F., 1984. *Mon. Not. R. astr. Soc.*, **210**, 813.
Heavens, A.F. & Drury, L.O'C., 1988. *Mon. Not. R. astr. Soc.*, submitted.
Heavens, A.F. & Meisenheimer, K., 1987. *Mon. Not. R. astr. Soc.*, **225**, 335.
Kirk, J.G. & Schneider, P., 1987. *Astrophys. J.*, **315**, 425.
Krimsky, G.F., 1977. *Dok. Akad. Nauk SSSR*, **234**, 1306.
Meisenheimer, K. & Heavens, A.F., 1986. *Nature*, **323**, 419.
Peacock, J.A., 1981. *Mon. Not. R. astr. Soc.*, **196**, 135.
Synge, J.L., 1957. *"The Relativistic Gas"*, North Holland Publishing Company (Amsterdam).

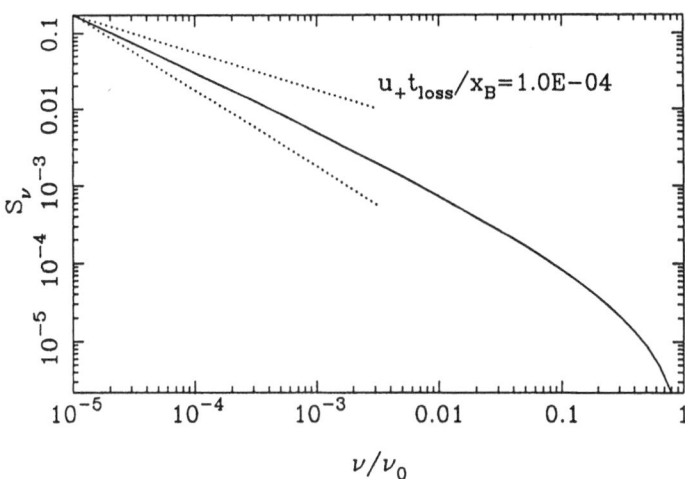

Synchrotron Spectrum: $B(x)=B_0(1+x/x_B)^{-2/3}$

$u_+t_{loss}/x_B=1.0E-04$

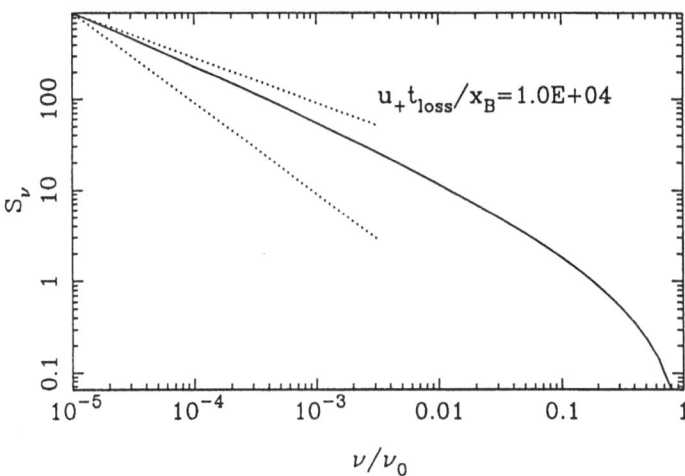

Synchrotron Spectrum: $B(x)=B_0(1+x/x_B)^{-2/3}$

$u_+t_{loss}/x_B=1.0E+04$

Fig.3 Approximate synchrotron flux from particles accelerated to $f(p) \propto p^{-4}$ up to a cutoff p_0 at a planar surface, losing energy *via* synchrotron emission as they are convected away at speed u_+. The field is assumed to decay spatially as $B(x) = B_0(1 + x/x_B)^{-2/3}$. τ_{loss} is the time for electrons at p_0 to lose half their energy in a field B_0. The dotted lines have $\alpha = 0.5$ and $\alpha = 1.0$. The spectral index is changing only slowly with frequency, and, for a wide range of $u_+\tau_{loss}/x_B$, the spectral index is neither very close to 0.5 or 1.0, over many decades near the cutoff.

Klaus Meisenheimer   Alan Heavens

SHOCK ACCELERATION THEORY APPLIED -
THE SPECTRA OF RADIO HOT SPOTS.

Klaus Meisenheimer

Max-Planck-Institut für Astronomie

D-6900 Heidelberg, West Germany

If synchrotron losses are taken into account, the theory of diffusive shock acceleration leads to electron energy distributions which agree perfectly with those inferred from the observed synchrotron spectra of radio hot spots. This agreement is used to derive the physical parameters of the acceleration process from observables.

The Synchrotron Spectra Of Radio Hot Spots

The detection of synchrotron radiation from radio hot spots at wavelengths below 1 cm constrains the overall spectra up to frequencies of 10^{15} Hz (Röser 1988, paper I). Radiative losses prohibit that the ultra-relativistic electrons responsible for the optical synchrotron light can be provided by the core (see Meisenheimer et al. 1988, paper II). They have to be accelerated in situ within the hot spots themselves. The high surface brightness of radio hot spots and their location well isolated from sources of strong thermal emission lead to extra-ordinary clean synchrotron spectra. In addition, one is able to resolve the hot spot emission regions with the VLA. This makes radio hot spots a promising laboratory to investigate the acceleration mechanism in extragalactic radio sources.

The Model

We assume a non-relativistic plasma jet to be decelerated in a strong collisionless shock at the hot spot. The relevant parameters on either side $i = 1,2$ of the shock are the flow speed u_i (in units of c) with respect to the shock at $x = 0$, the mean free path of the relativistic electrons $\lambda_i = \kappa_i/(3c)$ (κ = diffusion coefficient, see Drury's contribution) and the magnetic field B_i. It is convenient to parameterize the synchrotron losses in terms of a — dimensionless — loss rate $w_i = 1.65 \ 10^{-11} \ \lambda_i B_i^2/u_i^2$ (λ_i in pc, B_i in nT). The balance of synchrotron losses and acceleration gains leads to a maximum energy $\gamma_c = E_c/m_e c^2$ at which the canonical energy distribution $n(\gamma) \sim \gamma^{-q}$ cuts of rather steeply. Subsequent losses shift γ_c towards lower energies while the electrons are advected away from the shock to $x > 0$:

$$\gamma_c(x) = \frac{4/3 \ (r-1)}{r w_1 + w_2 \ (1 + x/x_o)} \qquad (1)$$

where $x_o = \lambda_2/(3u_2)$, $r = u_1/u_2$ and w_2 = constant is assumed (Heavens & Meisenheimer 1987). At the shock, $x = 0$, equ. (1) is equivalent to the energy cutoff derived by Webb et al. (1984). The integration of the electron spectra $n(\gamma,x)$ over a finite hot spot emission region $0 \leq x \leq L$ results in a total spectrum $N(\gamma,L)$ following the initial powerlaw $\sim \gamma^{-q}$ up to some "break" energy $\gamma_b = \gamma_c(L)$ where it steepens to γ^{-q-1}. The synchrotron spectra calculated from such model distributions show a $\Delta\alpha = 0.5$ break at ν_b and a high frequency cutoff at ν_c. Four free parameters characterize the model spectra:

- The low frequency spectral index $\alpha_o = (1-q)/2$ for $S_\nu \sim \nu^\alpha$,
- the cut off frequency $\nu_c = 42$ Hz $(B_2/nT) \ \gamma_c^2$, $\gamma_c = \gamma_c(0)$,
- the frequency ratio $\nu_c/\nu_b = (\gamma_c/\gamma_b)^2$,
- a normalization constant.

The model provides an excellent fit to the observed synchrotron spectra in the entire frequency range 10^9 to 10^{15} Hz (see Fig. 3 in paper I). Fig. 1 summarizes the results of the model fits: The low frequency spectral index α_o (Fig.1a) coincides around $\langle\alpha_o\rangle = -0.51\pm0.07$ (rms). Four out of five optically detected hot spots show a high frequency cutoff $\nu_c \approx 10^{14}$ Hz, just above the detection limit $\nu_c > 5 \ 10^{13}$ Hz (Fig. 1b). The majority of radio hot spots, however, cannot be detected in the optical. They are represented in our sample by 3C 123 east alone (shown by "<" in Fig.1b). So far, we have found only one example, the western hot spot of Pictor A, in which the synchrotron spectrum seems to run straight up to at least 10^{15} Hz ($\nu_c \geq 7 \ 10^{15}$ Hz, ">" in Fig.1b). The bi-modal distribution of the break frequency ν_b (Fig. 1c) indicates that one should differentiate between low loss hot spots having $\nu_b > 10^{12}$ Hz and high loss hot spots in which the spectral break occurs at radio frequencies, $\nu_b \leq 10$ GHz.

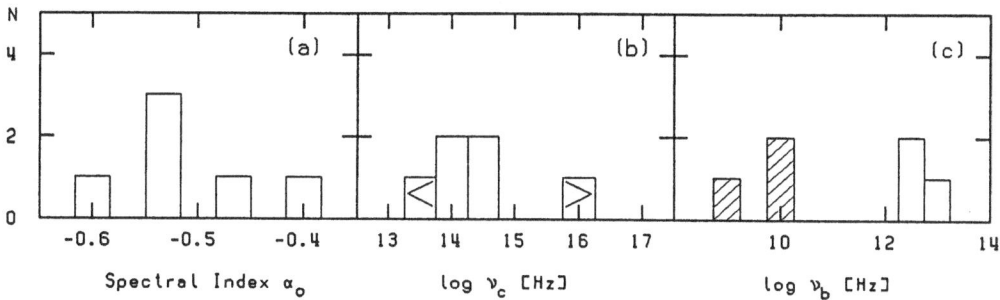

Figure 1: Parameters of the model fits to the synchrotron spectra of the hot spots in 3C 20 west, 3C 33 south, 3C 111 east, 3C 123 east, 3C 273 A and Pictor A west.
(a) Histogram of the low frequency spectral index α_0.

(b) Histogram of the cutoff frequency ν_c. Upper and lower limits are given for 3C 123 east (not detected at $\nu > 10^{14}$ Hz) and Pictor A west (no cutoff found), respectively.
(c) Histogram of the break frequency ν_b. High loss hot spots are hatched.

Magnetic Field Strength

The standard estimate of the magnetic field strength in synchrotron sources is based on the assumption that the relativistic particles and the field arrange themselves such that the total energy density is near its minimum ("minimum energy" field). Since neither this assumption nor the free parameter k (= ratio of energy stored in relativistic protons to that stored in electrons) can be accessed by direct observations it is important to have an independent clue to the field strength: Meisenheimer & Heavens (1986) showed that the downstream field B_2 can be calculated from the break frequency ν_b and the length L of the downstream emission region:

$$B_2 = 6.7 \text{ nT } [(r-1)u_2]^{2/3} (L/kpc)^{-2/3} (\gamma_c/\gamma_b-1)^{2/3} (\nu_c/10^{14}Hz)^{-1/3} \quad (2)$$

with $r = 1 - 3/(2\alpha_0)$ from the standard theory (Bell 1978). This field estimate again contains one unknown parameter, the downstream drift velocity u_2. However, by choosing $u_2 = 0.075$, i.e. $u_{jet} = r u_2 \approx 0.3$, B_2 should be correct to within a factor of 2 for a wide range of mildly relativistic jet speeds $0.1 < u_{jet} < 0.8$.

The field B_2 should be compared with the minimum energy field B_{me} in a cylindrical emission region of diameter D and length L (L and D have to be determined on well resolved radio maps). The uncertainty in the viewing angle can introduce large errors in L, especially when the emission region is thin, L << D (details are given in paper II). In Figure 2, I compare the fields $B_2(u_{jet}=0.3)$ and $B_{me}(k)$ for three different assumptions about k:

(i)   I assume a common value $k = 10$ (Δ in Fig. 2). $B_{me}(k=10)$ is correct to within a factor 2 for $0 < k < 100$. B_{me} and B_2 agree within a factor of 3 for all hot spots in our sample.

(ii) In our model of a finite emission region, k should increase from an initial value k_o produced at the shock to

$$k(L) \quad \approx \quad \frac{\ln(\gamma_c/\gamma_o)}{\ln(\gamma_b/\gamma_o)} \quad k_o$$

at the end of the region because energy losses of the protons are negligible. Here γ_o is the low energy limit of the electron distribution. Since $k \approx k(L)$ throughout most of the emission region, $k(L)$ should be used in calculating B_{me}. Indeed, I find a better agreement between B_2 and B_{me} by using $k(L)$ for $k_o = 2$ and $\gamma_o = 100$ ($\bullet$ in Fig. 2).

(iii) The best agreement between both field estimates is reached for $k_o = 2$ and a high injection energy $\gamma_o = 700$ (o). The synchrotron spectra would show a low frequency turn-over at ≈ 200 MHz.

From the agreement of B_2 and B_{me} I conclude that $B_{HS} = \sqrt{B_{me} B_2}$ for the intermediate case (ii) should be a very reliable estimate of the hot spot magnetic fields.

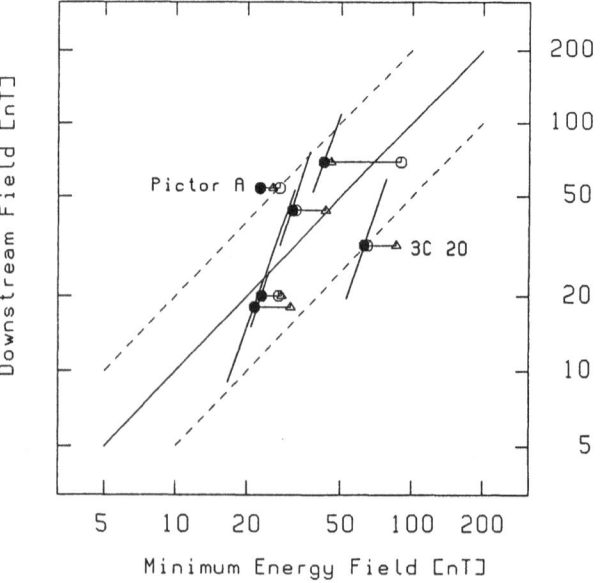

Figure 2: Comparison between two independent estimates of the hot spot magnetic field: Minimum energy estimate B_{me} and downstream field estimate B_2. Three different assumptions about the energy ratio k have been used (see text). The inclined error bars indicate how the uncertainties in the length of the emission regions affect both field estimates. The 45° lines show $B_2 = B_{me}$, $B_2 = 2 B_{me}$, and $B_2 = 1/2 B_{me}$.

Details of the Acceleration Process.

Knowing the field B_{HS} we are prepared to investigate the details of the acceleration process: The observed cutoff frequency ν_c directly yields the <u>maximum energy</u> (Fig. 3a) which is typically $\gamma_c \approx 3 \ 10^5$, i.e. $E_c = 150$ GeV in the optically detected hot spots. In Pictor A west the electrons are boosted to ≥ 7 times higher energies $\gamma_c \geq 2 \ 10^6$ (">" in Fig.3a). If synchrotron radiation dominates the losses the <u>acceleration time scale</u> at the highest energies $\gamma \approx \gamma_c$ is

$$\tau_{acc}(\gamma_c) = \tau_{loss} = 2.5 \ 10^{11} \ (B_{HS}/nT)^{-2} \ \gamma_c^{-1} \ \text{years,}$$

independent of the acceleration mechanism. I find time scales between < 100 years (Pic A) and 1000 years, at most 1/5 of the light traveling time accross the hot spot diameter. Within the framework of shock acceleration τ_{acc} corresponds to an electron <u>mean free path</u> of a few parsec (see Fig. 3b), that is 10^5 times the gyroradius of the relativistic electrons. With the exception of Pic A west ($\lambda_2 \leq 0.13$ pc) the mean free path of optically detected hot spots is in good agreement with theoretical considerations by Biermann & Strittmatter (1987, see also Biermann, this volume). However, the majority of optically <u>undetected</u> hot spots may have $\lambda_2 \gg 10$ pc.

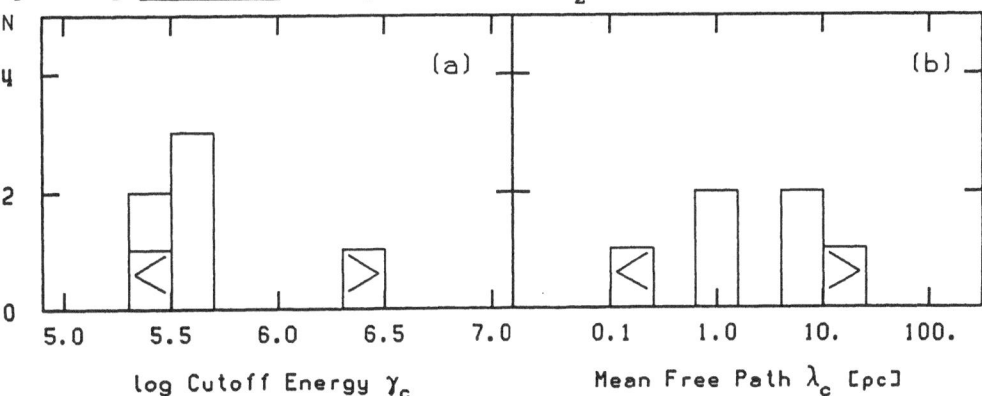

Figure 3: Parameters of the acceleration process (sample as in Fig. 1).
(a) Histogram of the maximum energy γ_c reached by the acceleration process. Upper and lower limits as in Fig. 1b.

(b) Histogram of the electron mean free path $\lambda_2(\gamma_c)$. An upper limit in γ_c corresponds to a lower limit in λ_2 and vice versa.

In the non-relativistic diffusion approximation (Bell 1978) the low frequency spectral index α_0 is a simple function of the shock compression ratio r. Our mean value $<\alpha_0> \approx -0.51$ (Fig. 1a) corresponds to $r \approx 4$. A detailed treatment of both the diffusion process and the downstream equation of state yields $\alpha_o \simeq -0.4$ for a wide variety of jet speeds $0.1 < u_{jet} < 0.8$ (see contributions by Kirk and by Heavens, this volume). Steeper spectra may be explained by smoothing the shock profile (Drury, this volume, Schneider & Kirk 1987). This idea is supported by the fact that we find the shortest acceleration time scale $\tau_{acc} \leq 100$ yrs (i.e. the highest acceleration efficiency) for Pictor A west which has the flattest spectrum, $\alpha_o = -0.4$. Finally, I mention the result of two consistency checks: Paper II shows that for a self-consistent parameter set u_2, λ_2, B_{HS} and an injection energy $\gamma_o \geq 100$ both the required injection rate of relativistic particles and the energy could easily be provided by the observed radio jets.

High Loss Hot Spots - Low Loss Hot Spots.

Three of the six hot spots in our sample show a $\Delta\alpha = 0.5$ break at $\nu_b \leq 10$ GHz (i.e. high losses: 3C 123 east, 3C 273 A, Pic A west). In the others (3C 20 west, 3C 33 south, 3C 111 east) any spectral break - if present - takes place above 10^{12} Hz (low losses). It is obvious from Fig. 4 that the main difference between both groups is due to the length L of the downstream emission region. The other parameters which could affect ν_b, namely B_{HS}, λ_2 and u_2 do not correlate with ν_b (paper II). That is, low loss hot spots are characterized by a short, disk-like emission region, $L << D$, while high loss hot spots show long, cylindrical emission regions $L \simeq D$.

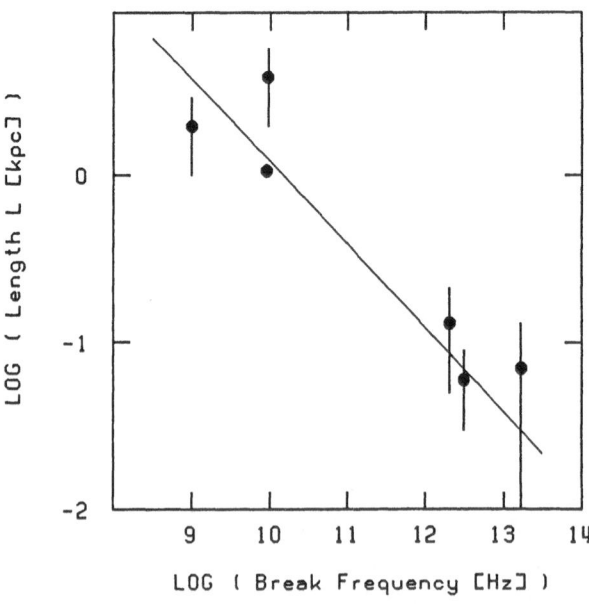

Figure 4: Correlation between the observed break frequency ν_b and the length L of the hot spot emission region. The straight line represents $\nu_b \sim L^{-2}$ which is expected if ν_b is determined by the length alone.

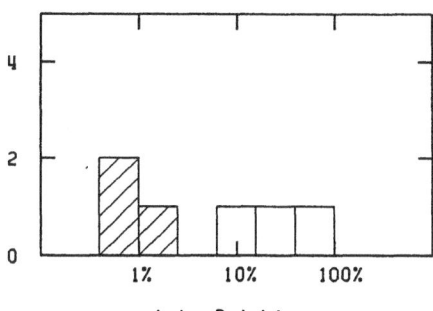

4

2

0

1%   10%   100%

Lobe Brightness

Figure 5: Histogram of the lobe brightness around the hot spot. The brightness is normalized to the peak hot spot flux in a 1 arcsec beam and refers to 5 GHz. High loss hot spots are hatched as in Fig. 1c.

One may use the bi-modal distribution of ν_b to test whether the electrons responsible for the lobe emission are provided by the particle accelerators in the hot spots: If true one would expect that only the low loss hot spots could power bright radio lobes – since

$$\nu_c(\text{lobe}) < \frac{B_{\text{lobe}}}{B_{\text{HS}}} \nu_b(\text{HS}) \ll \nu_b(\text{HS}).$$

Figure 5 shows that such a difference in the lobe brightness is in fact observed.

Based on the current sample of six radio hot spots with well determined overall spectra I conclude: Shock acceleration explains the observed spectra amazingly well. The acceleration regions are closely confined to a thin layer, but the emission regions may extend to a substantial length downstream. This can cause high synchrotron losses within the hot spots themselves. Consequently there are no energetic electrons left to power bright radio lobes around the high loss hot spots.

Acknowlegdement – The basic model has been developed in close collaboration with Alan Heavens. Discussions with P. Biermann, J. Kirk and P. Schneider deepened my insight into the subject.

Bell, A.R. 1978. Mon. Not. R. astr. Soc. 182, 147.

Biermann, P.L. and Strittmatter, P.A. 1987. Astrophys. J. 322, 643.

Biermann, P.L. 1988. This volume.

Drury L. O'C. 1988. This volume

Heavens, A.F. and Meisenheimer, K. 1987. Mon. Not. R. astr. Soc. 225, 335.

Heavens, A.F. 1988. This volume.

Kirk, J.G. 1988. This volume.

Meisenheimer, K. and Heavens, A.F. 1986, Nature 323, 419.

Meisenheimer, K., Röser, H.-J., Hiltner, P.R., Yates, M.G., Longair, M.S., Chini, R. and Perley, R.A. 1988. Submitted to Astron. Astrophys. (PAPER II)

Röser, H.-J. 1988. This volume. (PAPER I)

Schneider, P. and Kirk, J.G. 1987. Astrophys. J. (Letters) 323, L84.

Webb, G.M., Drury, L. O'C. and Biermann, P.L. 1984. Astron. Astrophys. 137, 185.

Peter Hiltner   Peter Biermann

Luke Drury

Friedrich Meyer

Reinhard Schlickeiser

SYNCHROTRON-EMISSION - PHOTONS AND NEUTRINOS - FROM SHOCKWAVE REGIONS IN ACTIVE GALACTIC NUCLEI, JET AND HOT SPOTS

Peter L. Biermann
Max-Planck-Institut für Radioastronomie,
Auf dem Hügel 69, D-5300 Bonn

ABSTRACT

Starting from the recently developed theory for the sharp cutoff in the nonthermal spectra of active galactic nuclei, we describe several new results, mostly based on the work of Pérez-Fournon, Klemens, Fritz and Meisenheimer. This work indicates that a) the shockwaves in jets and hot spots with a spectral cutoff in the near infrared/optical are near-relativistic or relativistic, b) the exact shape of the cutoff and its evolution along a jet can be understood, c) proton-proton collisions can produce strong X-ray synchrotron-emission via secondary electron-positron pairs, d) proton-photon collisions can result in a neutrino-luminosity of similar magnitude as the electromagnetic nonthermal luminosity, with a hard ($\sim E_{\bar{\nu}}^{-2}$) spectrum, and e) that the turbulent wave field is normally of Kolmogorov-character, even in radio sources.

1. INTRODUCTION

The sharp spectral cutoff observed in many active galactic nuclei, jets and spots near the frequency $\sim 3 \ 10^{14}$ Hz has recently found a lot of interest. Biermann and Strittmatter (1987) have formulated a theory how a cutoff at such a frequency can be understood, while Meisenheimer and Heavens (1986) have used more detailed observations to constrain diffusive shock acceleration models. Here we describe further recent developments in this area, based to a large part on the work by Pérez-Fournon (1985), Pérez-Fournon et al. (1988), Klemens (1987), Fritz (1987) and Meisenheimer (1988).

2.1 Shock velocities

The frequency ν^* of the cutoff in the non-thermal synchrotron emission is given by

$$\nu^* = 3 \ 10^{14} \ \text{Hz} \ \left[3b \ \left(\frac{U}{c}\right)^2\right] f(a) , \qquad\qquad (1)$$
$$\text{with } f(a) = (1+Aa)^{1/2}/(1+a)^{3/2} ,$$

(Biermann & Strittmatter 1987) in a simple model where electrons are accelerated up to an energy for which losses balance the gains. b is the fraction of magnetic field energy density in waves. a is the photon energy density relative to the magnetic field energy density, and A ≈200 depends only logarithmically on source parameters. An important assumption in the derivation of eq. (1) was the Kolmogorov-character of the turbulent wave spectrum, suggested by in-situ solar wind observations. There is circumstantial evidence that such a wave spectrum is a fair approximation at non-dissipative length scales in plasmas of the interstellar medium. It follows that all sources with a cutoff frequency not too far from $3 \ 10^{14}$ Hz have a shock velocity U (upstream velocity in the shock frame) not far from relativistic; this means that strong double radio sources with hot spots at or near their tips have a near-relativistic or relativistic flow all the way out to the hot spots. It follows (see Biermann & Strittmatter 1987) that at the magnetic field strength of B $\sim10^{-4}$ Gauss, typical for hot spots, protons can be accelerated to $\sim10^{12}$ GeV and have at most moderate adiabatic losses when they leak into intergalactic space. Thus, knot A in the M87 jet may be one important contributor to the very high energy cosmic rays observed near earth.

2.2 The exact shape of the cutoff

Webb et al. (1984) have calculated the exact form of the cutoff in the approximation, that the turbulent wave field has a k^{-2} dependence on wavenumber k which means a diffusion coefficient independent of energy. Perez-Fournon (1985) then calculated the radiative transfer in all four Stokes-parameters to study the structure of shock regions as seen by an observer. Fritz (1987), using a newly developed code, then calculated the shape of the cutoff spectrum for various functional forms of the wave spectrum, including the case of a Kolmogorov-law. He showed that the shape of the cutoff can be understood if proper account is taken of dust absorption which is likely to change the observed spectrum in some sources; this point merits further observations. Detailed observations of the evolution of the cutoff frequency along a jet (Pérez-Fournon et al. 1988) demonstrated that the cutoff frequency decreases outwards in the M87 jet. Again, this can be understood as a combination of decreasing relativistic boosting, decreasing shock velocity and decreasing photon intensity along the jet.

2.3 Secondary electron/positron synchrotron emission

The primary acceleration process produces a synchrotron cutoff near 3 10^{14} Hz or less. However, beyond this frequency, a number of other processes contribute to the emission, readily observable in X-rays, thermal emission from hot gas, inverse Compton emission (possibly increased by a strongly anisotropic radiation field), and synchrotron emission from secondary electron/positron pairs produced in proton-photon and proton-proton collisions. In order to properly describe this last process it is necessary to combine the acceleration of particles in a shockwave region and the production of new particles. Using a newly developed Monte-Carlo code, Klemens (1987) demonstrated that the FIR/optical/X-ray spectra of sources can readily be fitted by a combination of primary and secondary particle synchrotron emission, and has thus shown that this process may contribute significantly to the X-ray emission of the M87 jet.

2.4 Neutrino-emission

It is easy to show (see Biermann and Strittmatter 1987) that the neutrino-emission resulting from proton-photon collisions (limiting the proton-energies in strong radiation fields) produces a neutrino spectrum which is hard ($\sim E_{\overline{\nu}}^2$), extends to very high energies ($\sim 10^{19}$ eV in hot spots) and has a luminosity which can reach the entire non-thermal electromagnetic luminosity (10^{47} erg sec^{-1} for a fair number of quasars). Similarly, very high energy photons are produced, but are cut off by photon-photon pair-creation. Thus, very sensitive photon- and neutrino detectors will be required to observe this emission, with an effective area not too far below a km^2.

2.5 The wave spectrum

A critical assumption in the concept developed by Biermann and Strittmatter (1987) was the Kolmogorov character of the wave turbulence. The Kolmogorov-law was originally derived for incompressible, isotropic turbulence. Observations in the earth atmosphere and in the solar wind have clearly demonstrated that the Kolmogorov-law is a fair approximation at non-dissipative wavelengths; observations of the interstellar medium are consistent with the assumption of Kolmogorov-turbulence in the hot phase of the gas.

One can easily show that the observation of the integrated emission of a finite region downstream from a shock permits the determination of the turbulence powerlaw index assuming the flow velocity and the magnetic field strength to be constant throughout the region considered: The cutoff frequency decreases downstream from the shock with distance squared, hence the integrated emission has a $\nu^{-1/2}$ spectrum up to the local cutoff frequency ν_b at the end of the emission, then a ν^{-1} spectrum up to the cutoff frequency c_c near the shock (for strong shocks with a density ratio 4, in the test-particle approximation). With an observation of the length of this emission region contribution, the bend frequency ν_b and the cutoff frequency ν_c, and the assumption of minimum energy magnetic fields one can derive the appropriate turbulence wave spectral index. Meisenheimer (1988) has assembled such data for several sources. For all sources for which the shock strength is compatible with a non-relativistic equation of state (density ratio ≤ 4) the resulting turbulent wave spectral index is very close to the Kolmogorov value of 5/3. It thus appears that the Kolmogorov-law holds even in many radio sources.

3. CONCLUSION

Recent applications of the diffusive shock acceleration theory (e.g. Drury 1983) have shown remarkable success in explaining the observations.

ACKNOWLEDGEMENTS

I wish to thank Drs. L. O'C. Drury, K.-D. Fritz, Y. Klemens, W. Krülls, K. Meisenheimer, I. Pérez-Fournon, Th. Schmutzler, P. Strittmatter, H.J. Völk and G.M. Webb for their time and help to understand a little more the physics of shockwaves in active galaxies.

REFERENCES

Biermann, P.L., Strittmatter, P.A.: 1987, Ap.J. **322**, 643
Drury, L. O'C.: 1983, Rep. Progr. Phys. **46**, 973
Fritz, K.-O.: 1987, Ph.D. Thesis Bonn
Fritz, K.-O.: 1988, A&A (submitted)
Klemens, Y.: 1987, Ph.D. Thesis Bonn
Meisenheimer, K., Heavens, A.F.: 1986, Nature **323**, 419
Meisenheimer, K.: 1988, priv. comm.
Pérez-Fournon, I.: 1985, Ph.D. Thesis, Tenerife
Pérez-Fournon, I., Colina, L., Gonzalez-Serrano, J.I., Biermann, P.L.: 1988,
    Ap.J. Letters (in press)
Webb, G.M., Drury, L. O'C., Biermann, P.L.: 1984 A&A **137**, 185

REMARKS ABOUT DIFFUSIVE SHOCK WAVE ACCELERATION

Reinhard Schlickeiser
Max-Planck-Institut für Radioastronomie
Auf dem Hügel 69, 5300 Bonn 1, West Germany

1. Introduction

The theory of particle acceleration at astrophysical shocks has received much attention since this process is believed to be a universal mechanism to explain the origin of relativistic particles (for recent review see Blandford and Eichler 1987, Drury 1988). In the conventional picture of diffusive shock wave acceleration pitch-angle scattering of charged particles with purely magnetic Alfvén waves is used to confine particles near the shock, and particles gain energy by multiple crossings of the shock (hereafter referred to as the standard theory). Presumably the shock wave itself generates the Alfvénic turbulence in its vicinity. The transport equation then to be solved in the test particle limit is of first-order in momentum yielding simple power law particle spectra regardless of the momentum dependence of the allowed spatial diffusion, with a spectral index solely determined by the shock wave compression ratio.

Unfortunately, this preferential generation of power law particle spectra, with a spectral index being almost insensitive to the actual microscopic physical conditions (turbulence level, power spectrum of Alfvénic turbulence), is in conflict with astronomical observations where we witness ongoing particle acceleration. In-situ satellite measurements of energetic electrons and ions accelerated in solar flares (McGuire and von Rosenvinge 1984) indicated that the representation of the ion spectra as an exponential in rigidity and as a Bessel function in momentum per nucleon fit the data best, whereas power law spectra can be excluded. Another conflict of the standard theory has been the measured radio spectral index distributions of shell-type supernova remnants and bright spiral galaxies where it was not possible to account for the dispersion in spectral indices and independence from dynamical evolutionary effects (Lerche 1980).

In both these examples it has recently been shown by Dröge and Schlickeiser (1986) in case of solar flares and by Dröge, Lerche and Schlickeiser (1987) in case of supernova remnants and bright spiral galaxies that the inclusion of momentum diffusion of particles in the transport equation removes these discrepancies, and that the observations can be accounted for in a quantitative fashion. In both papers a space averaged transport equation has been employed which results if the scattering time method (Wang and Schlickeiser 1987; Schlickeiser, Sievers and Thiemann 1987) is applied to the problem. The momentum diffusion results from cyclotron damping of the electric field associated with each Alfvén wave and is an essential effect of diffusive shock wave acceleration. In this picture the shock wave is used as a source of turbulence and the acceleration is primarily due to cyclotron damping of this turbulence by the particles. Acceleration by multiple shock crossings is still there but a minor effect compared to cyclotron damping. This is a quite different view on the acceleration process near shocks as compared to the standard theory, but this view is in agreement with the acceleration of particles at least in solar flares, shell-type supernova remnants and bright spiral galaxies.

In the following we investigate in detail the topology of the particle transport equation. We will demonstrate that momentum diffusion indeed should be present in realistic physical models, whereas the widely used transport equation without the momentum diffusion term can only be reproduced under rather academic requirements on the Alfvén wave's propagation direction.

2. The cosmic ray transport equation

2.1 General equations

We consider the behaviour of energetic charged particles in an ordered uniform magnetic field $\vec{B}_o = B_o \, \vec{e}_z$ with zero electric field with N superposed non-dispersive transverse plane electromagnetic waves which propagate with respective phase velocities w_N along $\vec{B}_o$. The magnetic field is embedded in a cold background medium moving with the nonrelativistic bulk speed $\vec{U}$ parallel to $\vec{B}_o$. The total electromagnetic field then is

$$\vec{B} = (0,0,B_o) + \sum_{j=1}^{N} \int_o^{\infty} dk \, \left(B_{x_j}(k), \, B_{y_j}(k), \, 0\right) \, \exp\left\{ik(z-w_j t)\right\} \tag{2.1}$$

$$\vec{E} = \sum_{j=1}^{N} \frac{w_j}{c} \int_o^{\infty} dk \, \left(B_{y_j}(k), \, -B_{x_j}(k), \, 0\right) \, \exp\left\{ik(z-w_j t)\right\} \tag{2.2}$$

where Maxwell's induction law has been used to express the electric field in terms of the magnetic field components. The quasi-linear approximation of the collisionless Boltzmann equation, where the electromagnetic field (2.1) and (2.2) is used in the Coulomb-Lorentz force, describing the evolution of the gyrophase-averaged phase space density $f(z,p,\mu,t)$ yields the Fokker-Planck equation (for details see Schlickeiser 1988a)

$$\frac{\partial f}{\partial t} + v\mu \frac{\partial f}{\partial z} - \frac{S_0}{2} = \frac{\partial}{\partial \mu} D_{\mu\mu} \frac{\partial f}{\partial \mu} + \frac{\partial}{\partial \mu} D_{\mu p} \frac{\partial f}{\partial p}$$

$$+ \frac{1}{p^2} \frac{\partial}{\partial p} p^2 D_{\mu p} \frac{\partial f}{\partial \mu} + \frac{1}{p^2} \frac{\partial}{\partial p} p^2 D_{pp} \frac{\partial f}{\partial p} \qquad (2.3)$$

with the Fokker–Planck coefficients

$$D_{\mu\mu} = \sum_{j=1}^{N} \frac{\pi}{2} \frac{\Omega^2}{B_0^2} (1-\mu^2) \frac{\left[1 - \frac{\mu w_j}{v}\right]^2}{|v\mu - w_j|} \alpha_j(k) \qquad (2.4)$$

$$D_{\mu p} = \sum_{j=1}^{N} \frac{\pi}{2} \frac{\Omega^2}{B_0^2} (1-\mu^2) \frac{w_j p}{v} \frac{\left[1 - \frac{\mu w_j}{v}\right]}{|v\mu - w_j|} \alpha_j(k) \qquad (2.5)$$

$$D_{pp} = \sum_{j=1}^{N} \frac{\pi}{2} \frac{\Omega^2}{B_0^2} (1-\mu^2) \frac{w_j^2 p^2}{v^2} \frac{1}{|v\mu - w_j|} \alpha_j(k) \qquad (2.6)$$

where

$$\alpha_j(k) = I^L \left(k = \frac{-\Omega}{v\mu - w_j}\right) + I^R \left(k = \frac{\Omega}{v\mu - w_j}\right) \qquad (2.7)$$

refers to the power spectra of left-hand (L) and right-hand (R) polarized magnetic field fluctuations. S_0 in equation (2.3) represents sources and sinks of particles, $\Omega = \Omega_0/\gamma$ is the relativistic gyrofrequency, v the velocity, p the momentum and μ the cosine of the pitch angle of the particle. As one can see the value of the Fokker–Planck coefficients is determined by the directionality (values w_j) and the cross helicity of the waves, i.e. the intensity ratio of the left-hand ($B_L = B_X + iB_Y$) and right-hand ($B_R = B_X - iB_Y$) polarized waves.

Applying the diffusion approximation to equation (2.3) the fundamental transport equation for the isotropic part of the particle distribution function F(z,p,t) results (Schlickeiser 1988a)

$$\frac{\partial F}{\partial t} = \frac{\partial}{\partial z} \left[\kappa \frac{\partial F}{\partial z}\right] - \frac{1}{4p^2} \frac{\partial}{\partial p} (p^2 \, v \, a_1) \frac{\partial F}{\partial z} + \frac{1}{4} v \frac{\partial F}{\partial p} \frac{\partial a_1}{\partial z} + \frac{1}{p^2} \frac{\partial}{\partial p} p^2 a_2 \frac{\partial F}{\partial p} + S_0$$

$$(2.8)$$

where the spatial diffusion coefficient $\kappa(p,z)$, the rate of adiabatic deceleration or acceleration $a_1(p,z)$ — depending on the sign of $(\partial a_1/\partial z)$ — and the momentum diffusion coefficient $a_2(p,z)$ are related to the Fokker–Planck coefficients $D_{\mu\mu}$, $D_{\mu p}$, D_{pp} as

$$\kappa(p,z) = \frac{v^2}{8} \int_{-1}^{+1} d\mu \, (1-\mu^2)^2 / D_{\mu\mu} \qquad (2.9)$$

267

$$a_1(p,z) = \int_{-1}^{+1} d\mu \, (1-\mu^2) \, D_{\mu p}/D_{\mu\mu} \qquad (2.10)$$

$$a_2(p,z) = \frac{1}{2} \int_{-1}^{+1} d\mu \left[D_{pp} - \frac{D_{\mu p}^2}{D_{\mu\mu}} \right] \qquad (2.11)$$

Through the Fokker–Planck coefficients it is the cross helicity and directionality of the waves which determine the transport parameters κ, a_1 and a_2.

2.2 Momentum diffusion

We first consider the momentum diffusion coefficient a_2 in more detail. Inserting equations (2.4)–(2.6) in equation (2.11) we obtain

$$a_2(p,z) = \frac{\pi \, \Omega^2 \, p^2}{4 \, B_0^2 \, v^2} \int_{-1}^{+1} d\mu \, (1-\mu^2) \left\{ \left[\sum_{j=1}^{N} \frac{\alpha_j \, w_j^2}{|v\mu - w_j|} \cdot \sum_{j=1}^{N} \frac{\alpha_j \left(1 - \frac{\mu w_j}{v} \right)^2}{|v\mu - w_j|} \right. \right.$$

$$\left. \left. - \left[\sum_{j=1}^{N} \frac{\alpha_j \, w_j \left(1 - \frac{\mu w_j}{v} \right)}{|v\mu - w_j|} \right]^2 \right\} \middle/ \left\{ \sum_{j=1}^{N} \frac{\alpha_j \left(1 - \frac{\mu w_j}{v} \right)^2}{|v\mu - w_j|} \right\} \qquad (2.12)$$

Only in two cases the momentum diffusion coefficient a_2 is precisely zero:
(i)   all possible phase velocities are precisely zero. But that case would also yield a vanishing adiabatic deceleration rate $a_1 = 0$, and thus no acceleration of particles;

(ii)   only one phase speed of waves is present ($N = 1$). Then besides $a_2 = 0$ we obtain

$$a_1 = \frac{w_1 p}{v} \int_{-1}^{+1} d\mu \, (1-\mu^2) \left(1 - \frac{\mu w_1}{v} \right)^{-1}$$

$$= p \left[2 \frac{v}{w_1^2} - \left(\frac{v^2}{w_1^2} - 1 \right) \ln \frac{1 + \frac{w_1}{v}}{1 + \frac{w_1}{v}} \right] \propto \frac{4}{3} \frac{w_1}{v} \, p \qquad (2.13)$$

where the latter approximation holds for $w_1 \ll v$, and

$$\kappa = \frac{B_0^2 \, v^2}{4 \, \pi \, \Omega^2} \int_{-1}^{+1} \frac{d\mu \, (1-\mu^2) \, |v\mu - w_1|}{\alpha_1(k) \left(1 - \frac{\mu w_1}{v} \right)^2} \qquad (2.14)$$

which for power law type power spectra ($I(k) \propto k^{-q}$) exhibits the legendary resonance gap singularity at $\mu_R = w_1/v$ if $q \geq 2$.

Equation (2.8) in this case formally reduces to

$$\frac{\partial F}{\partial t} = \frac{\partial}{\partial z} k \frac{\partial F}{\partial z} - w_1 \frac{\partial F}{\partial z} + \frac{1}{3} \frac{dw_1}{dz} p \frac{\partial F}{\partial p} + S_0 \qquad (2.15)$$

which for $w_1 = U$ is the transport equation used in the standard theory.

In all other cases $(N > 1)$ the transport equation (2.8) contains nonvanishing terms representing spatial diffusion, spatial convection, adiabatic deceleration or acceleration (depending on the sign of $\partial a_1/\partial z$) and momentum diffusion.

2.3 The realistic transport equation

The simplest realistic case $(N = 2)$ has been discussed in detail by Schlickeiser (1988b). If the cold background medium consists of protons and electrons, there are two possible phase speeds of waves propagating along $\vec{B}_0$ at frequencies below the nonrelativistic proton gyrofrequency:

$$w_{1,2} = U \pm V_A \qquad (2.16)$$

referred to as forward (+) and backward (−) moving Alfvén waves. $V_A = B_0 (4\pi\, n_e \cdot [m_p + m_e])^{-1/2}$ is the Alfvén velocity. [As an aside, if the cold background medium would consist of electrons and positrons, we would obtain waves with $w_{1,2} = U \pm V_e$, where $V_e = (m_p/(2m_e))^{1/2} V_A = 30.3\, V_A$, with frequencies below the nonrelativistic electron gyrofrequency.] Assuming a slab-type power law turbulence spectrum $(\Gamma^{+,-}(k) = \Gamma_0^{+,-} k^{-q})$ with zero cross helicity for the waves Schlickeiser (1988b) determined

$$\kappa \propto \frac{v^{3-q} B_0^{\,2}}{2\pi\, (1+R)\, I_0^+\, \Omega^{2-q}} \left[\frac{2}{(2-q)(4-q)} + \frac{q-1}{q-2} \left(\frac{v}{U}\right)^{q-2} \right] \qquad (2.17)$$

$$a_1 \propto \frac{4}{3} \frac{p}{v} \left[U + \frac{1-R}{1+R} V_A \right] \qquad (2.18)$$

$$a_2 \propto \frac{2\pi\, \Omega^{2-q}\, R\, I_0^+\, v^{q-3}\, V_A^2\, p^2}{q(q+2)\, B_0^2}$$

$$= \frac{V_A^2\, p^2}{\kappa} \cdot \frac{R}{1+R} \frac{\left[\frac{2}{(2-q)(4-q)} + \frac{q-1}{q-2}\left(\frac{v}{U}\right)^{q-2}\right]}{q(q+2)} \qquad (2.19)$$

where $R = I_0^-/I_0^+$ refers to the intensity ratio of backward and forward moving waves.

Equation (2.19) expresses the close relation between spatial and momentum diffusion. Defining the time scale for particles to diffuse a distance L by pitch angle scattering along the ordered magnetic field as $\tau_D = L^2/\kappa$, and the time scale for a particle to increase its momentum by cyclotron damping as $\tau_F = p^2/a_2$, one finds that the product of these two time scales is

$$\tau_D(z,p) \cdot \tau_F(z,p) = \frac{1+R}{R} \quad \frac{q(q+2)}{\frac{2}{(2-q)(4-q)} + \frac{q-1}{q-2} \left(\frac{v}{U}\right)^{q-2}} \quad \left(\frac{L}{V_A}\right)^2 \tag{2.20}$$

which is a constant for relativistic particles ($v = c$) at all momenta. The value of the constant is independent of microscopic quantities (as $I_0^{+,-}$), and is given by the macroscopic properties (size L, Alfvén speed V_A, wave intensity ratio R) of the considered physical system. The significance of microscopic processes in the system, indicated by the respective time scales τ_D and τ_F, is determined at all momenta by macroscopic parameters. Random walk in spatial coordinates due to diffusion along magnetic field lines and momentum diffusion due to cyclotron damping of Alfvén waves are twin processes: once the time scale for one of them is known, the time scale of the corresponding process and thus its relevance can be estimated from equation (2.20). For solar flares we estimated $\tau_F \approx 1$ min, for bright spiral galaxies $\tau_F \approx 10^6 \, (p/GeVc^{-1})^{0.5}$ years (Schlickeiser 1988a) indicating the importance of momentum diffusion in these astrophysical systems.

With equations (2.17)–(2.19) the transport equation (2.8) is

$$\frac{\partial F}{\partial t} = \frac{\partial}{\partial z} \, \kappa \, \frac{\partial F}{\partial z} - \left[U + \frac{1-R}{1+R} \, V_A \right] \frac{\partial F}{\partial z} + \frac{1}{3} \frac{d}{dz} \left[U + \frac{1-R}{1+R} \, V_A \right] p \, \frac{\partial F}{\partial p} + \frac{1}{p^2} \frac{\partial}{\partial p} \, p^2 \, a_2 \, \frac{\partial F}{\partial p} + S_0 \tag{2.21}$$

which for the reasons given above is a much more realistic description of energetic charged particle transport.

2.4 Interlude

The popular and widely used particle transport equation without a momentum diffusion term ($a_2 = 0$) results rigorously from first principles if Alfvén waves of both polarization state moving with only one phase speed are present which would require a very fine-tuned wave generation process. However, as a consequence of the same assumption, a resonance gap point at $\mu_R = w_1/v$ occurs which leads to an infinitely large value of the spatial diffusion coefficient for power spectra spectral indices $q \gtrsim 2$. Within the confines of quasi-linear theory this transport equation appears to be inadequate.

In the context of diffusive shock wave acceleration there is another point worth mentioning. Even if we take for granted that the physics of wave generation in the upstream region of the shock wave can indeed be arranged that only forward waves are present, the work of McKenzie and Westphal (1969) and Morfill and Scholer (1977) has shown that the interaction of the shock and the forward Alfvén wave yields both forward and backward moving waves in the post shock region, a situation where momentum diffusion is unavoidable as we have demonstrated. Momentum diffusion has to be included for consistency.

There is no question that in a real turbulent medium as near a shock wave, plasma waves come from all directions with different phase speeds, so that $N \gg 1$. We therefore regard a particle transport equation without momentum diffusion as a serious misrepresentation of the acceleration process near shocks.

3. Diffusive shock wave acceleration with momentum diffusion

An argument put forward very often by advocates of the standard theory is that, even if momentum diffusion is present, it is a minor effect and negligible as compared to energy gain by multiple crossings of a strong shock. To assess the relative importance of these two energy gain processes, exact solutions of the transport equation (2.21) with and without momentum diffusion are required. Because of the drastically increased mathematical complexity in case $a_2 \neq 0$ such a solution has been obtained only lately.

Recently, Schlickeiser (1988b) succeeded to solve equation (2.21) in the case $R = 1$ for a momentum-independent ($q = 2$) spatial diffusion coefficient $\kappa(z)$ for any given velocity pattern $U(z)$ by adequately adjusting the spatial variation $\kappa(z)$. By using the scattering time method he showed that, in case of two free-escape boundaries located at distances $z = L_1$ and $z = -L_2$ from the shock ($z = 0$), the momentum spectrum is determined by

$$F(z,p) \;=\; \sum_{n=1}^{\infty} A_n(z)\, T_n(p) \tag{3.1}$$

an infinite sum of functions $T_n(p)$ with expansion coefficients $A_n(z)$ depending on the spatial variation of $\kappa(z)$ and the source distribution of particles $q(z)$. Each function $T_n(p)$ satisfies a leaky-box type equation of exactly the form considered by Dröge and Schlickeiser (1986), Dröge, Lerche and Schlickeiser (1987) and schlickeiser, Sievers and Thiemann (1987), where the corresponding eigenvalues (ψ_n, $n = 1,2,3,...$) of the spatial problem enter in the form of escape lifetimes. For the case considered of a momentum-independent $\kappa(z)$ each function $T_n(p) \propto p^{-\Gamma_n}$ yields a power law solution in momentum, where the spectral index Γ_n is related to the spatial eigenvalue ψ_n. Since the set of eigenvalues form an increasing series of numbers $\psi_1 < \psi_2 < \psi_3 \ldots$ the resulting momentum spectrum is a superposition of power law spectra with different spectral indices Γ_n. At large momenta far away from the injection momentum the smallest eigenvalue ψ_1, leading to the smallest spectral index Γ_1, dominates the spectrum.

In the illustrative example of an hyperbolic tangent type velocity distribution with a shock thickness z_c and upstream and downstream velocities of U_1 and U_2, respectively, and the variation $\kappa(z) = \kappa_0 \cosh^2(z/z_c)$ the smallest eigenvalue ψ_1 is calculated from variational methods (Rayleigh–Ritz and comparison method). It is shown that

$$\psi_1 = \frac{1}{2} + c_1 \, N_p^{-1} + c_2 \, N_p \qquad (3.2)$$

with the shock wave's Peclet number $N_p = z_c U_2/\kappa_0$ and c_1 and c_2 determined by the compression ratio $r = U_1/U_2$ and L_1, L_2, z_c, so that Γ_1 is a function of N_p and r. Figure 1 shows the variation of Γ_1 from N_p for various compression ratios in case of no momentum diffusion. As can be seen both from Figure 1 and equation (3.2) there exists an optimum value of $N_{p_{min}}$ for which Γ_1 attains its lowest value.

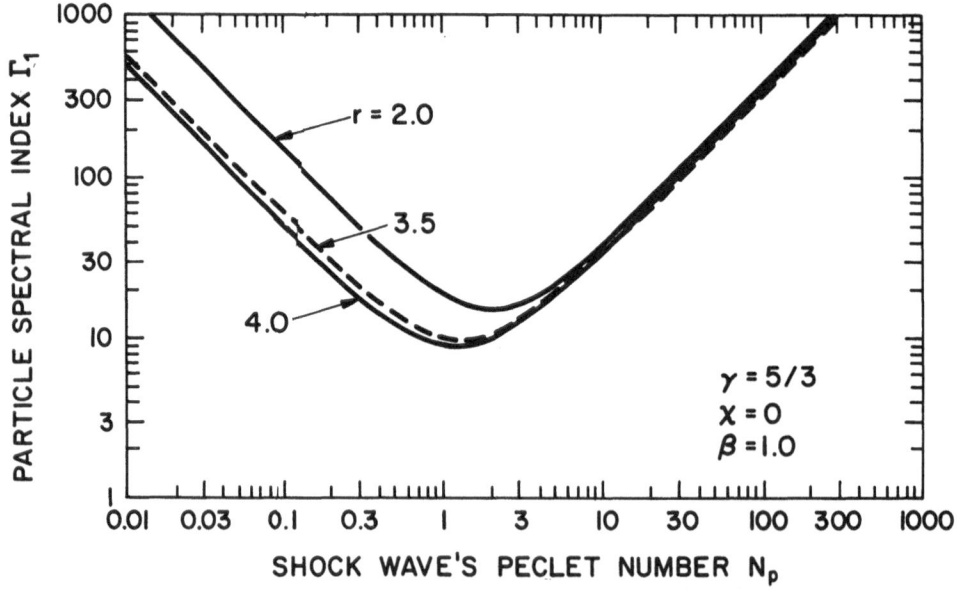

Figure 1: Power law spectral index Γ_1 as a function of the shock wave's Peclet number N_p for three different values of the compression ratio r in the case where momentum diffusion is neglected ($\chi = 0$) and the free-escape boundaries are located far away from the shock's transition region ($L_1 \gg z_c$, $L_2 \gg z_c$).

In Figure 2 we show the variation of Γ_1 if momentum diffusion is included. As can be seen a quite different dependence results. Whereas at low Peclet numbers the acceleration is rather ineffective, it becomes very effective at large Peclet numbers and is mainly due to momentum diffusion. In both calculations the free-escape boundaries have been assumed to be located far away ($L_1 \gg z_C$, $L_2 \gg z_C$) from the shock's transition region.

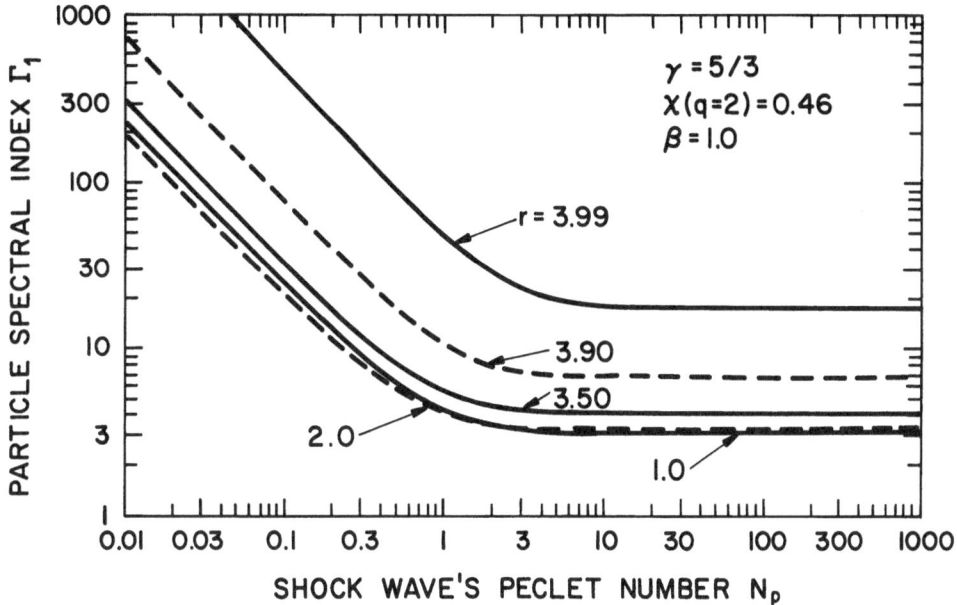

Figure 2:   _Power law spectral index_ Γ_1 _as a function of the shock wave's Peclet number_ N_p _for five different values of the compression ratio r in a nonrelativistic_ _($\gamma = 5/3$) medium with plasma beta_ $\beta = 1$ _in the case where momentum diffusion is included, and where the free-escape boundaries are located far away from the shock's transition region ($L_1 \gg z_C$, $L_2 \gg z_C$)._

We do not regard this particular example of the flow pattern U(z) and the spatial diffusion coefficient $\kappa(z)$ as an ideal representation of the physical situation near shocks. Our choice of the functional form of $\kappa(z)$ was dictated by the task to find an exact analytical solution, albeit it describes the concentration of the Alfvénic turbulence near the shock in the correct way. But this example serves one important purpose: it shows that inclusion of momentum diffusion leads to quantitatively different results for particle acceleration near shocks. Our results furthermore indicate that energy gain by momentum diffusion in shocks of large Peclet number N_p is much more efficient than the gain by multiple crossings, leading to a quite different view on the elementary acceleration process near shocks: it is mainly the cyclotron damping of the Alfvén waves produced by the shock which accelerates particles.

4. Summary

The topology of the particle transport equation used in particle acceleration theories at astrophysical shocks has been investigated. The popular and widely used transport equation without momentum diffusion results rigorously from first principles only if a drastic simplification of the Alfvén wave propagation directions is made: all waves move with the same phase speed. It is argued that this simplification is never realized in astrophysical objects and therefore is a severe misrepresentation of the physical acceleration process. Discarding this simplification leads to a situation where momentum diffusion of particles by cyclotron damping the Alfvénic turbulence is unavoidable. Solutions to the transport equation with and without the momentum diffusion term are presented which indicate the importance of momentum diffusion, leading to a quite different view on the elementary acceleration process. The acceleration of particles at shocks is primarily due to the cyclotron damping whereas the energy gain by multiple shock crossings is a minor effect. Astrophysical shock waves with large Peclet numbers accelerate particles very efficient.

Acknowledgement: I thank Ms. Ġ. Breuer for the careful typing of this manuscript.

References

Blandford, R., Eichler, D., 1987, Phys. Rep. 154, 1
Dröge, W., Schlickeiser, R., 1986, Astrophys. J. 305, 909
Dröge, W., Lerche, I., Schlickeiser, R., 1987, Astron. Astrophys. 178, 252
Drury, L.O.C., 1988, these proceedings
Lerche, I., 1980, Astron. Astrophys. 85, 141
McGuire, R.E., Von Rosenvinge, T.T., 1984, Adv. Space Res. 4, 117
McKenzie, J.K., Westphal, K.O., 1969, Planet. Space Sci. 17, 1029
Morfill, G.E., Scholer, M., 1977, Astrophys. Space Sci. 46, 73
Schlickeiser, R., 1988a, Astrophys. J., submitted
Schlickeiser, R., 1988b, Astrophys. J., submitted
Schlickeiser, R., Sievers, A., Thiemann, H., 1987, Astron. Astrophys. 182, 21
Wang, Y.-M., Schlickeiser, R., 1987, Astrophys. J. 313, 200

PARTICLE ACCELERATION IN HOTSPOTS

Wolfgang Kundt
Institut für Astrophysik der Universität Bonn
Auf dem Hügel 71, D-5300 Bonn / FRG

One can sometimes hear or read that in-situ particle acceleration via shocks was 'absolutely necessary' and even 'observationally well-confirmed' in certain astrophysical environments. It is the purpose of this short communication to point out that such statements are not correct. I am not aware of any astrophysical situation where in-situ acceleration would be required or even consistent with all the observed facts. The boosters to very high particle energies of which I am aware are rotating magnets and reconnecting magnetic fields [Kundt, 1984b, 1986b].

Shock acceleration is topical; this in itself explains why so many publications are devoted to it. Direct measurements of possible shock acceleration are restricted to the heliosphere for which Sarris & Krimigis [1985] and Krimigis et al [1986] have argued against its importance: The accelerations are highly directional and prompt - exceeding anisotropies of 10^{-4} - exponential in spectrum, and can be explained with locally ordered fields $\vec{E}, \vec{B}$ according to $\Delta E = e \int (\vec{E} + \vec{\beta} \times \vec{B}) \cdot d\vec{x}$; see also Pesses [1979].

In-situ acceleration has been invoked to explain the presence of cosmic rays, i.e. of ions and electrons with particle energies between 10^9 eV and 10^{20} eV which pervade the galactic disk. It is agreed that supernova shells - the key candidates for shock accelerators - cannot achieve energies in excess of some $10^{13\pm1}$ eV if they were turbulent. But the cosmic-ray spectrum is very smooth between 10^{11} eV and 10^{15} eV and shows no decline at the predicted upper cutoff. And it is hydrogen deficient. For these and several other reasons, shock acceleration cannot be the dominant CR booster [Kundt, 1983, 1984, 1986]. Better candidates are the (young) binary neutron stars, many of which have been identified as VHEγ-ray sources, with photon energies up to 10^{15} eV or even 10^{17} eV [e.g. Kundt, 1985].

In the case of the extragalactic radio sources and their hotspots, shock acceleration has been considered because of the short lifetimes of the synchrotron-emitting charges. Contrary to the cosmic-ray problem, here one wants to transfer the energy from the ions to the electrons. The alternative possibility of a quasi-lossfree supply - in the form of an $\vec{E} \times \vec{B}$-drift - is discussed in my other contribution to this workshop.

So why could all of the published calculations be misleading when it comes to applications? They are mostly performed in the test-particle limit in which particle interactions are elastic and wave- and particle-losses ignorable. Once the efficiencies are needed to be high ($\gtrsim 10^{-2}$), loss processes (from the ideal phase-space volume) become important.

More in detail, most shock calculations ignore the omnipresence of transverse magnetic fields which lock the particle motions [Pesses, 1979]. Further, it is known from the cosmic-ray problem that direct Fermi acceleration of electrons is much less efficient than acceleration of protons because of the electrons' shorter mean free path [Sikora et al., 1987]. Finally, the shocks invoked are in all cases minute in energy and speed compared with those likely to be present near the speed-of-light cylinder of the central rotating (super-) massive magnet.

Acknowledgement: I am thankful to Peter Scheuer for a discussion.

References

Krimigis, S.M., Sibeck, D.G., McEntire, R.W., 1986: Geophys. Res. Lett. 13, 1376
Kundt, W., 1983: Astrophys. Sp. Sci. 90, 59
Kundt, W., 1984a: Adv. Space Res. 4, 381
Kundt, W., 1984b: J. Astrophys. Astron. 5, 277
Kundt, W., 1985: Bull. Astron. Soc. India 13, 12
Kundt, W., 1986a, in: Cosmic Radiation in Contemporary Astrophysics, NATO ASI C 162, ed. M.M. Shapiro, Reidel, pp. 57, 67
Kundt, W., 1986b: Il Nuovo Cimento 9C, 469
Pesses, M.E., 1979: Proc. 16th ICRC at Kyoto II, OG 9-1-8, p. 33
Sarris, E.T., Krimigis, S.M., 1985: Astrophys. J. 298, 676
Sikora, M., Kirk, J.G., Begelman, M.C., Schneider, P., 1987: Astrophys. J. 320, L81

Wolfgang Kundt

Dieter Biskamp

Magnetic Reconnection and Particle Acceleration

Dieter Biskamp

Max-Planck-Institut für Plasmaphysik

8046 Garching bei München, Federal Republic of Germany

Abstract

The present status of the theory of magnetic reconnection is described. At high magnetic Reynolds numbers which are of particular astrophysical interest reconnecting systems are nonstationary, probably strongly turbulent, giving rise to turbulent resistivity and large parallel electric fields. These result in efficient electron acceleration due to the run-away effect. An estimate is given of the maximum electron energy that can be attained in hot spots by this mechanism.

I Introduction

The presence of high energy particles is a ubiquitous phenomenum in astrophysical systems. In addition to the cosmic ray back ground radiation there are various manifestations of local concentrations of energetic particles, in particular electrons, observed primarily by strong synchrotron radiation, e.g. in the hot spots of extragalatic radio sources. The origin of these particles is still the subject of lively discussions. The most favoured acceleration mechanism is diffuse shock acceleration or first order Fermi acceleration at shocks, discussed in detail by Drury[1]. The process is conceptually rather simple, does not require special properties of the shock except for the presence of extended magnetically turbulent upstream and wake regions, and yields a simple power law spectrum consistent with observations. The mechanism, however, requires certain minimum or "seed" particle energies which are rather easily provided for ions but cause some problems for electrons. Since the increase of energy is diffusive and hence an inherently slow process, electrons can easily loose their energy gain by synchrotron radiation which may limit attainable energies to rather low levels.

In contrast to Fermi acceleration which is mainly due to alternating $E_\perp$ induction fields, particle acceleration by d.c. $E_\parallel$ fields along the magnetic field is inherently more effective,

$$\frac{dW}{dt} = q \int E_\parallel \, ds \; . \tag{1}$$

Usually $E_\parallel = \eta j_\parallel$ is very small in a plasma due to the high electrical conductivity η^{-1}. Nevertheless high energies may be attained if the path $\int ds = L$ is sufficiently long. In a toroidal magnetic trap such as a tokamak electrons beyond a certain threshold energy may be freely accelerated by the weak toroidal electric field $E_\parallel \sim 0.1 V/m$ reaching energies up to $10^8 eV$, the total path length being $L \sim 10^9 m$ corresponding to 10^8 revolutions around the torus axis. Such perfectly confining magnetic configurations are scarcely realized in astrophysical systems. Hence efficient acceleration requires strong $E_\parallel$ fields. These can be set up in fast magnetic reconnection processes.

II Failure of Conventional Theory of Magnetic Reconnection

The term magnetic reconnection refers to the picture of magnetic field lines. These have a well-defined meaning in a highly conducting fluid, viz. thin magnetic flux tubes, which are carried along with the fluid, preserving their individuality, though they may be wound up in a very complex manner. Only due to finite electrical resistivity or some equivalent process may two field lines coming close together loose their identities being cut and reconnected in a different way. Though this is a local process, it leads to a change of field topology permitting new types of large scale plasma motions that would otherwise be inhibited. The change of the magnetic field is desribed by Faraday's law

$$\frac{\partial \vec{B}}{\partial t} = \nabla \times \left(\vec{v} \times \vec{B} \right) + \eta \nabla^2 \vec{B} \; . \tag{2}$$

Here the ratio of the diffusion term and the convection term

$$\frac{\left| \eta \nabla^2 \vec{B} \right|}{\left| \nabla \times \left(\vec{v} \times \vec{B} \right) \right|} \sim \frac{\eta}{vL} = \frac{1}{R_m} \tag{3}$$

is a convenient dimensionless measure of the resistivity, R_m being the magnetic Reynolds number. In practically all astrophysical plasmas R_m is large, essentially because of the large scales L. Hence magnetic diffusion is in general a very weak process. On the other hand magnetic processes such as solar flares always seem to require fast reconnection with time scales practically independent of R_m. The main theoretical problem therefore is to find models allowing sufficiently high reconnection rates.

Fast reconnection is not a diffuse process, but is strongly localized in current sheets. Such current sheets may arise at any point with non-vanishing magnetic shear and a gradient of the velocity along the direction of the shear perpendicular to the field, i.e. virtually everywhere in the plasma, as visualized in Fig. 1.

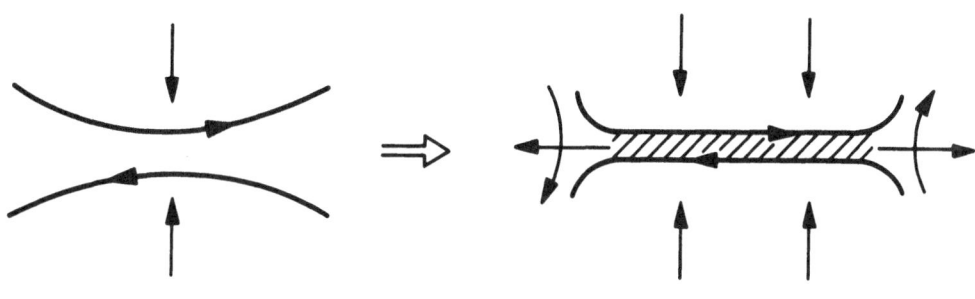

Fig.1 Current sheet formation. Magnetic field and velocity components in the plane perpendicular to and in the frame moving with the central field line.

While in the case of strong MHD turbulence there should be a random spatial distribution of rapidly varying current sheets, one can also think of quasi stationary configurations with one or only a few current sheets at well defined locations determined by the overall geometry. Such relatively simple configurations have been investigated in the conventional theory of magnetic reconnection. The basic assumption in these theoretical approaches is the existence of a two-dimensional subsystem around an x-type magnetic neutral point, which is small compared with the global magnetic configuration but large compared with the so called diffusion region around the neutral point, where the diffusion term in (1) is important. In this subsystem conditions adjust rapidly to changes in the global configuration, so that the evolution of the latter corresponds to a sequence of stationary states in the former, which are steady state solutions with the boundary conditions determined by the global system.

It has been taken for granted that steady state solutions with these properties exist in the limit $R_m \to \infty$, an example being Petschek's reconnection model[2] which became quite generally accepted. In fact much of the theoretical work on magnetic reconnection[3],[4] consists of modifications and refinements of this model, which is given schematically in Fig. 2. The theory is based on the effect that the motion of a plasma may be supersonic at any speed with respect to the slow mode for almost perpendicular propagation. Hence in analogy to a system of two supersonic gas jets colliding head-on, two pairs of slow shocks are generated standing back to back against the incoming plasma flow, diverting it into the outflow cone and accelerating it up to the Alfvén speed corresponding to the upstream magnetic field intensity. Petschek's configuration is characterized by a single parameter, the angle α of the outflow cone, which determines the ratio of inflow and outflow velocities, the so called reconnection rate $M = u/v_A$. The diffusion region is small with dimensions $O(\eta)$ and adjusts automatically to the external configuration. Since α is a free parameter, this class of solutions, considered as solutions of a small section out of the global magnetic configuration, seems to guaranty that reconnection and corresponding energy convertion rates depend only on the asymptotic plasma velocities, implying that M is essentially independent of η though the reconnection process of course requires magnetic diffusion.

Contrary to conventional wisdom, however, such solutions do not exist. The fundamental difference between the magnetized plasma and its gas dynamic analog is, that the plasma motion is not truly supersonic. Since plasma velocities are usually small compared with the compressional Alfvén mode - in fact the theory usually assumes incompressible plasma motions - information about the plasma behavior in front of the diffusion region can easily propagate upstream and affect the asymptotic inflow velocity. Mathematically speaking the inconsistency in the Petschek type solutions consists in essentially ignoring the boundary layer problem, i.e. the matching of the solution in the diffusion layer to that in the external region.

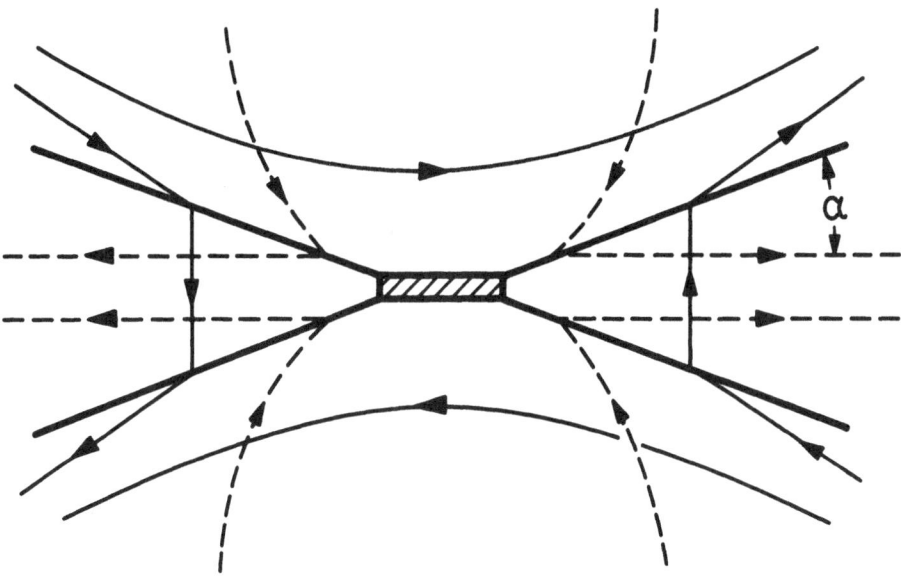

*Fig.2 Schematic representation of Petschek's recon-
nection configuration.*

283

The failure of Petschek's solution to apply for large Reynolds number has become evident only in recent years, when exact numerical solutions of the 2-dimensional resistive MHD equations revealed a completely different behavior. Figure 3 shows a set of three numerical solutions each computed from the same initial state in time until a stationary state had been reached, using the same boundary conditions but different values of the resistivity η. Here Φ is the stream function, $\vec{v} = \hat{z} \times \nabla\Phi$, and Ψ is the flux function, $\vec{B} = \hat{z} \times \nabla\Psi$. The conspicuous feature is that by reducing η the size of the diffusion region, i.e. the length of the current sheet, increases finally reaching the global system size.

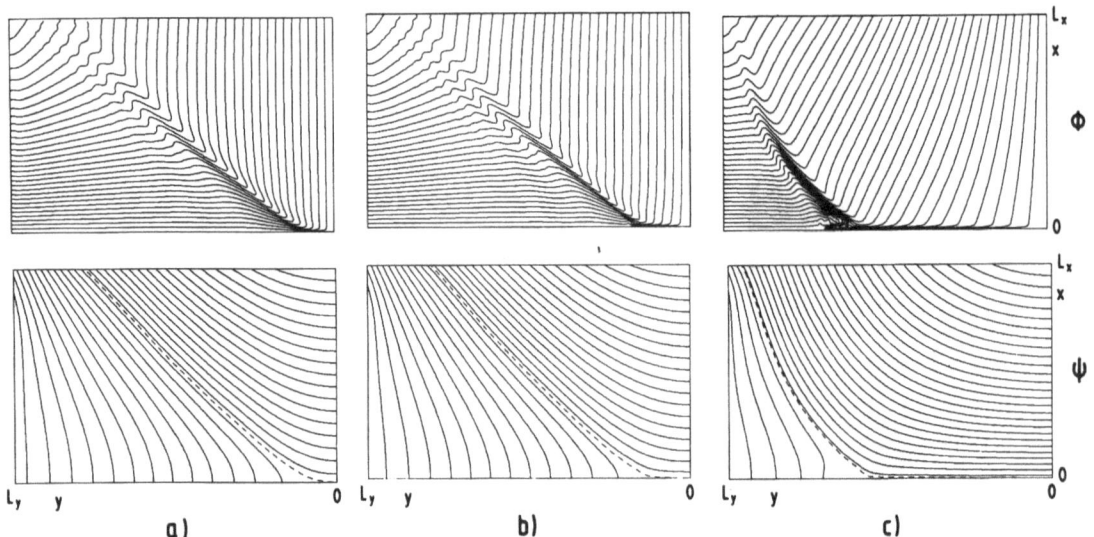

Fig.3 Steady state forced reconnection configurations with identical boundary conditions, differing only in the value of η: a) $\eta = \eta_0$, b) $\eta = \eta_0/2$, c) $\eta = \eta_0/4$. From Ref. 5.

Detailed scaling laws obtained from a series of numerical runs are given in Ref. 5. The physical picture is that for η falling below some natural magnetic diffusion rate $\eta_c \sim R_m^{-1/2}$ reconnection becomes increasingly inefficient. Consequently magnetic flux is piling up in front of the diffusion region with a corresponding slowing down of the upstream plasma flow compared with the prescribed boundary value of the inflow velocity. In addition to the increase in size the diffusion region develops an increasingly complex structure, for details see Refs. 5,6, such that a rigorous analytical treatment of the diffusion region appears to be practically impossible, not to speak of the matching problem mentioned above.

Since the choice of boundary conditions used in the numerical solutions has been the subject of some controversy and misunderstanding e.g. in Ref. 4, let me add a few remarks to the more detailed discussion given in Ref. 6. Conventionally consistency of boundary conditions is discussed for the nondissipative equations using the theory of characteristics. While this gives an unambiguous rule of the number of quantities that can be freely prescribed at the boundary for compressible fluid systems, conceptional difficulties arise in the incompressible case, which corresponds to a mixed hyperbolic-elliptic problem. Use of the equations for $\vec{B}$ and $\vec{v}$ as fundamental equations supplemented by the conditions $\nabla \cdot \vec{B} = \nabla \cdot \vec{v} = 0$ leads to integral constraints on the boundary values. Using equations for flux- and stream functions, which eliminate the divergence conditions in a convenient way, leads to higher order differential equations precluding application of the formalism of characteristics. In addition the mathematical theory has not been worked out rigorously for nonlinear multi-dimensional problems, where even in the compressible case numerical implication has not yet yielded convincing results. On the other hand computations such as those given in Ref. 5 show that diffusion coefficients far from only smoothening the behavior in boundary layers strongly affect the global configuration, in particular for large Reynolds numbers. In the dissipative system no real problem of boundary conditions arise, the only requirement for "natural" inflow and ouflow behavior is that in the limit of small dissipation no singular layers emerge at the boundaries of the numerical system. This still leaves considerable freedom in the choice of boundary conditions. The essential point made in Refs. 5,6 is that for sufficiently small resistivity the systems behavior depends far more strongly on the value of η than on the choice of the boundary conditions, however extreme.

III Nonstationary Reconnection

Stability investigations have shown[5] that current sheets arising during the process of magnetic reconnection (so called Sweet-Parker current sheets), which carry a strong inhomogeneous flow, are substantially more stable with respect to tearing modes than static current sheets. Only if the ratio of length over thickness exceeds $L/d > 10^2$, compared with $L/d > 10$ in the static case, tearing instability sets in. Since this threshold is passed for sufficiently small η, steady state reconnection does not exist in the limit $\eta \rightarrow 0$. On the contrary incompressible fluids always seem to develop fully three-dimensional turbulence at high Reynolds numbers, which is notoriously difficult to treat both analytically and numerically. Let us therefore first briefly discuss the nonstationary behavior at Reynolds numbers not very far above the transition point. The tearing instability leads to formation of plasmoids which are swept along the current sheet at high speed $v \sim v_A$ and ejected into the downstream region. Due to this nonsteady process reconnection is strongly enhanced compared with the unperturbed extended current sheet configuration, by generating secondary thin transient current sheets[6]. Further increase of the Reynolds number gives rise to additional dynamical fine structure. This behavior is indicated in Fig. 4, showing the break up of a current sheet, modelling events in the earth's magnetotail believed to cause magnetospheric substorms. The two cases displayed are for two values of η differing by a factor of 2 and otherwise identical conditions. The main process is the formation of a large plasmoid moving to the left with a long thin trailing current sheet. In the smaller η case this sheet is unstable leading to a secondary plasmoid following the main one. This increased dynamical activity compensates, at least partly, the reduction of the reconnection rate due to the smaller value of η.

Decreasing η even further a hierarchy of processes with smaller and smaller scales and higher irregularity is generated. They will certainly introduce some fine scale structures in the third dimension, even if the global configuration still remains quasi two-dimensional.

Such small scale MHD-turbulence can be treated on a statistical level. Averaging over small scales results in an effective resistivity independent of the local value of η, $\eta_{eff} \simeq \tilde{v}^2 \tau_{cor}$, for details see Ref. 7. In this sense a finite reconnection rate in the limit $\eta \to 0$ is in principle possible, in the same way as in hydrodynamic turbulence the energy dissipation rate is independent of the value of the viscosity.

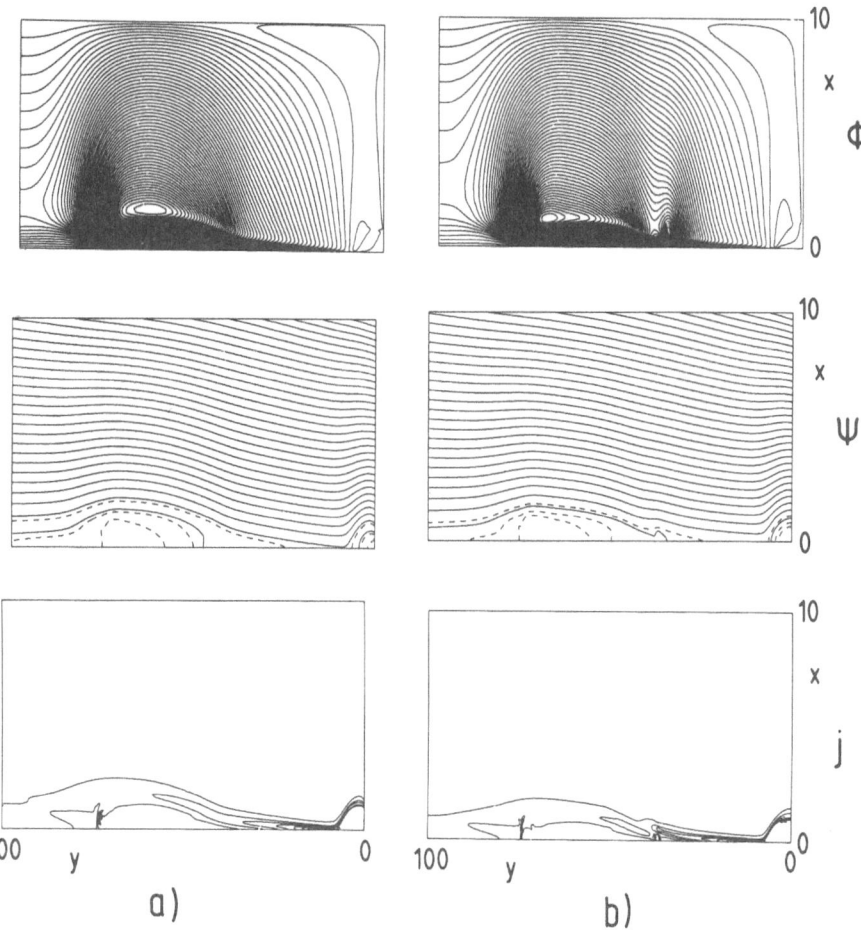

Fig.4 *Plasmoid formation in a current sheet with a)* $\eta = \eta_0$ *b)* $\eta = \eta_0/2$, *but otherwise identical initial and boundary conditions.*

Since astrophysical plasmas are often dilute and hence almost collisionless, MHD theory can only describe large scale features. If small scale structures are set up as expected in fast magnetic reconnection, kinetic or microscopic instabilities will be excited leading to micro-scale turbulence with wavelengths of the order of the Debye length e.g. the ion sound instability[8]. The picture of a single isolated process dominating the turbulence is, however, too simplistic. In strongly turbulent systems a rather smooth Kolmogorov type fluctuation spectrum extending over many decades in k-space, quite independently of the wavelength range of individual excitation mechanisms, seems to be a more realistic assumption. Since the high local current density j arising in magnetic reconnection processes, is probably the most effective source for excitation of turbulence, fluctuation levels and effective resistivity are often written phenomenologically in the form $\eta_{eff} = \gamma(j - j_0)^\nu$.

IV Runaway Electron Acceleration

Fast reconnection implies the presence of strong parallel electric fields $E_\| = \eta j \simeq uB$. In a resistive plasma following this simple form of Ohms law, the current density depends only on the instantaneous local values of η and $E_\|$, which implies that all electrons effectively randomize their directed energy gain. In reality, however, the Coulomb cross section is a strongly decreasing function of particle energy W. Hence electrons that happen to have substantially higher energies than their thermal companions are effectively collisionless and hence almost freely accelerated. This is called the runaway effect[9]. The effect is, however, not restricted to Coulomb collisions. In fact any kind of turbulent resistivity implies the scattering of electrons by electrostatic fields generated by charge clumps, i.e. local deviations from charge neutrality, which move at speeds given roughly by the phase velocities v_p of the turbulent waves. Except for special cases electron velocities exceed v_p, such that the collectively excited charge clumps act in the same way as individuel ions in a collisional plasma giving rise to the same runaway process for sufficiently energetic electrons. The number of runaway electrons depends on the strength of the electric field. For $E_\| \ll E_c = m_e v_e \nu_{eff}/e$, $v_e =$ electron thermal velocity, $\nu_{eff} =$ effective collision frequency, only a small fraction will be freely accelerated, while for $E_\| \gtrsim E_c$ the current will be mainly carried by runaways (implying that Ohms law would be dominated by electron inertia instead of resistivity). It should be noted that ion runaway is more difficult, so that this acceleration process primarily leads to high energy electrons in contrast to diffusive shock acceleration.

Application of these ideas to explain the observed synchrotron radiation in extragalactic radio sources, in particular hot spots, can only be very qualitative. The magnetic configuration is not known in any detail. The dynamical behavior of the jet and the embedded shocks including the hot spot at the jet's leading edge is presumably highly instationary[10], so that current sheets on all different spatial scales may be temporarily formed. For electrons of sufficiently high energy, say $W > 10^8 eV$, the details of the turbulent resistivity seem to be irrelevant, so that a crude estimate of the maximum electron energy $W_{max} \sim ce \int_0^{t_0} E_{||} dt$ can easily be obtained. Here $E_{||} \sim uB/c$, with u a typical plasma flow velocity, and t_0 is the acceleration time, which is given either by the lifetime of a major reconnection configuration of size R or by the extent $L_{||}$ of the current sheet parallel to the magnetic field, $t_0 \sim min(R/u, L_{||}/c)$. As the plasma velocity may be relativistic, $u \sim c$, both expressions are of the same order. To give a numerical example we choose $R \sim 10pc$ and $B \sim 10^{-5}G$, which yields $W_{max} \sim 10^{15} eV$. The energy spectrum depends on the rate at which particles are extracted from the bulk distribution of low energy scattered particles to become freely accelerated, and on the particle confinement in the region of large $E_{||}$. Because of the inherent uncertainty in these effects a definite spectrum cannot be predicted. The spectrum should be relatively flat, not inconsistent with the observed powerlaw $\propto W^{-2}$.

References

1) L.O'C. Drury, this conference

2) H.E. Petschek, in AAS/NASA Symposium on the Physics of Solar Flares, edited by W.N. Hess (NASA, Washington, DC, 1964), p. 425

3) V.M. Vasyliunas, Rev. Geophys. 13, 303 (1975)

4) T. G. Forbes, E.R. Priest, Rev. Geophys. 25, 1583 (1987)

5) D. Biskamp, Phys. Fluids 29, 1520 (1986)

6) D. Biskamp, Magnetic Fields and Extragalactic Objects, Proc. Cargèse Workshop, June 1987, E. Asseo and D. Grésillon editors, p. 37

7) F. Krause, K.H. Raedler, Mean Field Magnetohydrodynamics and Dynamo Theory, Pergamon, 1981

8) see e.g. C.T. Dum, R. Chodura, D. Biskamp, Phys. Rev. Lett. <u>32</u>, 1231 (1974)

9) H. Dreicer, Phys. Rev. <u>115</u>, 242 (1959)

10) M. Norman, this conference

STOCHASTIC ACCELERATION OF RELATIVISTIC ELECTRONS IN
SYNCHROTRON SOURCES WITH TURBULENTLY
RECONNECTING MAGNETIC FIELDS

Wayne A. Christiansen
Department of Physics and Astronomy
University of North Carolina
Chapel Hill, NC 27514, USA

Introduction:

Although magnetic reconnection as a mechanism for particle acceleration has
received considerable attention in solar and space physics, it has not been discussed
too often as an in situ acceleration mechanism for synchrotron emitting extragalactic
sources. Nevertheless, in the context of the physics of extragalactic synchrotron
sources, reconnection does have some distinct advantages over the more popular Fermi
type mechanisms which were discussed this morning (Drury et al. this volume). There
are three distinct areas of extragalactic radio astrophysics where I believe that
reconnection mechanisms exhibit distinctive advantages.

Direct Particle Acceleration:

By this, I mean that the basic outcome of the magnetic reconnection process, as
discussed this afternoon by Biskamp, is the generation of electric fields in current
sheets and (or) diffusion zones. There are two important aspects relating these
reconnection generated electric fields to particle acceleration in synchrotron
sources: First, direct electric field acceleration does not discriminate against
electrons, which is why high energy laboratory electron accelerators are LINAC's.
Second, in a reconnecting plasma, the acceleration of electrons takes place in regions
where the magnetic field is weak, so the synchrotron radiation drag on the accelerating
electrons is minimized.

Source Morphology:

With the advent of high resolution, high dynamic range radio imaging, it is
becoming increasingly clear that in extragalactic sources observed at high frequencies,
the radio lobe morphology is dominated by hot spots and filamentation. One can clearly
see this in the beautiful images presented by Perley and Carilli (CygA) and Owen (M87)
as well as the optical observations of Röser and Mesenheimer at this workshop. Such
synchrotron radiating features clearly represent order of magnitude enhancements in the
volume emissivity and may present some difficulties in standard models of lobe emission
relating to inferred overpressures (as mentioned at this meeting by Scheuer). In a
vigorously reconnecting plasma, however, which is observed via synchrotron radiation,
filamentation is a natural by-product of the process itself. Since synchrotron

emissivity is proportional to the square of the magnetic field intensity, the observed filaments delineate regions of unreconnected magnetic fields, which are illuminated by relativistic electrons which, in turn, have diffused in from the weak field regions where they were accelerated.

Seed Particle Production:

It would be a mistake to assume that particle acceleration mechanisms are mutually exclusive. In particular the ubiquitous Fermi mechanisms almost certainly contribute to the maintenance of relativistic electrons in radio sources. However, as mentioned in several papers this morning, there is consensus that Fermi mechanisms for accelerating electrons generally require that the initial energy of the electrons be already relativistic (Lorentz factors of 1000 or so) before Fermi mechanisms can efficiently bring about further acceleration (up to Lorentz factors of 10^7, as observed). By way of contrast, reconnection generated electric fields have no low energy cutoff where they cannot accelerate electrons; so they can and probably do serve as injectors for other complementary mechanisms such as Fermi.

Having outlined the reasons for suspecting that reconnection may play an important role in radio source physics, let me now examine specific areas in which reconnection may contribute to our understanding of synchrotron sources.

Energetics:

The total energy generated by reconnecting an initial magnetic field, B_0, is

$$\Delta E_{RC} = \text{Volume} \; \bar{\underline{X}}(\frac{B_0^2 - B_{RC}^2}{8\pi})$$

where B_{RC} is the residual field after reconnection and "Volume" is the relevant radiating source volume. Biskamp (this workshop) has discussed various ways in which reconnection occurs (e.g. via current sheet formation, tearing modes, etc.). The bottom line is that the magnitude of the residual field scales rather simply with the field line merger velocity, V_c, i.e.

$$B_{RC} \sim (\frac{V_c}{V_A})B_0$$

where V_A is the Alfven velocity associated with the initial field, B_0. Thus, the total energy obtained by anhilating B_0 is

$$\Delta E_{RC} = E_o (1-(V_c/V_A)^2)$$

where $E_o = \text{Volume} \; \bar{\underline{X}} \; (B_0^2/8\pi)$ is the magnetic energy originally existing in the source. Our observations are instantaneous, however, so we are more interested in the instantaneous power available in reconnecting sources which may, in principle, be channeled into in situ particle acceleration, thus maintaining the source's synchrotron

292

emission. If the scale size of the reconnecting volume is R, then the average time scale for complete reconnection is R/V_c. Thus, the time average power released by reconnection is

$$P_{RC} \sim E_o \, (1-(V_c/V_A)^2)(V_c/R).$$

The application of reconnection models to particle acceleration in solar flares must meet the severe constraint that the above power must be released in a short time. Hence, in solar physics a great deal of attention has been directed toward studies of and mechanisms for "fast reconnection". In the extragalactic case, however, "fast reconnection" is not necessarily required since, at least in extended sources, variability is not observed.

In Table 1, I list "estimated" reconnection "power" for typical radio emitting regions of interest, where I have optimistically chosen $V_A \sim 0.1c$ and $V_c \sim 0.1 \, V_A$ (For cautionary remarks concerning V_c see Biskamp's contribution to this volume). If the reconnecting field is actually already very "filamentary," or becomes so through tearing mode development, all of the above powers may rise (or fall) by a factor $f_f(R/r_f)$ where f_f is the filament's volume-filling factor and r_f is the filament scale size.

Table 1

	LOBES	JETS	HOT SPOTS
B_0:	10^{-5} gauss	10^{-5} gauss	10^{-3} gauss
R:	10 kpc	1 kpc	1 kpc
		L 100 kpc	
P_{rc}:	10^{43-44} ergs/s	10^{43-44} ergs/s	10^{45} ergs/s

It must be noted that in all cases the power available is impressively large (10^{43}-10^{45} ergs/s) and comparable to the radio luminosities of the strongest sources. However, power release "does not a synchrotron emitter make" and the energy released by reconnection may be channeled into "useless" forms such as thermal heat which do not result in synchrotron emission. Therefore, we need to examine the process in terms of its ability to accelerate synchrotron emitting electrons.

Electron Acceleration:

In the current sheet or reconnection zone, electrons with small pitch angles

293

relative to the electric field direction may get a very strong kick, i.e.

$$\Delta \varepsilon \sim 2eEl_s$$

where $E = (V_c/c)B_0$ is the strength of the electric field in the current sheet and l_s is the coherence length over which the electron continues to remain in the current sheet. Naively, if we take $V_c/c = 0.01$, $B_0 = 10^{-5}$ gauss and assume that l_s is in parsecs, which is allowed by high resolution mapping of surface brightness variations, then we find that

$$\Delta \varepsilon_{naive} \sim 10^{13} l_s(pc) \text{ (in eV per encounter)}$$

This would imply that the electron Lorentz factor could easily reach 10^7 in one encounter with a reconnection zone and, if true, would satisfy even the requirements for optical synchrotron emitters pointed out by Meisenheimer and Röser at this meeting. That seems too good to be true, and it is, because of streaming instabilities, etc. which will limit the acceleration path to a length, l_s, considerably less than the longitudinal scale of the electric field itself.

On the other hand, in a turbulent or semi-turbulent reconnecting plasma, a number of current sheets with electric fields will coexist at any time. Thus, a statistical process involving multiple encounters with reconnection generated electric fields may dominate electron acceleration. Here again, particle acceleration by reconnection in extragalactic radio sources differs from the models for acceleration by reconnection which are invoked for solar flares, since acceleration takes place in a spatially extended region rather than at a single point.

In considering multiple encounters with electric fields as a mechanism for electron acceleration, it must be recognized that if the electric field vectors are truly random in direction and intensity, no net acceleration can occur because deceleration is equally likely. However, in synchrotron emitting sources (both optical and radio) we know that true randomness does not exist. The emergent synchrotron radiation is polarized! This means that the source magnetic field and, hence, the electric field vectors in the reconnection zones do have net directionality.

In traveling from one reconnection zone (or current sheet) to the next the relativistic electrons will preferentially remain in the weak, reconnected residual magnetic field. Because of streaming instabilities, the accelerated electrons can only diffuse through the residual field at the local Alfven speed, V_{AR}. As a result, if l_{MF}

is the mean free path between electric field encounters, the net energy gain rate per electron is

$$\left\langle \frac{d\varepsilon}{dt} \right\rangle \sim \Delta\varepsilon \ F(\frac{V_{AR}}{l_{MF}}),$$

where $F \leq 1$ is the fractional excess of energy-gaining encounters.

The parameter, F, is thus a measure of the net polarization in the reconnection generated electric field. Lacking detailed information on the morphology of the reconnecting magnetic field, we can only estimate a lower bound for F. In a uniform magnetic field, the net polarization of synchrotron radiation emitted by a power-law distribution of relativistic electrons is about 70%, providing the source is optically thin and Faraday rotation is negligible. If the magnetic field has a turbulent component the net polarization is reduced with the result that (for the doubly thin case)

$$p = (0.7)((N_+ - N_-)/(N_+ + N_-))$$

where N_+ represents the number of cells in the telescope beam possessing the dominant magnetic field orientation and N_- represents the number of cells with the opposite orientation. Since the net polarization in the electric fields generated by reconnection must reflect the net polarization of the original magnetic field we find that

$$F \geq p/(0.7)$$

This is a lower limit because we have not included the enhancing effects of tearing modes. Tearing modes will generate a series of magnetic islands and current sheets in which all of the electric fields have the same orientation, thus allowing the accelerating electrons to interact with a series of parallel electric fields. In any case, it may be expected that since p is usually observed to be a few percent, then $F \sim 0.1$.

Finally, it is clear that a crucial parameter in determining the effectiveness of reconnection as an acceleration mechanism is l_{MF}. If l_{MF} is too large the mechanism will fail.

Competing with the energy gain discussed above are the synchrotron losses induced by the weak residual magnetic field between reconnection zones. The result is that in such a reconnecting plasma there is a natural cutoff in the relativistic electrons' energy spectrum corresponding to the energy at which synchrotron losses balance the energy gains from reconnection, i.e.

$$\langle\frac{d\epsilon}{dt}\rangle = \Delta\epsilon F(\frac{V_A}{l_{MF}}) = \frac{d\epsilon}{dt}\Big|_{sync} = 10^{-15}B_{RC}^2\ \gamma_{co}^{\ 2}$$

where γ_{co} is the Lorentz factor of electrons having the cutoff energy. The result is
that

$$\gamma_{co}^{\ 2} = 1.5 \times 10^{20}(\frac{Fl_s}{l_{MF}})$$

We may now set limits on the connection between l_s and l_{MF} which are required by
observations, as shown in Table 2.

Table 2

Optical Synchrotron Radiation	Fermi Seeds
$\gamma_{co} \sim 10^7$	$\gamma_{co} \sim 10^3$
$l_{MF} \sim 10^6\ Fl_s$	$l_{MF} \sim 10^{14}\ Fl_s$

Both l_{MF} and l_s in Table 2 are highly uncertain but an order of magnitude estimate
of l_s is, $l_s \sim 10^{12} - 10^{14}$cm, which would require a very fine scale in the reconnecting
field (i.e. $l_{MF} \lesssim 1$ pc) if reconnection is to boost the electron cutoff energy high
enough to allow optical synchrotron radiation. On the other hand, if one is to only
ask that reconnection accelerate seed particles for the Fermi mechanism it is clear
that this can be accomplished easily.

Because l_s is likely to be quite short, i.e. $\ll$ a parsec, it seems unlikely that
reconnection generated electric fields can supply the high energy electrons required to
generate optical synchrotron radiation as the result of single encounters. However, if
the reconnecting magnetic field in the synchrotron source has a turbulent component, a
large number of current sheets or diffusion zones containing electric fields may
coexist at any moment. Thus, relativistic electrons may reach high energies as a
result of a stocastic acceleration in the resultant partially polarized electric fields
associated with reconnection zones.

Luke Drury   Rick Perley   Frazer Owen

Instead of a summary talk, the concluding session was devoted to a general discussion of three topics which had been selected by means of an opinion poll about "the most important question" after the end of the regular sessions:
(1) Hot spots and emission line observations,
(2) the flow into / out of hot spots,
(3) spectra, particle acceleration and magnetic fields
   (introduced by Bob Fosbury, Peter Scheuer and myself, respectively). It was not possible to include the discussion in these proceedings. But the questions themselves may stimulate future work. Therefore, some of them are given below.

Klaus Meisenheimer

What is the connection between radio emitting plasma and emission line gas in radio galaxies ?

How can we get dense (10^0 cm^{-3}) gas and strong (>1μGauss) magnetic fields 100 kpc out in a galactic atmosphere ?

How reliable are numerical simulations ?

Where are the superhot knots (see contribution by Colin Lonsdale) in the computer simulations ?

What is the composition of the jet / hot spot material ? What is the flow velocity into and out of the hot spots ?

How can a jet be recollimated (after passing trough the primary hot spot) to produce a bright "splatter spot".

Could accurate multi-frequency observations (I,Q,U) be used to determine the flow pattern in radio hot spots ?

How important are magnetic fields ? Are there observational indicators to estimate the ratio of Pointing to mass energy flux ?

Are the ultra-relativistic particles accelerated within the hot spots or transported into the hot spots ?

Why are there so few published maps which resolve hot spots ?

Object list

Cen A (1322-427)	127, 130, 131, 143, 145
Coma A (3C 277.3, 1251+278)	**101**, 127, 128, *129*, 130, 131, 132, 138, 141, 143, 145, 146
Cygnus A (3C 405, 1957+405)	2, 5, 6, 7, 14, 15, 22, 24, 33, **51**, *55-60*, 92, **105**, 123, 127, 128, 164, 180, 183, 187, 188, 189, 194, 202, 210, 291
Fornax A	146
Hercules A (3C 348)	5
Minkowski's Object (PKS 0123-016A, see also NGC 541)	*131*, 145
M 51 (1327+474)	127, 139, 141
M 81	180
M 84 (3C 272.1)	123
M 87 (3C 274, 1228+126)	**77**, *81*, **82**, 124, 127, 128, 143, 187, 188, 291
jet	77, *80*, **82**, **89**, 181, 187, 189, 262, 263
M 124	180
NGC 541 (see Minkowski's Object)	124, 127, 130, 131, 143, 145
NGC 1068 (0240-002)	127, 139, 141, 142
NGC 5929 (1524-418)	127, 132, 139, 141
NGC 7385 (2247+111)	127, 145
OR-017	see 1510-089
Pictor A (0518-458)	7, **97**, *112*, 187, 194, 210
east	*100*, 104
west	**97**, *100*, 107, 108, 109, 146, 165, 180, 255, 257, 258
Sgr A*	180
3C 9	162
3C 20 (0040+517)	*111*
east	30
west	**94**, 107, 108, 255, 258
3C 33 (0106+130)	7, *8*, **61**, *62*, 123
north	**63**
south	**63**, **94**, *95*, 107, 108, *111*, 115, *116*, 146, 255, 258
3C 41	*39*
3C 63	125
3C 68.1	162
3C 75	123, 189
3C 79 (0308+169)	106
3C 89	123
3C 98	124, *125*

299

Italic page numbers refer to maps.

Bold page numbers indicate more extensive coverage.

Lecture Notes in Mathematics

Lecture Notes in Physics

G. L. Verschuur, K. I. Kellermann, National Radio Astronomy Observatory, Charlottesville, VA, USA (Eds.)
With the Assistance of E. Bouton

Galactic and Extragalactic Radio Astronomy

2nd edition. 1988. 207 figures. XXII, 694 pages. ISBN 3-540-96575-0

Contents: Galactic Nonthermal Continuum Emission. – HII Regions and Radio Recombination Lines. – Neutral Hydrogen and the Diffuse Interstellar Medium. – Molecules as Probes of the Interstellar Medium and of Star Formation. – Interstellar Molecules and Astrochemistry. – Astronomical Masers. – The Structure of Our Galaxy Derived from Observations of Neutral Hydrogen. – The Galactic Center. – Radio Stars. – Supernova Remnants. – Pulsars. – Extragalactic Neutral Hydrogen. – Radio Galaxies and Quasars. – The Microwave Background Radiation. – Radio Sources and Cosmology. – Index.

M. H. Soffel, University of Tübingen, FRG

Relativity in Astronomy, Celestial Mechanics and Geodesy

1989. 32 figures. XIV, 208 pages. ISBN 3-540-18906-8

Contents: Relativity in Astrometry, Celestial Mechanics and Geodesy. – Newtonian and Relativistic Space-Time. – Reference Frames and Astrometry. – Celestial Mechanics. – Geodesy. – Appendix. – References. – Subject Index.

K. Rohlfs, Ruhr-University of Bochum, FRG

Tools of Radio Astronomy

1986. 127 figures. XII, 319 pages. ISBN 3-540-16188-0

Contents: Radio Astronomical Fundamentals. – Electromagnetic Wave Propagation Fundamentals. – Wave Polarization. – Fundamentals of Antenna Theory. – Filled Aperture Antennas. – Interferometers and Aperture Synthesis. – Receivers. – Emission Mechanisms of Continuous Radiation. – Some Examples of Thermal and Nonthermal Radio Sources. – Line Radiation Fundamentals. – Line Radiation of Neutral Hydrogen. – Recombination Lines. – Interstellar Molecules and Their Line Radiation. – Appendices A – E. – List of Symbols. – References. – Subject Index.

Springer-Verlag
Berlin Heidelberg New York
London Paris Tokyo Hong Kong

Springer